Progress in Mathematical Physics
Volume 32

G.F. Torres del Castillo

3-D Spinors, Spin-Weighted Functions and their Applications

Springer Science+Business Media, LLC

G.F. Torres del Castillo
Universidad Autonoma de Puebla
Instituto de Ciencias
Avenue San Claudio y Rio Verde
Puebla, Puebla 72570
Mexico

Library of Congress Cataloging-in-Publication Data

Torres del Castillo, G. F. (Gerardo Francisco), 1956-
3-D spinors, spin-weighted functions and their applications / G.F. Torres del Castillo.
p. cm. – (Progress in mathematical physics ; v. 20)
Includes bibliographical references and index.
ISBN 978-0-8176-3249-6 ISBN 978-0-8176-8146-3 (eBook)
DOI 10.1007/978-0-8176-8146-3
1. Spinor analysis. I. Title. II. Series.

QA433.T67 2003
515'.63–dc21

2003051982
CIP

AMS Subject Classifications: 15A66, 22E70, 33E30, 53B20, 83C60, 35Q40, 35Q75, 83C05

Printed on acid-free paper.

Originally published by Birkhäuser Boston in 2003

ISBN 978-0-8176-3249-6 SPIN 10933817

Reformatted from the authors' files by TEXniques, Inc., Cambridge, MA.

9 8 7 6 5 4 3 2 1

Contents

Preface

The spinor calculus employed in general relativity is a very useful tool; many expressions and computations are considerably simplified if one makes use of spinors instead of tensors. Some advantages of the spinor formalism applied in the four-dimensional space-time of general relativity come from the fact that each spinor index takes two values only, which simplifies the algebraic manipulations.

Spinors for spaces of any dimension can be defined in connection with representations of orthogonal groups and in the case of spaces of dimension three, the spinor indices also take two values only, which allows us to apply some of the results found in the two-component spinor formalism of four-dimensional space-time. The spinor formalism for three-dimensional spaces has been partially developed, mainly for spaces with a definite metric, also in connection with general relativity (*e.g.*, in space-plus-time decompositions of space-time), defining the spinors of three-dimensional space from those corresponding to four-dimensional space-time, but the spinor formalism for three-dimensional spaces considered on their own is not widely known or employed.

One of the aims of this book is to give an account of the spinor formalism for three-dimensional spaces, with definite or indefinite metric, and its applications in physics and differential geometry. Another is to give an elementary treatment of the spin-weighted functions and their various applications in mathematical physics. The best-known example of the spin-weighted functions are the spin-weighted spherical harmonics, which are a generalization of the ordinary spherical harmonics and, as the latter, are very useful in the solution by separation of variables of partial differential equations. By means of the spin-weighted spherical harmonics one can give a unified treatment of fields of any spin, without requiring definitions of the vector, tensor and spinor spherical harmonics employed in electrodynamics, quantum mechanics and general relativity.

Apart from Chapter 1, which is intended to be an elementary introduction to the spinors of three-dimensional space, the book is divided into two somewhat independent parts; three chapters are devoted to the properties and applications of spin-weighted functions and the last three chapters deal with spinors in three-dimensional space and their applications. Among the topics not included in this book are the global aspects related to the existence of spinor structures and the relationship of spinors to Clifford algebras.

It is assumed that the reader has some familiarity with tensor calculus, linear algebra, elementary group theory, Riemannian manifolds and special functions. The examples considered in the book are taken from classical mechanics, electrodynamics, quantum mechanics, general relativity, elasticity and differential geometry. The Dirac equation is considered at several places in the book, starting from the standard form of the equation as given in quantum mechanics books like Schiff (1968), without assuming a detailed knowledge about the Dirac four-component spinors and their transformation properties. For the last chapter, it is convenient to have some knowledge of general relativity and of the corresponding two-component spinor formalism.

I would like to acknowledge my indebtedness to Professor Jerzy F. Plebański and to Sir Roger Penrose for their influence. I also thank one of the reviewers of this book for many valuable suggestions.

3-D Spinors,
Spin-Weighted Functions
and their Applications

1
Rotations and Spinors

It is a well-known fact that rotations in three-dimensional Euclidean space can be represented by means of complex 2×2 matrices and that this representation is related to the stereographic projection of complex numbers (see, *e.g.*, Goldstein 1980, Penrose and Rindler 1984, Burn 1985, Sattinger and Weaver 1986, Stillwell 1992). However, the form in which these results are usually established is somewhat indirect and, therefore, it is difficult to appreciate the naturalness of these relationships and their geometric origin (*cf.* also Misner, Thorne and Wheeler 1973).

In this chapter we employ the correspondence between points of the sphere and points of the complex plane to show that rotations in three dimensions can be represented by a certain class of functions of a complex variable and by 2×2 matrices, which are related to spinors. In Section 1.1 it is shown that the system of differential equations that determine the movement of a vector under rotations can be transformed, by means of the stereographic projection, into a single differential equation. It is shown that the solution of such an equation can be represented by a unitary 2×2 matrix and the matrix corresponding to a rotation about a given axis through an arbitrary angle is obtained. In Section 1.2, spinors are introduced and the relationship between the results of Section 1.1 and the treatment given in other works (*e.g.*, Payne 1952, Goldstein 1980) is established. It is shown that each point of the space can be represented by means of a spinor and that a spinor also represents a triad of vectors of the same magnitude orthogonal to each other. The properties of the 3×3 real matrices that represent rotations are obtained, as well as their explicit form. In Section 1.4 it is shown that, in a three-dimensional space with an indefinite metric, spinors can also be defined in a geometrical way.

1.1 Representations of rotations

The stereographic projection establishes a correspondence between the points of the sphere $S^2 = \{(x, y, z) \in \mathbb{R}^3 \mid x^2 + y^2 + z^2 = 1\}$ and those of the complex plane in the following way. The straight line joining the "north pole" of the sphere, represented by (0,0,1), with an arbitrary point $(x, y, z) \neq (0, 0, 1)$ of the sphere, intersects the xy plane at some point with coordinates (X, Y), or at $X + \mathrm{i}Y$, regarding the xy plane as the complex plane (see Fig. 1). Thus, the point (x, y, z) of the sphere is associated with the complex number $\zeta \equiv X + \mathrm{i}Y$. The points of the straight line joining the points (0,0,1) and (x, y, z) are of the form $(0, 0, 1) + t[(x, y, z) - (0, 0, 1)] = (tx, ty, 1 + t(z - 1))$, hence this line intersects the xy plane for $t = 1/(1 - z)$ at the point $(x, y, 0)/(1 - z)$, which corresponds to the complex number

$$\zeta = \frac{x + \mathrm{i}y}{1 - z}, \tag{1.1}$$

by identifying the xy plane with the complex plane. Thus, $\zeta\bar{\zeta} = \dfrac{x^2 + y^2}{(1 - z)^2} = \dfrac{1 - z^2}{(1 - z)^2} = \dfrac{1 + z}{1 - z}$, which implies that $z = \dfrac{\zeta\bar{\zeta} - 1}{\zeta\bar{\zeta} + 1}$ and from (1.1) it follows that, $x + \mathrm{i}y = \dfrac{2\zeta}{\zeta\bar{\zeta} + 1}$. Hence, the inverse relation to (1.1) is given by

$$x = \frac{\zeta + \bar{\zeta}}{\zeta\bar{\zeta} + 1}, \qquad y = \frac{\zeta - \bar{\zeta}}{\mathrm{i}(\zeta\bar{\zeta} + 1)}, \qquad z = \frac{\zeta\bar{\zeta} - 1}{\zeta\bar{\zeta} + 1}. \tag{1.2}$$

In terms of spherical coordinates, $(x, y, z) = (\sin\theta\ \cos\phi, \sin\theta\ \sin\phi, \cos\theta)$, and from (1.1) one finds the equivalent expression

$$\zeta = \mathrm{e}^{\mathrm{i}\phi} \cot \tfrac{1}{2}\theta.$$

The point (0,0,1), which corresponds to $\theta = 0$, can be associated with the point at infinity and, in this manner, there is a one-to-one correspondence between the points of the sphere and the points of the extended complex plane.

Under an arbitrary rotation about the origin, each point of the sphere is mapped into another point of the sphere and since, by means of the stereographic projection, there exists a one-to-one correspondence between the points of the sphere and the points of the extended complex plane, a rotation determines a transformation of the extended complex plane onto itself. It will be shown that this transformation can be easily obtained for any rotation.

Under counterclockwise rotations about the axis defined by a unit vector $\mathbf{n}$, the position vector $\mathbf{r}$ of an arbitrary point of the three-dimensional space rotates

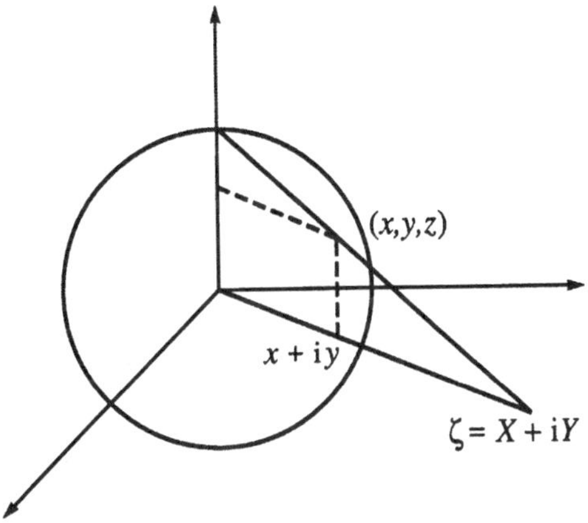

Figure 1: Stereographic projection of the unit sphere on the complex plane.

according to

$$\frac{d\mathbf{r}}{d\alpha} = \mathbf{n} \times \mathbf{r}, \tag{1.3}$$

where α is the angle of rotation about $\mathbf{n}$. Equation (1.3) constitutes a system of three linear differential equations that can be integrated directly (for instance, by writing the right-hand side of (1.3) in matrix form and making use of the exponential); however, (1.3) can be transformed into a simpler equation by means of (1.1) and (1.2), showing, at the same time, several connections with other areas.

Under rotations about the axis $\mathbf{n}$, the complex number ζ, corresponding to a point $\mathbf{r} \equiv (x, y, z)$ of the sphere S^2, varies according to [see (1.1) and (1.3)]

$$\begin{aligned} \frac{d\zeta}{d\alpha} &= \frac{1}{1-z}\left(\frac{dx}{d\alpha} + i\frac{dy}{d\alpha}\right) + \frac{x+iy}{(1-z)^2}\frac{dz}{d\alpha} \\ &= \frac{1}{1-z}[n_2 z - n_3 y + i(n_3 x - n_1 z)] + \frac{x+iy}{(1-z)^2}(n_1 y - n_2 x), \end{aligned}$$

where $\mathbf{n} = (n_1, n_2, n_3)$. Making use of (1.1) and (1.2) one finds that

$$\begin{aligned} \frac{d\zeta}{d\alpha} &= -\frac{i}{2}(n_1 - in_2)\left[\zeta^2 - \frac{2n_3}{n_1 - in_2}\zeta - \frac{n_1 + in_2}{n_1 - in_2}\right] \\ &= -\frac{i}{2}(n_1 - in_2)(\zeta - \zeta_1)(\zeta - \zeta_2), \end{aligned} \tag{1.4}$$

where

$$\zeta_1 \equiv \frac{n_3 + 1}{n_1 - in_2}, \qquad \zeta_2 \equiv \frac{n_3 - 1}{n_1 - in_2}. \tag{1.5}$$

The fact that $\overline{\zeta}$ does not appear in (1.4) means that the functions of the extended complex plane onto itself that represent rotations are analytic.

By contrast with (1.3), (1.4) can be integrated in an elementary way; indeed, making use of (1.4) and (1.5), one finds that

$$-\mathrm{i}\int_0^\alpha \mathrm{d}\alpha = \frac{2}{n_1 - \mathrm{i}n_2}\int_\zeta^{\zeta'} \frac{\mathrm{d}\zeta}{(\zeta-\zeta_1)(\zeta-\zeta_2)} = \int_\zeta^{\zeta'} \frac{\mathrm{d}\zeta}{\zeta-\zeta_1} - \int_\zeta^{\zeta'} \frac{\mathrm{d}\zeta}{\zeta-\zeta_2},$$

where ζ' denotes the image of ζ under the rotation about $\mathbf{n}$ through the angle α. Hence,

$$\ln\left[\frac{\zeta'-\zeta_1}{\zeta-\zeta_1}\frac{\zeta-\zeta_2}{\zeta'-\zeta_2}\right] = -\mathrm{i}\alpha,$$

which implies that

$$\zeta' = \frac{\mathrm{e}^{-\mathrm{i}\alpha}\zeta_2(\zeta-\zeta_1) - \zeta_1(\zeta-\zeta_2)}{\mathrm{e}^{-\mathrm{i}\alpha}(\zeta-\zeta_1) - (\zeta-\zeta_2)} = \frac{\mathrm{e}^{-\mathrm{i}\alpha/2}\zeta_2(\zeta-\zeta_1) - \mathrm{e}^{\mathrm{i}\alpha/2}\zeta_1(\zeta-\zeta_2)}{\mathrm{e}^{-\mathrm{i}\alpha/2}(\zeta-\zeta_1) - \mathrm{e}^{\mathrm{i}\alpha/2}(\zeta-\zeta_2)}$$

and, substituting expressions (1.5), one obtains

$$\zeta' = \frac{(\cos\frac{1}{2}\alpha + \mathrm{i}n_3\sin\frac{1}{2}\alpha)\zeta + (\mathrm{i}n_1 - n_2)\sin\frac{1}{2}\alpha}{((\mathrm{i}n_1 + n_2)\sin\frac{1}{2}\alpha)\zeta + \cos\frac{1}{2}\alpha - \mathrm{i}n_3\sin\frac{1}{2}\alpha}. \tag{1.6}$$

In place of the variable ζ it is convenient to employ

$$\xi \equiv \overline{\zeta} = \frac{x - \mathrm{i}y}{1-z} = \mathrm{e}^{-\mathrm{i}\phi}\cot\tfrac{1}{2}\theta. \tag{1.7}$$

By contrast with ζ, the variable ξ defines an orientation on the complex plane that coincides with that induced by the orientation of the sphere under the stereographic projection (this is equivalent to the condition that the Jacobian determinant $\left|\frac{\partial(\mathrm{Re}\,\xi, \mathrm{Im}\,\xi)}{\partial(\theta,\phi)}\right|$ be positive at all points of the sphere). Then, from (1.6) and (1.7) it follows that

$$\xi' = \frac{(\cos\frac{1}{2}\alpha - \mathrm{i}n_3\sin\frac{1}{2}\alpha)\xi - (\mathrm{i}n_1 + n_2)\sin\frac{1}{2}\alpha}{((-\mathrm{i}n_1 + n_2)\sin\frac{1}{2}\alpha)\xi + \cos\frac{1}{2}\alpha + \mathrm{i}n_3\sin\frac{1}{2}\alpha}. \tag{1.8}$$

Equation (1.8) is of the form

$$\xi' = \frac{\beta\xi + \gamma}{\delta\xi + \epsilon}, \tag{1.9}$$

where β, γ, δ, and ϵ are complex numbers and we can associate with this transformation the 2×2 complex matrix formed by these coefficients,

$$\begin{pmatrix} \beta & \gamma \\ \delta & \epsilon \end{pmatrix}. \tag{1.10}$$

(The functions of the form (1.9) are known as Möbius transformations (see, *e.g.*, Knopp 1952, Burn 1985).) If we make a second rotation after the one given by (1.9), this will be represented by an expression of the form

$$\xi'' = \frac{\eta\xi' + \kappa}{\mu\xi' + \nu}, \tag{1.11}$$

where η, κ, μ, and ν are complex numbers, which corresponds to the matrix

$$\begin{pmatrix} \eta & \kappa \\ \mu & \nu \end{pmatrix}.$$

The effect of the composition of these two rotations is obtained by substituting (1.9) into (1.11), in order to obtain ξ'' in terms of ξ,

$$\xi'' = \frac{\eta\dfrac{\beta\xi + \gamma}{\delta\xi + \epsilon} + \kappa}{\mu\dfrac{\beta\xi + \gamma}{\delta\xi + \epsilon} + \nu} = \frac{(\eta\beta + \kappa\delta)\xi + (\eta\gamma + \kappa\epsilon)}{(\mu\beta + \nu\delta)\xi + (\mu\gamma + \nu\epsilon)}.$$

This transformation corresponds to the matrix

$$\begin{pmatrix} \eta\beta + \kappa\delta & \eta\gamma + \kappa\epsilon \\ \mu\beta + \nu\delta & \mu\gamma + \nu\epsilon \end{pmatrix},$$

which is just the usual matrix product

$$\begin{pmatrix} \eta & \kappa \\ \mu & \nu \end{pmatrix}\begin{pmatrix} \beta & \gamma \\ \delta & \epsilon \end{pmatrix}.$$

This means that a transformation of the form (1.9) can be represented by the matrix (1.10), so that the composition of transformations corresponds to the matrix product. However, the matrix associated to the transformation (1.9) is not uniquely defined, since it can also be written as

$$\xi' = \frac{\lambda\beta\xi + \lambda\gamma}{\lambda\delta\xi + \lambda\epsilon},$$

for any complex number $\lambda \neq 0$; therefore all the matrices $\begin{pmatrix} \lambda\beta & \lambda\gamma \\ \lambda\delta & \lambda\epsilon \end{pmatrix} = \lambda\begin{pmatrix} \beta & \gamma \\ \delta & \epsilon \end{pmatrix}$ must be considered as equivalent to $\begin{pmatrix} \beta & \gamma \\ \delta & \epsilon \end{pmatrix}$. In order to reduce this ambiguity, one can impose the condition that the determinant of $\begin{pmatrix} \beta & \gamma \\ \delta & \epsilon \end{pmatrix}$ be equal to 1; in this manner only two matrices will correspond to the transformation

(1.9) (with one of these matrices being the negative of the other). In what follows it will be assumed that the matrices (1.10) have determinant equal to 1.

The matrix

$$Q \equiv \begin{pmatrix} \cos\frac{1}{2}\alpha - in_3 \sin\frac{1}{2}\alpha & -(in_1 + n_2)\sin\frac{1}{2}\alpha \\ (-in_1 + n_2)\sin\frac{1}{2}\alpha & \cos\frac{1}{2}\alpha + in_3 \sin\frac{1}{2}\alpha \end{pmatrix}, \tag{1.12}$$

formed by the coefficients appearing in (1.8), in addition to having determinant equal to 1, is unitary, *i.e.*,

$$QQ^\dagger = I, \tag{1.13}$$

where $Q^\dagger$ is the conjugate transpose of Q and I denotes the identity 2×2 matrix. This means that $Q^\dagger$ is the inverse of Q; in fact, the inverse of Q can be obtained by replacing α by $-\alpha$ in (1.12), which coincides with the result of transposing the complex conjugate of Q. Thus, the matrices (1.12), which represent rotations in three-dimensional space, belong to the group SU(2) formed by the unitary 2×2 matrices with determinant equal to 1. From (1.12) it follows that the rotation angle α is related to the trace of Q by means of the expression

$$\operatorname{tr} Q = 2\cos\tfrac{1}{2}\alpha. \tag{1.14}$$

Equation (1.12) is equivalent to

$$Q = \cos\tfrac{1}{2}\alpha\, I - \mathrm{i}\sin\tfrac{1}{2}\alpha\,(n_1\sigma_1 + n_2\sigma_2 + n_3\sigma_3), \tag{1.15}$$

where

$$\sigma_1 \equiv \begin{pmatrix} 0 & 1 \\ 1 & 0 \end{pmatrix}, \qquad \sigma_2 \equiv \begin{pmatrix} 0 & -\mathrm{i} \\ \mathrm{i} & 0 \end{pmatrix}, \qquad \sigma_3 \equiv \begin{pmatrix} 1 & 0 \\ 0 & -1 \end{pmatrix}, \tag{1.16}$$

are the Pauli matrices, which satisfy the relations

$$\sigma_j\sigma_k = \delta_{jk} I + \mathrm{i}\varepsilon_{jkm}\sigma_m; \tag{1.17}$$

ε_{jkm} is the Levi-Civita symbol and, as in what follows, we sum over repeated indices. Then, from (1.15) and (1.17) one finds that

$$\begin{aligned} Q\sigma_k &= \left[\cos\tfrac{1}{2}\alpha\, I - \mathrm{i}\sin\tfrac{1}{2}\alpha\, n_j\sigma_j\right]\sigma_k \\ &= \cos\tfrac{1}{2}\alpha\,\sigma_k - \mathrm{i}\sin\tfrac{1}{2}\alpha\, n_k\, I + \sin\tfrac{1}{2}\alpha\,\varepsilon_{jkm}n_j\sigma_m, \end{aligned}$$

and taking into account that the trace of the Pauli matrices is equal to zero and $\operatorname{tr} I = 2$,

$$\operatorname{tr} Q\sigma_k = -2\mathrm{i}\sin\tfrac{1}{2}\alpha\, n_k. \tag{1.18}$$

Hence, for a given matrix $Q \in \mathrm{SU}(2)$, (1.14) and (1.18) allow us to find the axis and the angle of the rotation represented by Q. From (1.18) it follows that any rotation in three dimensions is equivalent to a rotation about some axis [see also (1.42)].

1.2 Spinors

By expressing ξ as the quotient of two complex quantities

$$\xi = \frac{u}{v} \tag{1.19}$$

and, similarly, $\xi' = u'/v'$, the transformation (1.9) can be written in the form

$$\frac{u'}{v'} = \frac{\beta(u/v) + \gamma}{\delta(u/v) + \epsilon} = \frac{\beta u + \gamma v}{\delta u + \epsilon v},$$

which holds if the equations $u' = \beta u + \gamma v$, $v' = \delta u + \epsilon v$, are satisfied. These equations can be written in the matrix form

$$\begin{pmatrix} u' \\ v' \end{pmatrix} = \begin{pmatrix} \beta & \gamma \\ \delta & \epsilon \end{pmatrix} \begin{pmatrix} u \\ v \end{pmatrix}. \tag{1.20}$$

The complex numbers u and v can be regarded as the components of a complex vector ψ which transforms under rotations according to (1.20); such vectors are called spinors. Writing $\psi = \begin{pmatrix} u \\ v \end{pmatrix}$ and analogously $\psi' = \begin{pmatrix} u' \\ v' \end{pmatrix}$, the "transformation law" (1.20) can be abbreviated as

$$\psi' = Q\psi. \tag{1.21}$$

The fact that the matrices Q are unitary [(1.13)] implies that $\psi^\dagger\psi$ is invariant under rotations, since $(\psi')^\dagger\psi' = (Q\psi)^\dagger Q\psi = \psi^\dagger Q^\dagger Q\psi = \psi^\dagger\psi$.

According to the preceding results, each point of the sphere S^2 corresponds to a complex number ζ or ξ and the latter can be associated with a two-component spinor ψ. Under rotations about the origin, ξ transforms by means of the linear fractional transformation (1.9), while the spinor ψ transforms according to the linear transformation (1.21). Spinors not only can be employed to represent points of space, but they have other applications; they are frequently used in the description of spin-1/2 particles. It will be shown in the following chapters that, starting from the spinors defined above, one can construct higher rank spinors and one obtains an alternative formalism to tensor analysis in three dimensions (Torres del Castillo 1990a, 1992a, 1994a,b).

The unitary matrix Q corresponding to an arbitrary rotation [(1.12)] has the property that, after increasing the rotation angle by 2π, one does not obtain again the matrix Q, but $-Q$. This means that by rotating a spinor through 2π about any axis, the spinor is multiplied by -1 and only after a rotation through 4π does one obtain the original spinor; nevertheless, the matrices Q and $-Q$ produce the same rotation of points of the space [see, *e.g.*, (1.8) and (1.19)]. (A discussion about the relationship between this behavior of spinors and the Pauli exclusion principle can be found in Feynman 1987.)

Roughly speaking, the definition of the notion of spinors on a Riemannian manifold requires the possibility of assigning consistently the change of sign of a spinor under rotations through 2π (see, *e.g.*, Wald 1984, Penrose and Rindler 1984, Lawson and Michelsohn, 1989). It turns out that any orientable three-dimensional manifold admits a spinor structure which, however, may not be unique.

According to (1.7) and (1.19), the components of a spinor ψ, corresponding to a point of the sphere, are given by

$$u = \lambda\, e^{-i\phi/2} \cos \tfrac{1}{2}\theta, \qquad v = \lambda\, e^{i\phi/2} \sin \tfrac{1}{2}\theta, \tag{1.22}$$

where λ is an arbitrary nonzero complex number. On the other hand, from (1.7) and (1.19) we have $\zeta = \overline{u}/\overline{v}$ and, substituting this expression into (1.2), making use of (1.6), it follows that the Cartesian coordinates of the point of the sphere corresponding to the complex number ζ can be written as

$$x = \frac{\overline{u}v + \overline{v}u}{\overline{u}u + \overline{v}v}, \qquad y = \frac{i\overline{v}u - i\overline{u}v}{\overline{u}u + \overline{v}v}, \qquad z = \frac{\overline{u}u - \overline{v}v}{\overline{u}u + \overline{v}v} \tag{1.23}$$

or, equivalently,

$$x_i = \frac{\psi^\dagger \sigma_i \psi}{\psi^\dagger \psi}, \tag{1.24}$$

where $(x_1, x_2, x_3) \equiv (x, y, z)$. If now (x_1, x_2, x_3) are the Cartesian coordinates of any point of the space different from (0,0,0) and $r = \sqrt{x_1^2 + x_2^2 + x_3^2}$ denotes the usual radial coordinate, the point with coordinates x_i/r belongs to the sphere and therefore the coordinates x_i/r can be expressed in the form (1.24)

$$\frac{x_i}{r} = \frac{\psi^\dagger \sigma_i \psi}{\psi^\dagger \psi}. \tag{1.25}$$

Being arbitrary in the choice of the factor λ appearing in (1.22), we find it convenient to impose the condition $|\lambda|^2 = r$, *i.e.*, $\psi^\dagger \psi = r$, which still leaves the phase of λ undetermined. Then from (1.25) we have

$$x_i = \psi^\dagger \sigma_i \psi \tag{1.26}$$

and writing $\lambda = \sqrt{r}\, e^{-i\chi/2}$, where χ is some real number (the factor $-1/2$ is introduced for later convenience), from (1.22) we obtain

$$\psi = \sqrt{r}\, e^{-i\chi/2} \begin{pmatrix} e^{-i\phi/2} \cos \frac{1}{2}\theta \\ e^{i\phi/2} \sin \frac{1}{2}\theta \end{pmatrix}. \tag{1.27}$$

(Substitution of (1.27) into (1.26) yields the standard expression for Cartesian coordinates in terms of spherical coordinates.)

Since $\phi^\dagger \psi = \operatorname{tr} \psi\phi^\dagger$, for any pair of two-component spinors, the relation (1.26) can be written as

$$x_i = \operatorname{tr} \sigma_i \psi\psi^\dagger. \tag{1.28}$$

The product $\psi\psi^\dagger$ is a 2×2 matrix and therefore it can be expressed as a linear combination of the Pauli matrices (1.16) and the unit matrix, which form a basis for the complex 2×2 matrices. Then, writing $\psi\psi^\dagger = a_k\sigma_k + bI$, and making use of (1.17) we see that $\operatorname{tr} \sigma_i \psi\psi^\dagger = \operatorname{tr} \sigma_i (a_k\sigma_k + bI) = 2a_i$, which compared with (1.28) gives $a_i = x_i/2$. Similarly, $\operatorname{tr} \psi\psi^\dagger = 2b$; thus

$$\psi\psi^\dagger = \tfrac{1}{2}x_k\sigma_k + \tfrac{1}{2}(\psi^\dagger\psi)\, I$$

or, equivalently, defining the traceless Hermitian 2×2 matrix

$$P \equiv x_k\sigma_k, \tag{1.29}$$

we obtain

$$P = 2\psi\psi^\dagger - (\psi^\dagger\psi)\, I. \tag{1.30}$$

According to (1.16) and (1.29), the matrix P that corresponds to the point (x, y, z) is explicitly given by

$$P = \begin{pmatrix} z & x - \mathrm{i}y \\ x + \mathrm{i}y & -z \end{pmatrix}. \tag{1.31}$$

Under the rotation represented by a matrix $Q \in \mathrm{SU}(2)$, by (1.30), (1.21), and (1.13), the matrix P transforms into

$$\begin{aligned} P' &= 2Q\psi(Q\psi)^\dagger - [(Q\psi)^\dagger Q\psi]\, I \\ &= 2Q\psi\psi^\dagger Q^\dagger - (\psi^\dagger Q^\dagger Q\psi)I \\ &= 2Q\psi\psi^\dagger Q^\dagger - (\psi^\dagger\psi)I \\ &= Q[2\psi\psi^\dagger - (\psi^\dagger\psi)I]Q^\dagger = QPQ^\dagger. \end{aligned} \tag{1.32}$$

Writing P' in an analogous form to (1.31) with (x', y', z') in place of (x, y, z) and using (1.32), one can obtain the Cartesian coordinates of a point after making any rotation (*cf.* Goldstein 1980).

Relation with quaternions

A quaternion can be defined as a "hypercomplex" number of the form $a + b\,\mathrm{i} + c\,\mathrm{j} + d\,\mathrm{k}$, where a, b, c, and d are real numbers and the units i, j, k, satisfy the relations

$$\mathrm{i}^2 = \mathrm{j}^2 = \mathrm{k}^2 = -1,$$
$$\mathrm{ij} = \mathrm{k} = -\mathrm{ji}, \qquad \mathrm{jk} = \mathrm{i} = -\mathrm{kj}, \qquad \mathrm{ki} = \mathrm{j} = -\mathrm{ik}.$$

As in the case of matrix multiplication, the product of quaternions is associative, is distributive over the sum and is not commutative. The conjugate quaternion of $q = a + b\mathrm{i} + c\mathrm{j} + d\mathrm{k}$ is defined as $\overline{q} = a - b\mathrm{i} - c\mathrm{j} - d\mathrm{k}$. It can be verified that, because of the relations (1.17), the matrices I, $-i\sigma_1$, $-i\sigma_2$, $-i\sigma_3$ satisfy the same multiplication rules as 1, i, j, k, therefore an arbitrary quaternion $a+b\mathrm{i}+c\mathrm{j}+d\mathrm{k}$ can be represented by the matrix $aI - ib\sigma_1 - ic\sigma_2 - id\sigma_3 = \begin{pmatrix} a - id & -c - ib \\ c - ib & a + id \end{pmatrix}$. In this manner, the matrix Q given by (1.12) is associated with the quaternion

$$q = \cos \tfrac{1}{2}\alpha + \sin \tfrac{1}{2}\alpha\,(n_1\mathrm{i} + n_2\mathrm{j} + n_3\mathrm{k}),$$

while $Q^\dagger$ is associated with $\overline{q}$. The condition $QQ^\dagger = I$ [(1.13)] amounts to $q\overline{q} = 1$.

To any point (x, y, z) of the space one can associate the pure quaternion or vector quaternion

$$p \equiv x\mathrm{i} + y\mathrm{j} + z\mathrm{k}$$

[*cf.* (1.29)]. Then, from the previous results it follows that the product

$$p' = qp\overline{q}$$

[*cf.* (1.32)], which turns out to be also a pure quaternion, corresponds to the image of (x, y, z) under the rotation represented by q (see also Misner, Thorne and Wheeler 1973, Penrose and Rindler 1984). (In fact, quaternions were introduced by W.R. Hamilton in 1843 in order to describe rotations in three dimensions.)

The induced SO(3) transformations

The relationship between the coordinates (x', y', z') and (x, y, z) can be given in an explicit form using (1.26), which yields $x'_i = \psi'^\dagger \sigma_i \psi'$; then from (1.21) we obtain

$$x'_i = (Q\psi)^\dagger \sigma_i Q\psi = \psi^\dagger Q^\dagger \sigma_i Q\psi. \tag{1.33}$$

Making use of (1.15) and (1.17), a straightforward computation gives

$$Q^\dagger \sigma_i Q = \cos\alpha\,\sigma_i + (1 - \cos\alpha) n_i n_j \sigma_j - \sin\alpha\,\varepsilon_{ijk} n_k \sigma_j, \tag{1.34}$$

which can be written as

$$Q^\dagger \sigma_i Q = a_{ij}\sigma_j \tag{1.35}$$

with

$$a_{ij} = \cos\alpha\ \delta_{ij} + (1-\cos\alpha)n_i n_j - \sin\alpha\ \varepsilon_{ijk}n_k. \tag{1.36}$$

Substituting (1.35) into (1.33) we find that $x'_i = \psi^\dagger a_{ij}\sigma_j\psi = a_{ij}\psi^\dagger\sigma_j\psi = a_{ij}x_j$, that is

$$x'_i = a_{ij}x_j. \tag{1.37}$$

Thus, $A \equiv (a_{jk})$ is a real 3×3 matrix that represents a rotation about the axis $\mathbf{n}$ through an angle α. Equation (1.37) gives directly the desired relation between (x', y', z') and (x, y, z). By a straightforward computation it can be verified that (1.37) and (1.36) constitute a solution of (1.3) (that is, $\mathrm{d}x'_j/\mathrm{d}\alpha = \varepsilon_{jkm}n_k x'_m$).

The entries of the inverse of the matrix A can be obtained by replacing α by $-\alpha$ in (1.36) and the effect of this substitution is equivalent to interchanging the indices i and j. Therefore

$$A^{-1} = A^{\mathrm{t}}, \tag{1.38}$$

where the superscript t denotes transposition. This means that A is an orthogonal matrix and, since $\det A = 1$, A belongs to the group SO(3) formed by the orthogonal 3×3 matrices with determinant equal to $+1$, where the group operation is the usual matrix product.

Some properties of the matrix A can be derived directly from (1.35), without using the explicit form (1.36). By (1.13) and (1.17), we have

$$\begin{aligned} Q^\dagger\sigma_j Q Q^\dagger \sigma_k Q &= Q^\dagger \sigma_j\sigma_k Q = Q^\dagger(\delta_{jk}I + \mathrm{i}\varepsilon_{jkm}\sigma_m)Q \\ &= \delta_{jk}I + \mathrm{i}\varepsilon_{jkm}Q^\dagger\sigma_m Q, \end{aligned}$$

hence, using (1.35) and (1.17) again,

$$a_{ji}\sigma_i a_{km}\sigma_m = \delta_{jk}I + \mathrm{i}\varepsilon_{jkm}a_{ms}\sigma_s$$

or

$$a_{ji}a_{km}(\delta_{im}I + \mathrm{i}\varepsilon_{ims}\sigma_s) = \delta_{jk}I + \mathrm{i}\varepsilon_{jkm}a_{ms}\sigma_s.$$

Then, the linear independence of $\{I, \sigma_1, \sigma_2, \sigma_3\}$ implies that

$$a_{ji}a_{km}\delta_{im} = \delta_{jk}, \tag{1.39}$$

which means that A is orthogonal, and

$$\varepsilon_{ims}a_{ji}a_{km} = \varepsilon_{jkm}a_{ms}.$$

From this last equation and (1.39) it follows that

$$\varepsilon_{ims} a_{ji} a_{km} a_{ps} = \varepsilon_{jkp}, \tag{1.40}$$

which means that the determinant of A is equal to 1.

Making use of the explicit expression (1.36), we find that

$$\operatorname{tr} A = 1 + 2\cos\alpha, \tag{1.41}$$

and

$$a_{ij}\varepsilon_{ijm} = -2\sin\alpha\, n_m. \tag{1.42}$$

Thus, given a matrix $A \in \mathrm{SO}(3)$, (1.41) and (1.42) allow us to find the angle and the axis of the rotation represented by A, except in the case where the rotation is through 0 or π [*cf.* (1.14) and (1.18)].

Equation (1.42) can be written in a form almost identical to that of (1.18) by defining the three 3×3 pure imaginary matrices S_k, with entries given by $(S_k)_{lm} \equiv -\mathrm{i}\varepsilon_{klm}$. Explicitly,

$$S_1 = \begin{pmatrix} 0 & 0 & 0 \\ 0 & 0 & -\mathrm{i} \\ 0 & \mathrm{i} & 0 \end{pmatrix}, \quad S_2 = \begin{pmatrix} 0 & 0 & \mathrm{i} \\ 0 & 0 & 0 \\ -\mathrm{i} & 0 & 0 \end{pmatrix}, \quad S_3 = \begin{pmatrix} 0 & -\mathrm{i} & 0 \\ \mathrm{i} & 0 & 0 \\ 0 & 0 & 0 \end{pmatrix}.$$

The matrices S_k satisfy the same commutation relations as the matrices $\frac{1}{2}\sigma_k$ (namely, $[S_j, S_k] = \mathrm{i}\varepsilon_{jkm}S_m$) and, just as the Pauli matrices, the matrices S_k are hermitian and have vanishing trace. Then, (1.42) amounts to

$$\operatorname{tr} AS_k = -2\mathrm{i}\sin\alpha\, n_k$$

[*cf.* (1.18)]. It may be noted that, written in terms of Cartesian components, (1.3) is equivalent to $\mathrm{d}x_i/\mathrm{d}\alpha = -\mathrm{i}(n_k S_k)_{ij} x_j$.

The group manifold SU(2), being homeomorphic to S^3, the unit sphere in $\mathbb{R}^4$, is simply connected, while SO(3) is not (see, *e.g.*, Penrose and Rindler 1984, Sattinger and Weaver 1986). The existence of the continuous homomorphism of SU(2) onto SO(3) given by (1.35), which is locally one-to-one, implies that SU(2) is the universal covering group of SO(3).

Apart from the vector $\mathbf{r} = (x_1, x_2, x_3)$ defined by the spinor ψ according to (1.26), there is a complex vector, $\mathbf{M}$, that can be constructed with a spinor ψ, which will allow us to give a geometrical meaning to the factor $e^{-\mathrm{i}\chi/2}$ in (1.27). The Cartesian components of the vector $\mathbf{M}$ will be defined by (Payne 1952, Torres del Castillo 1990a)

$$M_j \equiv \psi^{\mathrm{t}} \varepsilon \sigma_j \psi, \tag{1.43}$$

where

$$\varepsilon \equiv \begin{pmatrix} 0 & 1 \\ -1 & 0 \end{pmatrix}. \tag{1.44}$$

As a consequence of the general formula

$$Q^{-1} = -\frac{1}{\det Q}\, \varepsilon Q^{\mathrm{t}} \varepsilon, \tag{1.45}$$

applicable to any nonsingular 2×2 matrix, and of the fact that $\varepsilon^2 = -I$, any matrix belonging to SU(2) satisfies $Q^{\mathrm{t}}\varepsilon = \varepsilon Q^{-1} = \varepsilon Q^{\dagger}$. Therefore, under the rotation represented by Q, the components of $\mathbf{M}$ transform as

$$M'_j = (Q\psi)^{\mathrm{t}} \varepsilon \sigma_j Q\psi = \psi^{\mathrm{t}} Q^{\mathrm{t}} \varepsilon \sigma_j Q\psi = \psi^{\mathrm{t}} \varepsilon Q^{\dagger} \sigma_j Q\psi,$$

hence, by (1.35) and (1.43),

$$M'_j = \psi^{\mathrm{t}} \varepsilon a_{jk} \sigma_k \psi = a_{jk} M_k, \tag{1.46}$$

as required for any vector [*cf.* (1.37)]. Substituting (1.16), (1.27), and (1.44) into (1.43) one finds that

$$\begin{aligned} \mathbf{M} &= r e^{-\mathrm{i}\chi} [(\cos\theta\, \cos\phi, \cos\theta\, \sin\phi, -\sin\theta) + \mathrm{i}(-\sin\phi, \cos\phi, 0)] \\ &= r e^{-\mathrm{i}\chi} (\mathbf{e}_\theta + \mathrm{i}\mathbf{e}_\phi) \\ &= r[(\cos\chi\, \mathbf{e}_\theta + \sin\chi\, \mathbf{e}_\phi) + \mathrm{i}(-\sin\chi\, \mathbf{e}_\theta + \cos\chi\, \mathbf{e}_\phi)], \end{aligned} \tag{1.47}$$

where $\{\mathbf{e}_r, \mathbf{e}_\theta, \mathbf{e}_\phi\}$ is the orthonormal basis induced by the spherical coordinates. The explicit expression (1.47) shows that the real and imaginary parts of $\mathbf{M}$, $\mathrm{Re}\,\mathbf{M}$ and $\mathrm{Im}\,\mathbf{M}$, are orthogonal to each other and are obtained by rotating the vectors $r\mathbf{e}_\theta$ and $r\mathbf{e}_\phi$ about $\mathbf{e}_r$ through the angle χ. Furthermore, $\{\mathrm{Re}\,\mathbf{M}, \mathrm{Im}\,\mathbf{M}, \mathbf{r}\}$ is an orthogonal set such that $|\mathrm{Re}\,\mathbf{M}| = |\mathrm{Im}\,\mathbf{M}| = |\mathbf{r}| = r = \psi^{\dagger}\psi$.

It may be noticed that the components x_i of the vector $\mathbf{r}$, given by (1.26), can also be written in a form analogous to (1.43). Indeed, since $\varepsilon^{\mathrm{t}} = -\varepsilon$ and $\varepsilon^2 = -I$, from (1.26) we find that

$$x_i = \psi^{\dagger} \sigma_i \psi = \psi^{\dagger} \varepsilon^{\mathrm{t}} \varepsilon \sigma_i \psi = (\varepsilon \overline{\psi})^{\mathrm{t}} \varepsilon \sigma_i \psi, \tag{1.48}$$

which is similar to (1.43), with $\varepsilon\overline{\psi}$ in place of one of the spinors ψ appearing in (1.43). The product $\varepsilon\overline{\psi}$ is a spinor with the same transformation properties as ψ; in fact, with the aid of (1.45), we find that under the transformation (1.21), $\varepsilon\overline{\psi}$ transforms according to

$$\begin{aligned} \varepsilon\overline{\psi} \mapsto \varepsilon(\overline{Q\psi}) &= \varepsilon(Q^{\dagger})^{\mathrm{t}} \overline{\psi} = -(Q^{\dagger}\varepsilon)^{\mathrm{t}} \overline{\psi} = -(Q^{-1}\varepsilon)^{\mathrm{t}} \overline{\psi} = (\varepsilon Q^{\mathrm{t}} \varepsilon\varepsilon)^{\mathrm{t}} \overline{\psi} \\ &= -(\varepsilon Q^{\mathrm{t}})^{\mathrm{t}} \overline{\psi} = Q(\varepsilon\overline{\psi}). \end{aligned} \tag{1.49}$$

The spinor $-\varepsilon\overline{\psi}$ will be referred to as the *mate* of ψ and it will be denoted by $\widehat{\psi}$, that is

$$\widehat{\psi} \equiv -\varepsilon\overline{\psi}. \tag{1.50}$$

The mate of a spinor ψ is also called the conjugate (Cartan 1966, Ch. III) or the adjoint of ψ and is often denoted as $\psi^\dagger$. In order to avoid confusion, we shall continue using $\psi^\dagger$ to denote the conjugate transpose of ψ.

In this manner, (1.48) takes the form $x_i = -\widehat{\psi}^{\mathrm{t}}\varepsilon\sigma_i\psi$. The definition (1.50) implies that $\widehat{\widehat{\psi}} = -\psi$. If $\psi \neq 0$, then the set $\{\psi, \widehat{\psi}\}$ is linearly independent and hence a basis for the two-component spinors. Under the substitution of ψ by $\widehat{\psi}$, the vectors Re **M**, Im **M**, and **r** are replaced by −Re **M**, Im **M**, and −**r**, respectively.

The condition that Re **M** and Im **M** be orthogonal to each other and of the same magnitude is equivalent to

$$\mathbf{M} \cdot \mathbf{M} = 0. \tag{1.51}$$

(A vector satisfying this condition is called null or *isotropic* (Cartan 1966).) It may be verified directly that (1.51) holds by noting that the definition (1.43) gives

$$M_1 = u^2 - v^2, \qquad M_2 = \mathrm{i}(u^2 + v^2), \qquad M_3 = -2uv, \tag{1.52}$$

where u and v are the components of ψ. Making use of the explicit relations (1.52) it can be seen that given a null vector **M**, there exists a spinor ψ, defined up to sign, such that (1.43) holds.

If the two-component spinor ψ is an eigenspinor of $Q \in \mathrm{SU}(2)$, then so is its mate; writing $Q\psi = \mathrm{e}^{-\mathrm{i}\alpha/2}\psi$, for some $\alpha \in \mathbb{R}$, where we have taken into account the fact that the eigenvalues of a unitary matrix have modulus equal to 1, we obtain [see (1.49)]

$$Q\widehat{\psi} = Q(-\varepsilon\overline{\psi}) = -\varepsilon(\overline{Q\psi}) = -\varepsilon\mathrm{e}^{\mathrm{i}\alpha/2}\overline{\psi} = \mathrm{e}^{\mathrm{i}\alpha/2}\widehat{\psi}.$$

Then, if ψ is normalized in the sense that $\psi^\dagger\psi = 1$, we also have $\widehat{\psi}^\dagger\widehat{\psi} = 1$ and since $\psi^\dagger\widehat{\psi} = 0$, we can write

$$Q = \mathrm{e}^{-\mathrm{i}\alpha/2}\psi\psi^\dagger + \mathrm{e}^{\mathrm{i}\alpha/2}\widehat{\psi}\widehat{\psi}^\dagger. \tag{1.53}$$

According to (1.18) the Cartesian components of the axis of the rotation represented by Q are proportional to $\operatorname{tr}\sigma_k Q$. On the other hand, from (1.28) we have $\operatorname{tr}\sigma_k\psi\psi^\dagger = x_k$, where the x_k are the components of the vector **r** associated with ψ, therefore $\operatorname{tr}\sigma_k\widehat{\psi}\widehat{\psi}^\dagger = -x_k$, and

$$\operatorname{tr}\sigma_k Q = \mathrm{e}^{-\mathrm{i}\alpha/2}x_k - \mathrm{e}^{\mathrm{i}\alpha/2}x_k = -2\mathrm{i}\sin\tfrac{1}{2}\alpha\, x_k,$$

showing that Q corresponds to a rotation about **r** through an angle α [*cf.* (1.18)].

Euler angles

As we have seen, any rotation in three-dimensional Euclidean space can be represented by a matrix $Q \in \mathrm{SU}(2)$. In some applications, the rotations are parametrized by Euler angles (see, *e.g.*, Goldstein 1980, Davydov 1988). Following the "y convention", according to the terminology employed in Goldstein (1980), the rotation with Euler angles ϕ, θ, χ is obtained by composing a rotation about the z axis through ϕ, followed by a rotation through θ about the resulting y' axis and by a rotation about the new z'' axis through χ. Thus, if $Q_{\mathbf{n}}(\alpha)$ denotes the SU(2) matrix corresponding to a rotation through the angle α about the axis $\mathbf{n}$ [(1.12)], then the SU(2) matrix $Q(\phi, \theta, \chi)$ that represents the rotation with Euler angles ϕ, θ, χ is the product $Q_{\mathbf{c}}(\chi) Q_{\mathbf{b}}(\theta) Q_{\mathbf{a}}(\phi)$, where $\mathbf{a} = (0, 0, 1) = \mathbf{e}_z$, $\mathbf{b}$ is the image of $\mathbf{e}_y = (0, 1, 0)$ under the rotation defined by $Q_{\mathbf{a}}(\phi)$, and $\mathbf{c}$ is the image of (0,0,1) under the rotation $Q_{\mathbf{b}}(\theta) Q_{\mathbf{a}}(\phi)$.

Since $\mathbf{b}$ is the image of (0,1,0) under the rotation $Q_{\mathbf{a}}(\phi)$, from (1.32) it follows that $b_j \sigma_j = Q_{\mathbf{a}}(\phi) \sigma_2 [Q_{\mathbf{a}}(\phi)]^{\dagger}$, therefore, using (1.15) and (1.13),

$$\begin{aligned} Q_{\mathbf{b}}(\theta) &= \cos \tfrac{1}{2}\theta \; I - \mathrm{i} \sin \tfrac{1}{2}\theta \; Q_{\mathbf{a}}(\phi) \sigma_2 [Q_{\mathbf{a}}(\phi)]^{\dagger} \\ &= Q_{\mathbf{a}}(\phi) \left(\cos \tfrac{1}{2}\theta \; I - \mathrm{i} \sin \tfrac{1}{2}\theta \; \sigma_2 \right) [Q_{\mathbf{a}}(\phi)]^{\dagger} = Q_{\mathbf{a}}(\phi) Q_{\mathbf{e}_y}(\theta) [Q_{\mathbf{a}}(\phi)]^{\dagger}, \end{aligned}$$

and

$$Q_{\mathbf{b}}(\theta) Q_{\mathbf{a}}(\phi) = Q_{\mathbf{a}}(\phi) Q_{\mathbf{e}_y}(\theta) = Q_{\mathbf{e}_z}(\phi) Q_{\mathbf{e}_y}(\theta).$$

Similarly, since $\mathbf{c}$ is the image of (0,0,1) under the rotation $Q_{\mathbf{b}}(\theta) Q_{\mathbf{a}}(\phi)$, we have, $c_j \sigma_j = Q_{\mathbf{b}}(\theta) Q_{\mathbf{a}}(\phi) \sigma_3 [Q_{\mathbf{b}}(\theta) Q_{\mathbf{a}}(\phi)]^{\dagger}$; hence, from (1.15),

$$\begin{aligned} Q_{\mathbf{c}}(\chi) &= \cos \tfrac{1}{2}\chi \; I - \mathrm{i} \sin \tfrac{1}{2}\chi \; Q_{\mathbf{b}}(\theta) Q_{\mathbf{a}}(\phi) \sigma_3 [Q_{\mathbf{b}}(\theta) Q_{\mathbf{a}}(\phi)]^{\dagger} \\ &= Q_{\mathbf{b}}(\theta) Q_{\mathbf{a}}(\phi) \left(\cos \tfrac{1}{2}\chi \; I - \mathrm{i} \sin \tfrac{1}{2}\chi \; \sigma_3 \right) [Q_{\mathbf{b}}(\theta) Q_{\mathbf{a}}(\phi)]^{\dagger} \\ &= Q_{\mathbf{b}}(\theta) Q_{\mathbf{a}}(\phi) Q_{\mathbf{e}_z}(\chi) [Q_{\mathbf{b}}(\theta) Q_{\mathbf{a}}(\phi)]^{\dagger}. \end{aligned}$$

Thus, from the relations derived above and (1.13), one obtains

$$\begin{aligned} Q(\phi, \theta, \chi) &= Q_{\mathbf{c}}(\chi) Q_{\mathbf{b}}(\theta) Q_{\mathbf{a}}(\phi) = Q_{\mathbf{b}}(\theta) Q_{\mathbf{a}}(\phi) Q_{\mathbf{e}_z}(\chi) \\ &= Q_{\mathbf{e}_z}(\phi) Q_{\mathbf{e}_y}(\theta) Q_{\mathbf{e}_z}(\chi) \end{aligned} \tag{1.54}$$

or, in explicit form, making use of (1.12),

$$Q(\phi, \theta, \chi) = \begin{pmatrix} \mathrm{e}^{-\mathrm{i}(\phi+\chi)/2} \cos \tfrac{1}{2}\theta & -\mathrm{e}^{-\mathrm{i}(\phi-\chi)/2} \sin \tfrac{1}{2}\theta \\ \mathrm{e}^{\mathrm{i}(\phi-\chi)/2} \sin \tfrac{1}{2}\theta & \mathrm{e}^{\mathrm{i}(\phi+\chi)/2} \cos \tfrac{1}{2}\theta \end{pmatrix}. \tag{1.55}$$

This last expression shows that the spinor (1.27), with $r = 1$, is given by

$$\psi = Q(\phi, \theta, \chi) \begin{pmatrix} 1 \\ 0 \end{pmatrix}. \tag{1.56}$$

Substituting the spinor $\begin{pmatrix} 1 \\ 0 \end{pmatrix}$ into (1.26) and (1.43) one finds that $\{\mathrm{Re}\,\mathbf{M}, \mathrm{Im}\,\mathbf{M}, \mathbf{r}\} = \{\mathbf{e}_x, \mathbf{e}_y, \mathbf{e}_z\}$. Therefore, from (1.37), (1.46), and (1.56) one concludes that the orthonormal basis $\{\mathrm{Re}\,\mathbf{M}, \mathrm{Im}\,\mathbf{M}, \mathbf{r}\}$ defined by the spinor (1.27) is obtained from $\{\mathbf{e}_x, \mathbf{e}_y, \mathbf{e}_z\}$ by the rotation represented by $Q(\phi, \theta, \chi)$.

For a normalized spinor ψ, there exists a unique matrix $Q \in \mathrm{SU}(2)$ satisfying (1.56). In fact, if $\psi = \begin{pmatrix} u \\ v \end{pmatrix}$, with $|u|^2 + |v|^2 = 1$, the matrix $Q \in \mathrm{SU}(2)$ that satisfies (1.56) is $\begin{pmatrix} u & -\overline{v} \\ v & \overline{u} \end{pmatrix}$. Hence, (1.56) establishes a one-to-one relationship between normalized spinors and matrices belonging to SU(2), in this manner each normalized spinor represents a rotation (but this relationship is two-to-one since the spinors ψ and $-\psi$ represent the same rotation).

Geometrical representation of a spinor

A spinor ψ can be represented geometrically, making use of the vectors $\mathbf{r}$ and $\mathbf{M}$ defined by ψ according to (1.26) and (1.43). For instance, ψ can be represented by a flag (or an ax, Payne 1952); the flagpole is the vector $\mathbf{r}$ and the flag lies in the plane spanned by $\mathbf{r}$ and $\mathrm{Re}\,\mathbf{M}$, pointing in the direction of $\mathrm{Re}\,\mathbf{M}$ (see Fig. 2). However, the spinors ψ and $-\psi$ correspond to the same flag, which is related to the fact that under a rotation through 2π a spinor is transformed into its negative, while any geometrical object representing the spinor is left unchanged. If the

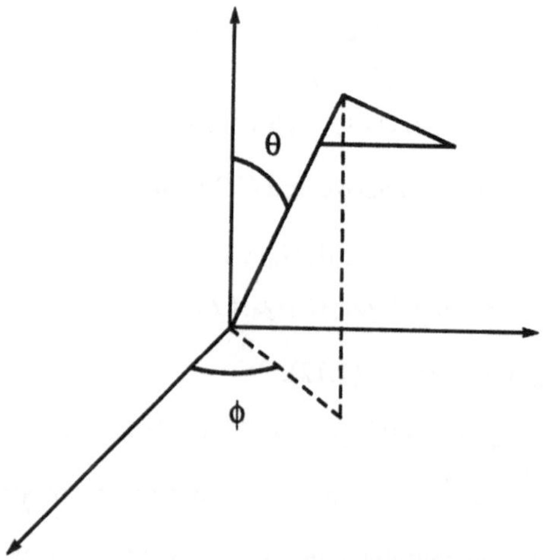

Figure 2: Geometrical representation of a spinor.

components of ψ are parametrized in the form (1.27), θ and ϕ are the usual polar and azimuth angles of the flagpole, r is the length of the flagpole and the flag makes an angle χ with the vector $\mathbf{e}_\theta$.

Equivalently, the spinor ψ can be represented by the (right-handed) orthogonal triad $\{\mathbf{r}, \operatorname{Re}\mathbf{M}, \operatorname{Im}\mathbf{M}\}$ and, again, the spinors ψ and $-\psi$ lead to the same triad.

Spinor indices and connection symbols

In what follows it will be convenient to label the components of a spinor by means of indices $A, B, \ldots$, which take the values 1 and 2 and the summation convention will apply whenever there is a repeated spinor index appearing as a subscript and as a superscript. From (1.45) we see that, for any 2×2 matrix Q with determinant equal to 1, $Q^{\mathrm{t}} \varepsilon Q = \varepsilon$, which is analogous to (1.28); thus, in the same way as the δ_{ij}, which are the components of the metric tensor in Cartesian coordinates, are employed to lower or raise the tensor indices, the spinor indices will be raised or lowered by means of the matrix ε given in (1.44),

$$(\varepsilon_{AB}) = \begin{pmatrix} 0 & 1 \\ -1 & 0 \end{pmatrix} = (\varepsilon^{AB}),$$

following the convention

$$\psi_A = \varepsilon_{AB} \psi^B, \tag{1.57}$$

that is, $\psi_1 = \psi^2$ and $\psi_2 = -\psi^1$. Since $\varepsilon_{AB}\varepsilon^{BC} = -\delta_A^C$, the inverse relation to (1.57) is

$$\psi^A = -\varepsilon^{AB} \psi_B = \varepsilon^{BA} \psi_B. \tag{1.58}$$

It should be remarked that many authors (*e.g.*, Penrose and Rindler 1984) follow the convention according to which $\psi^A = \varepsilon^{AB} \psi_B$. The antisymmetry of ε_{AB} implies that, for any pair of spinors with components ψ^A and ϕ^A,

$$\psi^A \phi_A = -\psi_A \phi^A,$$

since, by (1.57), $\psi^A \phi_A = \psi^A \varepsilon_{AB} \phi^B = -\varepsilon_{BA} \psi^A \phi^B = -\psi_B \phi^B = -\psi_A \phi^A$.

If the entries of the Pauli matrix σ_i are denoted by ${\sigma_i}^A{}_B$ (with the superscript labeling the rows and the subscript labeling the columns), then, following the convention (1.57), the entries of the matrix product $\varepsilon \sigma_i$, which are $\varepsilon_{AB} {\sigma_i}^B{}_C$, will be denoted by σ_{iAC} and the components of $\mathbf{M}$, defined by (1.43), can be written as

$$M_i = \sigma_{iAB} \psi^A \psi^B. \tag{1.59}$$

(The position of the spinor indices of the entries of the Pauli matrices is chosen in such a way that each spinor index appearing as a superscript on the right-hand side of (1.59) is contracted with a spinor index appearing as a subscript.) Similarly, (1.48) amounts to

$$x_i = -\sigma_{iAB}\widehat{\psi}^A\psi^B, \tag{1.60}$$

where, according to (1.50), $\widehat{\psi}^A = -\varepsilon^{AB}\overline{\psi^B}$ (note that, because of (1.21) and (1.13), $\overline{\psi'} = \overline{Q\psi} = (Q^{-1})^{\mathrm{t}}\overline{\psi}$, *i.e.*, the components of $\overline{\psi}$ transform under (1.21) by means of the matrix Q^{-1}, therefore, $\overline{\psi^B}$ transforms as if the spinor index B was a subscript) hence,

$$\widehat{\psi}_A = \overline{\psi^A} \qquad \text{or} \qquad \widehat{\psi}^A = -\overline{\psi_A}. \tag{1.61}$$

The products of ε with the Pauli matrices are

$$\varepsilon\sigma_1 = \begin{pmatrix} 1 & 0 \\ 0 & -1 \end{pmatrix}, \quad \varepsilon\sigma_2 = \begin{pmatrix} \mathrm{i} & 0 \\ 0 & \mathrm{i} \end{pmatrix}, \quad \varepsilon\sigma_3 = \begin{pmatrix} 0 & -1 \\ -1 & 0 \end{pmatrix}, \tag{1.62}$$

which are symmetric matrices,

$$\sigma_{iAB} = \sigma_{iBA}. \tag{1.63}$$

Furthermore, from (1.17) it follows that

$$\sigma_i\sigma_j + \sigma_j\sigma_i = 2\delta_{ij}I, \tag{1.64}$$

hence, $(\varepsilon\sigma_i)\sigma_j + (\varepsilon\sigma_j)\sigma_i = 2\delta_{ij}\varepsilon$ or, equivalently,

$$\sigma_{iAB}\sigma_j{}^B{}_C + \sigma_{jAB}\sigma_i{}^B{}_C = 2\delta_{ij}\varepsilon_{AC}. \tag{1.65}$$

By contracting both sides of this last equation with ε^{AC}, making use of (1.58), we find that

$$\sigma_{iAB}\sigma_j{}^{AB} = -2\delta_{ij}. \tag{1.66}$$

Equation (1.64) means that the matrices σ_i form a representation of the generators of the Clifford algebra of $\mathbb{R}^3$. Another relationship satisfied by the connection symbols σ_{iAB}, which can be regarded as the inverse of (1.66), is

$$\delta^{ij}\sigma_{iAB}\sigma_{jCD} = -(\varepsilon_{AC}\varepsilon_{BD} + \varepsilon_{AD}\varepsilon_{BC}). \tag{1.67}$$

The correctness of (1.67) can be demonstrated by noting that each side is symmetric in the pairs of indices AB and CD and under the interchange of AB with CD, therefore, it suffices to show that (1.67) holds for six independent combinations of the values of the indices. From (1.62) it also follows that, under complex conjugation, the connection symbols σ_{iAB} satisfy

$$\overline{\sigma_{i11}} = -\sigma_{i22}, \qquad \overline{\sigma_{i12}} = \sigma_{i21},$$

which can be expressed as [*cf.* (1.61)]

$$\overline{\sigma_{iAB}} = -\sigma_i{}^{AB}. \tag{1.68}$$

It must be stressed that, in the same manner as the components of a vector with respect to some basis are just a representation of a geometrical object, the components of a spinor correspond to an invariantly defined object. Analogously, the SU(2) matrices, which act on the spinor components, form a concrete representation of the group of linear transformations that preserve the Hermitian (positive definite) inner product between spinors (given by $\langle \phi, \psi \rangle = \overline{\phi^1}\psi^1 + \overline{\phi^2}\psi^2$ in terms of the spinor components with respect to one of the bases considered here). The expression for the components of the mate of a spinor depends on the spinor basis employed; the appropriate expression for an arbitrary basis would be obtained from the relation $\langle \phi, \psi \rangle = \widehat{\phi}_A \psi^A$.

1.3 Elementary applications

Let us consider the motion of a rigid body with a fixed point in the framework of classical mechanics. The configuration of the body at time t can be represented by the matrix $(a_{ij})(t) \in$ SO(3) that corresponds to the rotation leading from the configuration of the body at $t = 0$ to the configuration at time t (hence, $(a_{ij})(0) = I$). According to (1.15), a rotation through an infinitesimal angle $\mathrm{d}\alpha$ about the axis defined by a unit vector $\mathbf{n}$ has the form $I - \frac{1}{2}\mathrm{i}n_j\sigma_j\,\mathrm{d}\alpha$; therefore, if $Q(t)$ is one of the two SU(2) matrices corresponding to $(a_{ij})(t)$,

$$\begin{aligned} Q(t+\mathrm{d}t) &\simeq [I - \tfrac{1}{2}\mathrm{i}n_j(t)\sigma_j\omega(t)\,\mathrm{d}t]Q(t) \\ &= Q(t) - \tfrac{1}{2}\mathrm{i}\omega(t)n_j(t)\sigma_j Q(t)\,\mathrm{d}t, \end{aligned}$$

where $\omega(t)$ is the angular velocity of the rigid body at time t and $n_i(t)$ are the Cartesian components of the instantaneous axis of rotation at time t. Hence, if $\boldsymbol{\omega} \equiv \omega\mathbf{n}$, one finds that

$$\frac{\mathrm{d}Q(t)}{\mathrm{d}t}[Q(t)]^{-1} = -\frac{\mathrm{i}}{2}\omega_j(t)\sigma_j. \tag{1.69}$$

(Note that the left-hand side of the last equation is invariant under the replacement of Q by $-Q$.)

For instance, if $Q(t)$ is parametrized by Euler angles as in (1.54), using (1.34) we have

$$\frac{\mathrm{d}Q(t)}{\mathrm{d}t}[Q(t)]^{-1} = \frac{\mathrm{d}Q_{\mathbf{e}_z}(\phi)}{\mathrm{d}t}[Q_{\mathbf{e}_z}(\phi)]^\dagger + Q_{\mathbf{e}_z}(\phi)\frac{\mathrm{d}Q_{\mathbf{e}_y}(\theta)}{\mathrm{d}t}[Q_{\mathbf{e}_y}(\theta)]^\dagger[Q_{\mathbf{e}_z}(\phi)]^\dagger$$

$$+ Q_{\mathbf{e}_z}(\phi)Q_{\mathbf{e}_y}(\theta)\frac{\mathrm{d}Q_{\mathbf{e}_z}(\chi)}{\mathrm{d}t}[Q_{\mathbf{e}_z}(\chi)]^\dagger[Q_{\mathbf{e}_y}(\theta)]^\dagger[Q_{\mathbf{e}_z}(\phi)]^\dagger$$

$$= -\frac{\mathrm{i}}{2}\left[\sigma_3\frac{\mathrm{d}\phi}{\mathrm{d}t} + (\cos\phi\,\sigma_2 - \sin\phi\,\sigma_1)\frac{\mathrm{d}\theta}{\mathrm{d}t}\right.$$

$$\left. + (\cos\theta\,\sigma_3 + \sin\theta\cos\phi\,\sigma_1 + \sin\theta\sin\phi\,\sigma_2)\frac{\mathrm{d}\chi}{\mathrm{d}t}\right].$$

A comparison of this expression with (1.69) gives the Cartesian components of the angular velocity in terms of the derivatives of the Euler angles.

Spin-1/2 particles

The Pauli matrices were introduced in order to describe the spin of the electron. If $\mathbf{N}$ is a fixed unit vector, the 2×2 matrix $\frac{1}{2}\hbar\mathbf{N}\cdot\boldsymbol{\sigma} = \frac{1}{2}\hbar N_i\sigma_i$ represents the component of the spin along $\mathbf{N}$. Writing $N_i = -\sigma_{iAB}\widehat{o}^A o^B$, where o^A is a spinor such that $o^A\widehat{o}_A = 1$ [see (1.60)], the entries of the matrix $\mathbf{N}\cdot\boldsymbol{\sigma}$ are [see (1.67)]

$$\begin{aligned}(N_i\sigma_i)^A{}_B &= N_i\sigma_i{}^A{}_B = -\sigma_{iCD}\widehat{o}^C o^D\sigma_i{}^A{}_B = \widehat{o}^C o^D(\delta^A_C\varepsilon_{BD} + \delta^A_D\varepsilon_{BC})\\ &= \widehat{o}^A o_B + o^A\widehat{o}_B.\end{aligned}$$

This expression shows that the spinors o^A and $\widehat{o}^A$ are the eigenspinors of $\mathbf{N}\cdot\boldsymbol{\sigma}$, with eigenvalue 1 and -1, respectively. Thus, by expressing a unit vector $\mathbf{N}$ in the form $N_i = -\sigma_{iAB}\widehat{o}^A o^B$, one has at once the eigenspinors of $\mathbf{N}\cdot\boldsymbol{\sigma}$. If the polar and azimuth angles of $\mathbf{N}$ are known, the components o^A are given by (1.22). The Pauli matrices are defined in such a way that the spinor $\begin{pmatrix}1\\0\end{pmatrix}$ and its mate, $\begin{pmatrix}0\\1\end{pmatrix}$, are eigenspinors of σ_3.

For instance, the eigenspinors of $\sigma_1 = \mathbf{e}_x\cdot\boldsymbol{\sigma}$ can be obtained by noting that the direction of the unit vector $\mathbf{e}_x$ has the angles $\theta = \pi/2$, $\phi = 0$; hence, according to (1.22), the (normalized) spinors corresponding to this direction and, therefore, the normalized eigenspinors of σ_1 with eigenvalue 1, are of the form $(\mathrm{e}^{\mathrm{i}\chi/2}/\sqrt{2})\begin{pmatrix}1\\1\end{pmatrix}$. The eigenspinors of σ_1 with eigenvalue -1 can be obtained by finding the mates of those with eigenvalue 1 or, equivalently, by finding the spinors corresponding to the direction $-\mathbf{e}_x$, which has the angles $\theta = \pi/2$, $\phi = \pi$.

The Dirac equation

The Dirac equation is

$$\mathrm{i}\hbar\partial_t\psi = H\psi \equiv -\mathrm{i}\hbar c\boldsymbol{\alpha}\cdot\nabla\psi + Mc^2\beta\psi, \tag{1.70}$$

where ψ is a four-component column and the 4×4 matrices α_i and β satisfy the relations $\alpha_i\alpha_j + \alpha_j\alpha_i = \delta_{ij}I$, $\alpha_i\beta + \beta\alpha_i = 0$, $\beta^2 = I$ (see, *e.g.*, Messiah 1962,

Bjorken and Drell 1964, Schiff 1968). By choosing the matrices α_i and β in the standard form

$$\alpha_i = \begin{pmatrix} 0 & \sigma_i \\ \sigma_i & 0 \end{pmatrix}, \qquad \beta = \begin{pmatrix} I & 0 \\ 0 & -I \end{pmatrix},$$

the four-component column ψ can be expressed as $\psi = \begin{pmatrix} u \\ v \end{pmatrix}$, where u and v are two-component spinors, and the Dirac equation is given by

$$\begin{aligned} \mathrm{i}\hbar\partial_t u &= -\mathrm{i}\hbar c\boldsymbol{\sigma}\cdot\nabla v + Mc^2 u, \\ \mathrm{i}\hbar\partial_t v &= -\mathrm{i}\hbar c\boldsymbol{\sigma}\cdot\nabla u - Mc^2 v. \end{aligned} \tag{1.71}$$

The two-component spinor $\boldsymbol{\sigma}\cdot\nabla u$ transforms under rotations in the same manner as the spinor u and, therefore, the Dirac equation is invariant under rotations (actually, is also invariant under the Lorentz transformations, see, *e.g.*, Rose 1961, Messiah 1962, Bjorken and Drell 1964). In effect, under the rotation defined by a matrix $Q \in \mathrm{SU}(2)$, the Cartesian coordinates of the points of the space transform as $x'_i = a_{ij}x_j$ [see (1.37)] or, equivalently, since (a_{ij}) is orthogonal, $x_i = a_{ji}x'_j$, hence, $\partial'_i = (\partial x_j/\partial x'_i)\partial_j = a_{ij}\partial_j$ and using the fact that $u' = Qu$, (1.13), (1.35), and (1.39) we have

$$\boldsymbol{\sigma}\cdot\nabla' u' = \sigma_i\partial'_i u' = \sigma_i a_{ij}\partial_j Qu = a_{ij}Qa_{ik}\sigma_k\partial_j u = Q\boldsymbol{\sigma}\cdot\nabla u.$$

The Dirac equation admits plane wave solutions

$$u = u_0 \mathrm{e}^{\mathrm{i}(\mathbf{k}\cdot\mathbf{r}-\omega t)}, \qquad v = v_0 \mathrm{e}^{\mathrm{i}(\mathbf{k}\cdot\mathbf{r}-\omega t)}, \tag{1.72}$$

where u_0 and v_0 are constant two-component spinors. Substituting (1.72) into (1.71) one obtains

$$(\mathbf{k}\cdot\boldsymbol{\sigma})v_0 = \frac{E - Mc^2}{\hbar c}u_0, \qquad (\mathbf{k}\cdot\boldsymbol{\sigma})u_0 = \frac{E + Mc^2}{\hbar c}v_0,$$

where $E = \hbar\omega$. Since $(\mathbf{k}\cdot\boldsymbol{\sigma})^2 = |\mathbf{k}|^2 I$, by combining these equations one finds that in order to have a nontrivial solution, $\hbar^2c^2k^2 = E^2 - M^2c^4$. Thus, for a constant two-component spinor u_0 and a real vector $\mathbf{k}$, a plane wave solution of the Dirac equation is given by

$$\begin{pmatrix} u \\ v \end{pmatrix} = \begin{pmatrix} u_0 \\ \frac{\hbar c}{E+Mc^2}(\mathbf{k}\cdot\boldsymbol{\sigma})u_0 \end{pmatrix} \mathrm{e}^{\mathrm{i}(\mathbf{p}\cdot\mathbf{r}-Et)/\hbar}$$

with $E = \pm(M^2c^4 + p^2c^2)^{1/2}$ and $\mathbf{p} = \hbar\mathbf{k}$.

If $\mathbf{k} \neq 0$, then the Cartesian components of $\mathbf{k}$ can be expressed in the form $k_i = -\sigma_{iAB}\widehat{\kappa}^A\kappa^B$, and the spinors κ and $\widehat{\kappa}$ can be used as a basis. Writing $u_0 = a_-\kappa + a_+\widehat{\kappa}$, we have $(\mathbf{k}\cdot\boldsymbol{\sigma})u_0 = k(a_-\kappa - a_+\widehat{\kappa})$. Therefore, the plane wave solutions of the Dirac equation with nonvanishing wave vector $\mathbf{k}$ are given by

$$\begin{pmatrix} u \\ v \end{pmatrix} = \begin{pmatrix} a_-\kappa + a_+\widehat{\kappa} \\ \frac{pc}{E+Mc^2}(a_-\kappa - a_+\widehat{\kappa}) \end{pmatrix} \mathrm{e}^{\mathrm{i}(\mathbf{p}\cdot\mathbf{r}-Et)/\hbar}.$$

The Weyl equation

The Weyl equation for the massless neutrino can be written as

$$\mathrm{i}\boldsymbol{\sigma}\cdot\nabla\psi = \frac{\mathrm{i}}{c}\partial_t\psi, \tag{1.73}$$

where ψ is a two-component spinor field. As shown above, the left-hand side of (1.73) transforms under rotations in the same manner as the spinor ψ and, therefore, the Weyl equation is invariant under rotations (actually, is also invariant under the Lorentz transformations, but this invariance will not be considered here, see, *e.g.*, Rose 1961).

The Weyl equation admits plane wave solutions, *i.e.*, solutions of the form

$$\psi = \psi_0\,\mathrm{e}^{\mathrm{i}(\mathbf{k}\cdot\mathbf{r}-\omega t)}, \tag{1.74}$$

where ψ_0 is a constant two-component spinor, $\mathbf{k}$ is a constant vector and ω is a real constant. Substituting (1.74) into (1.73) one obtains

$$-\mathbf{k}\cdot\boldsymbol{\sigma}\,\psi_0 = \frac{\omega}{c}\psi_0, \tag{1.75}$$

which implies that $\omega/c = |\mathbf{k}| \equiv k$ (assuming $\omega > 0$) and if we express the Cartesian components of $\mathbf{k}$ in the form $k_i = -\sigma_{iAB}\widehat{\kappa}^A\kappa^B$, it follows that ψ_0 is proportional to $\widehat{\kappa}$. The minus sign appearing in (1.75) means that the spin of the neutrino is in the opposite direction to its momentum.

Dynamical symmetries of the two-dimensional harmonic oscillator

The Hamiltonian of a two-dimensional isotropic harmonic oscillator,

$$H = \frac{1}{2m}(p_x^2 + p_y^2) + \frac{m\omega^2}{2}(x^2 + y^2), \tag{1.76}$$

can be written in the form

$$H = \frac{1}{2m}\psi^\dagger\psi = \frac{1}{2m}\psi^A\widehat{\psi}_A$$

with

$$\psi = \begin{pmatrix} \psi^1 \\ \psi^2 \end{pmatrix} = \begin{pmatrix} ip_x + m\omega x \\ ip_y + m\omega y \end{pmatrix}.$$

The Poisson brackets between the components ψ^A and $\widehat{\psi}^A$ are

$$\{\psi^A, \psi^B\} = 0, \qquad \{\widehat{\psi}^A, \widehat{\psi}^B\} = 0, \qquad \{\psi^A, \widehat{\psi}^B\} = -2im\omega\varepsilon^{AB}. \quad (1.77)$$

Since $Q^t\varepsilon Q = \varepsilon$, for any 2×2 matrix Q with determinant equal to 1, the Poisson brackets (1.77) are invariant under the transformations $\psi \mapsto Q\psi$ for $Q \in \mathrm{SU}(2)$ [see also (5.36)]. $\psi^A\widehat{\psi}_A$ is also invariant under these transformations, which are, therefore, canonical and leave the Hamiltonian invariant, *i.e.*, are dynamical symmetries of the two-dimensional isotropic harmonic oscillator.

With each one-parameter group of canonical transformations that leave the Hamiltonian invariant there is associated, at least locally, a constant of the motion. Under the one-parameter group generated by a function G, the rate of change of any function f, defined on the phase space, is given by

$$\frac{df}{ds} = \{f, G\}.$$

The rate of change of ψ^A under the SU(2) transformation $\psi \mapsto Q\psi$, with $Q = \cos\frac{1}{2}\alpha\, I - i\sin\frac{1}{2}\alpha\, n_j\sigma_j$ [see (1.15)] is given by $d\psi/d\alpha = -\frac{1}{2}in_j\sigma_j\psi$, thus

$$\begin{aligned}\frac{d\psi^A}{d\alpha} &= -\frac{i}{2}n_j\sigma_j{}^A{}_C\psi^C = \frac{i}{2}n_j\sigma_{jBC}\varepsilon^{AB}\psi^C = -\frac{1}{4m\omega}n_j\sigma_{jBC}\{\psi^A, \widehat{\psi}^B\}\psi^C \\ &= -\frac{1}{4m\omega}n_j\{\psi^A, \sigma_{jBC}\widehat{\psi}^B\psi^C\},\end{aligned}$$

therefore, the (real-valued) functions

$$S_j \equiv -\frac{1}{4m\omega}\sigma_{jAB}\widehat{\psi}^A\psi^B$$

are constants of the motion that generate the action of SU(2) on the phase space. S_2 is essentially the angular momentum, $S_2 = \frac{1}{2}(xp_y - yp_x)$, and its conservation is a consequence of the rotational symmetry of the Hamiltonian, which is the only obvious symmetry. The Poisson brackets between the generating functions S_i are given by

$$\begin{aligned}\{S_i, S_j\} &= \frac{1}{(4m\omega)^2}\{\sigma_{iAB}\widehat{\psi}^A\psi^B, \sigma_{jCD}\widehat{\psi}^C\psi^D\} \\ &= \frac{1}{(4m\omega)^2}\sigma_{iAB}\sigma_{jCD}2im\omega(\varepsilon^{AD}\psi^B\widehat{\psi}^C - \varepsilon^{BC}\psi^D\widehat{\psi}^A) \\ &= \frac{i}{8m\omega}(\sigma_{iBC}\sigma_j{}^C{}_A - \sigma_{jBC}\sigma_i{}^C{}_A)\widehat{\psi}^A\psi^B\end{aligned}$$

$$= \frac{\mathrm{i}}{8m\omega} 2\mathrm{i}\varepsilon_{ijk}\sigma_{kBA}\widehat{\psi}^A\psi^B$$
$$= \varepsilon_{ijk}S_k.$$

According to our previous results, under the SU(2) transformation, $\psi \mapsto Q\psi$, the functions S_i transform linearly by means of the SO(3) matrix (a_{ij}) given by (1.36).

The two-dimensional isotropic harmonic oscillator is related to the Kepler problem with negative energy in two dimensions in the following manner. The Hamiltonian of the Kepler problem in two dimensions written in Cartesian coordinates is

$$H = \frac{1}{2m}(p_x^2 + p_y^2) - \frac{k}{\sqrt{x^2+y^2}},$$

where k is a constant. In terms of the parabolic coordinates, u, v, defined by $x = \frac{1}{2}(u^2 - v^2)$, $y = uv$, we have

$$H = \frac{1}{u^2+v^2}\left[\frac{1}{2m}(p_u^2 + p_v^2) - 2k\right], \tag{1.78}$$

where p_u and p_v are the canonical momenta conjugate to u and v, respectively ($p_u = up_x + vp_y$, $p_v = -vp_x + up_y$). Hence, the hypersurface in phase space $H = E$ corresponds to $h_{\mathrm{E}} = 2k$, where

$$h_{\mathrm{E}} \equiv \frac{1}{2m}(p_u^2 + p_v^2) - E(u^2+v^2),$$

which is of the form (1.76) with $E = -\frac{1}{2}m\omega^2 (< 0)$. Since the hypersurface $h_{\mathrm{E}} = 2k$ is invariant under the canonical transformations $\psi \mapsto Q\psi$, with $Q \in \mathrm{SU(2)}$ and

$$\psi = \begin{pmatrix} \mathrm{i}p_u + m\omega u \\ \mathrm{i}p_v + m\omega v \end{pmatrix},$$

so is the hypersurface $H = E$. Taking into account that (u, v, p_u, p_v) and $(-u, -v, -p_u, -p_v)$ correspond to the same point (x, y, p_x, p_y), it follows that SO(3) acts on the phase space as a dynamical symmetry group of the two-dimensional Kepler problem with negative energy.

1.4 Spinors in spaces with indefinite metric

In the case of three-dimensional spaces with indefinite metric, we can also define the corresponding spinors starting from geometrical considerations, making use again of the stereographic projection, but this time of the sheet $z \geqslant 1$ of the hyperboloid $x^2 + y^2 - z^2 = -1$ in the space $\mathbb{R}^3$ with the indefinite metric

$$\mathrm{d}x^2 + \mathrm{d}y^2 - \mathrm{d}z^2 \equiv g_{ij}\mathrm{d}x^i\mathrm{d}x^j, \tag{1.79}$$

onto the open disk $|\zeta| < 1$ of the complex plane. The straight line joining the point $(0,0,-1)$ with an arbitrary point $(x,y,z) \in M \equiv \{(x,y,z) \in \mathbb{R}^3 \mid x^2+y^2-z^2 = -1, z \geqslant 1\}$ intersects the xy plane at the point $(x,y,0)/(1+z)$, therefore, the point $(x,y,z) \in M$ can be associated with the complex number

$$\zeta = \frac{x+\mathrm{i}y}{1+z}, \tag{1.80}$$

which satisfies the condition $|\zeta| < 1$. From (1.80) it follows that

$$x = \frac{\zeta+\overline{\zeta}}{1-\zeta\overline{\zeta}}, \qquad y = \frac{\zeta-\overline{\zeta}}{\mathrm{i}(1-\zeta\overline{\zeta})}, \qquad z = \frac{1+\zeta\overline{\zeta}}{1-\zeta\overline{\zeta}}. \tag{1.81}$$

Equation (1.3) can be replaced by

$$\frac{\mathrm{d}x^i}{\mathrm{d}\alpha} = g^{ij}\varepsilon_{jkl}n^k x^l, \tag{1.82}$$

where (g^{ij}) is the inverse of (g_{ij}), hence $(g^{ij}) = \mathrm{diag}(1,1,-1) = (g_{ij})$ and the n^k are the components of a constant (real) vector. (In other words, any Killing vector field of (1.79) is of the form $g^{ij}\varepsilon_{jkl}n^k x^l(\partial/\partial x^i)$; therefore (1.82) gives the one-parameter groups of isometries of the metric (1.79) and M is invariant under these transformations.) Making use of (1.80)–(1.82) one finds that

$$\begin{aligned}\frac{\mathrm{d}\zeta}{\mathrm{d}\alpha} &= -\frac{\mathrm{i}}{2}(n^1-\mathrm{i}n^2)\left[\zeta^2 - \frac{2n^3}{n^1-\mathrm{i}n^2}\zeta + \frac{n^1+\mathrm{i}n^2}{n^1-\mathrm{i}n^2}\right]\\ &= -\frac{\mathrm{i}}{2}(n^1-\mathrm{i}n^2)(\zeta-\zeta_1)(\zeta-\zeta_2),\end{aligned} \tag{1.83}$$

where

$$\zeta_1 \equiv \frac{n^3+\sqrt{-n^k n_k}}{n^1-\mathrm{i}n^2}, \qquad \zeta_2 \equiv \frac{n^3-\sqrt{-n^k n_k}}{n^1-\mathrm{i}n^2} \tag{1.84}$$

and the tensor indices are raised or lowered in the usual way, with the aid of the metric tensor (*e.g.*, $n_k = g_{ki}n^i$).

In the present case the value of $n^k n_k$ can be positive, negative, or zero. When $n^k n_k$ is different from zero, the vector n^k can be normalized in such a way that $n^k n_k$ is equal to 1 or -1; then, from (1.83) and (1.84) we find that ζ transforms by means of the linear fractional transformation

$$\zeta' = \frac{\beta\zeta+\gamma}{\delta\zeta+\epsilon}, \tag{1.85}$$

with the matrix $\begin{pmatrix} \beta & \gamma \\ \delta & \epsilon \end{pmatrix}$ given by

$$\begin{cases} \begin{pmatrix} \cos\frac{1}{2}\alpha + in^3 \sin\frac{1}{2}\alpha & (n^2 - in^1)\sin\frac{1}{2}\alpha \\ (n^2 + in^1)\sin\frac{1}{2}\alpha & \cos\frac{1}{2}\alpha - in^3\sin\frac{1}{2}\alpha \end{pmatrix}, & \text{if } n^k n_k = -1, \\ \begin{pmatrix} 1 + in^3\frac{1}{2}\alpha & (n^2 - in^1)\frac{1}{2}\alpha \\ (n^2 + in^1)\frac{1}{2}\alpha & 1 - in^3\frac{1}{2}\alpha \end{pmatrix}, & \text{if } n^k n_k = 0, \\ \begin{pmatrix} \cosh\frac{1}{2}\alpha + in^3 \sinh\frac{1}{2}\alpha & (n^2 - in^1)\sinh\frac{1}{2}\alpha \\ (n^2 + in^1)\sinh\frac{1}{2}\alpha & \cosh\frac{1}{2}\alpha - in^3\sinh\frac{1}{2}\alpha \end{pmatrix}, & \text{if } n^k n_k = 1. \end{cases} \tag{1.86}$$

It may be noticed that in all cases, $\epsilon = \overline{\beta}$, $\delta = \overline{\gamma}$ and $\beta\epsilon - \gamma\delta = 1$; this means that the matrix $Q \equiv \begin{pmatrix} \beta & \gamma \\ \delta & \epsilon \end{pmatrix}$ satisfies the conditions

$$Q\eta Q^\dagger = \eta, \tag{1.87}$$

where

$$\eta = \begin{pmatrix} 1 & 0 \\ 0 & -1 \end{pmatrix}, \tag{1.88}$$

and $\det Q = 1$; hence, Q belongs to the group SU(1,1). Introducing the matrices

$$\tilde{\sigma}_1 \equiv \begin{pmatrix} 0 & -\mathrm{i} \\ \mathrm{i} & 0 \end{pmatrix}, \quad \tilde{\sigma}_2 \equiv \begin{pmatrix} 0 & 1 \\ 1 & 0 \end{pmatrix}, \quad \tilde{\sigma}_3 \equiv \begin{pmatrix} \mathrm{i} & 0 \\ 0 & -\mathrm{i} \end{pmatrix} \tag{1.89}$$

[*cf.* (1.16)], (1.86) can be written as

$$Q = \begin{cases} \cos\frac{1}{2}\alpha\, I + \sin\frac{1}{2}\alpha\, n^k\tilde{\sigma}_k, & \text{if } n^k n_k = -1, \\ I + \frac{1}{2}\alpha\, n^k\tilde{\sigma}_k, & \text{if } n^k n_k = 0, \\ \cosh\frac{1}{2}\alpha\, I + \sinh\frac{1}{2}\alpha\, n^k\tilde{\sigma}_k, & \text{if } n^k n_k = 1 \end{cases} \tag{1.90}$$

[*cf.* (1.15)]. The matrices (1.89) have vanishing trace and satisfy the relations

$$\tilde{\sigma}_i\eta + \eta\tilde{\sigma}_i^\dagger = 0 \tag{1.91}$$

[which imply that the matrices $\tilde{\sigma}_i$ form a basis for the Lie algebra of SU(1,1)] and

$$\tilde{\sigma}_i\tilde{\sigma}_j = g_{ij}I + \varepsilon_{ijk}\tilde{\sigma}^k. \tag{1.92}$$

Equivalently, we have, $\tilde{\sigma}^i\tilde{\sigma}^j = g^{ij}I + \varepsilon^{ijk}\tilde{\sigma}_k$, where ε^{ijk} is defined by $\varepsilon^{ijk} = g^{il}g^{jm}g^{kn}\varepsilon_{lmn}$. This definition implies that ε^{ijk} is also totally antisymmetric with $\varepsilon^{123} = -1$.

Writing $\zeta = u/v$, where u and v are two complex numbers with $|v| > |u|$, from (1.81) we have

$$x = \frac{u\bar{v} + v\bar{u}}{v\bar{v} - u\bar{u}}, \qquad y = \frac{\mathrm{i}(v\bar{u} - u\bar{v})}{v\bar{v} - u\bar{u}}, \qquad z = \frac{v\bar{v} + u\bar{u}}{v\bar{v} - u\bar{u}}. \tag{1.93}$$

With the aid of the matrices (1.88) and (1.89), these expressions can be written as

$$x^i = -\frac{\mathrm{i}\psi^\dagger \eta \tilde{\sigma}^i \psi}{\psi^\dagger \eta \psi}, \qquad (x^1, x^2, x^3) \in M, \tag{1.94}$$

where

$$\psi \equiv \begin{pmatrix} u \\ v \end{pmatrix}.$$

Then, the transformation (1.85) follows from the linear transformation

$$\psi' = Q\psi. \tag{1.95}$$

If now (x^1, x^2, x^3) is a point such that $x^k x_k < 0$ and $x^3 > 0$, then the point with coordinates $x^i/\sqrt{-x^k x_k}$ belongs to M and according to (1.94), these coordinates can be expressed as

$$\frac{x^i}{\sqrt{-x^k x_k}} = -\frac{\mathrm{i}\psi^\dagger \eta \tilde{\sigma}^i \psi}{\psi^\dagger \eta \psi}.$$

Making use of the ambiguity in the definition of u and v, we can impose the condition $\psi^\dagger \eta \psi = -\sqrt{-x^k x_k}$ [note that, because of (1.87), $\psi^\dagger \eta \psi$ is invariant under the transformations (1.95)], then

$$x^i = \mathrm{i}\psi^\dagger \eta \tilde{\sigma}^i \psi = \mathrm{i}\,\mathrm{tr}\, \tilde{\sigma}^i \psi \psi^\dagger \eta. \tag{1.96}$$

The 2×2 matrix $\psi\psi^\dagger\eta$ can be expressed as a linear combination of the matrices $\tilde{\sigma}^j$ and the unit matrix. Writing $\psi\psi^\dagger\eta = a_j \tilde{\sigma}^j + bI$, from (1.89) and (1.92), we find that $\mathrm{tr}\,\psi\psi^\dagger\eta = \psi^\dagger\eta\psi = 2b$ and $\mathrm{tr}\,\tilde{\sigma}^i \psi\psi^\dagger\eta = \mathrm{tr}\,\tilde{\sigma}^i(a_j\tilde{\sigma}^j + bI) = 2g^{ij}a_j = 2a^i$. Therefore,

$$\psi\psi^\dagger\eta = -\tfrac{1}{2}\mathrm{i}x^j\tilde{\sigma}_j + \tfrac{1}{2}(\psi^\dagger\eta\psi)I \tag{1.97}$$

and by defining the traceless 2×2 matrix

$$P \equiv -\mathrm{i}x^j\tilde{\sigma}_j,$$

we have

$$P = 2\psi\psi^\dagger\eta - (\psi^\dagger\eta\psi)I.$$

Using the fact that for any SU(1,1) matrix Q, $Q^\dagger\eta Q = \eta$, it follows that under a transformation of the form (1.95),

$$P \mapsto QPQ^{-1}.$$

From (1.89) one finds that P is given explicitly by

$$P = \begin{pmatrix} x^3 & -x^1 - \mathrm{i}x^2 \\ x^1 - \mathrm{i}x^2 & -x^3 \end{pmatrix}.$$

Using (1.96) we find that, under the transformation (1.95),

$$\begin{aligned} x'^i &= \mathrm{i}(Q\psi)^\dagger \eta \tilde{\sigma}^i (Q\psi) = \mathrm{i}\psi^\dagger Q^\dagger \eta \tilde{\sigma}^i Q\psi \\ &= \mathrm{i}\psi^\dagger \eta Q^{-1} \tilde{\sigma}^i Q\psi. \end{aligned}$$

Each matrix $\rho^i \equiv Q^{-1}\tilde{\sigma}^i Q$ is traceless and satisfies the condition $\rho^i\eta + \eta\rho^{i\dagger} = 0$ [see (1.91)] (in effect, using (1.87) and (1.91), $\rho^i\eta = Q^{-1}\tilde{\sigma}^i Q\eta = Q^{-1}\tilde{\sigma}^i \eta (Q^{-1})^\dagger = -Q^{-1}\eta\tilde{\sigma}^{i\dagger}(Q^{-1})^\dagger = -\eta Q^\dagger \tilde{\sigma}^{i\dagger}(Q^{-1})^\dagger = -\eta(Q^{-1}\tilde{\sigma}^i Q)^\dagger = -\eta\rho^{i\dagger}$), which implies that ρ^i is a linear combination of the matrices $\tilde{\sigma}^k$ with real coefficients, *i.e.*,

$$Q^{-1}\tilde{\sigma}^i Q = a^i{}_j \tilde{\sigma}^j, \tag{1.98}$$

where $(a^i{}_j)$ is some real 3×3 matrix. It will be shown that the matrix $(a^i{}_j)$ belongs to the group $\mathrm{SO}_0(2,1)$, formed by the 3×3 real matrices $(a^i{}_j)$ such that $\det(a^i{}_j) = 1$, $a^i{}_k a^j{}_l g_{ij} = g_{kl}$ and $a^3{}_3 > 0$. Indeed, from (1.92) one obtains

$$Q^{-1}\tilde{\sigma}^i Q Q^{-1}\tilde{\sigma}^j Q = Q^{-1}\tilde{\sigma}^i\tilde{\sigma}^j Q = Q^{-1}(g^{ij} I + \varepsilon^{ijk} g_{km}\tilde{\sigma}^m)Q,$$

therefore, using (1.98),

$$a^i{}_m \tilde{\sigma}^m a^j{}_k \tilde{\sigma}^k = g^{ij} I + \varepsilon^{ijk} g_{km} a^m{}_l \tilde{\sigma}^l$$

or

$$a^i{}_m a^j{}_k (g^{mk} I + \varepsilon^{mkl} g_{lp}\tilde{\sigma}^p) = g^{ij} I + \varepsilon^{ijk} g_{km} a^m{}_l \tilde{\sigma}^l.$$

Then, using the linear independence of $\{I, \tilde{\sigma}^1, \tilde{\sigma}^2, \tilde{\sigma}^3\}$, we find

$$a^i{}_m a^j{}_k g^{mk} = g^{ij}, \tag{1.99}$$

which is equivalent to $a^i{}_k a^j{}_l g_{ij} = g_{kl}$, and

$$\varepsilon^{mkl} a^i{}_m a^j{}_k g_{lp} = \varepsilon^{ijk} g_{km} a^m{}_p. \tag{1.100}$$

By combining (1.99) and (1.100) we obtain

$$\varepsilon^{mkl} a^i{}_m a^j{}_k a^q{}_l = g^{lp} \varepsilon^{ijk} g_{km} a^m{}_p a^q{}_l = \varepsilon^{ijk} g_{km} g^{mq} = \varepsilon^{ijq},$$

showing that $\det(a^i{}_j) = 1$. Finally, from (1.98) and (1.92), noting that $\tilde{\sigma}_3 = \mathrm{i}\eta$ and using (1.87), we have $a^3{}_3 = \frac{1}{2}\operatorname{tr} Q^{-1}\tilde{\sigma}^3 Q\tilde{\sigma}_3 = \frac{1}{2}\operatorname{tr} Q^{-1}\eta Q\eta = \frac{1}{2}\operatorname{tr} Q^{-1}(Q^{-1})^\dagger > 0$. Hence, for a given matrix $Q \in \mathrm{SU}(1,1)$, (1.98) yields a matrix $(a^i{}_j) \in \mathrm{SO}_0(2, 1)$; this mapping is two-to-one, since Q and $-Q$ give rise to the same matrix $(a^i{}_j)$, and is a group homomorphism. (The group SU(1,1) is not simply connected and therefore is not the universal covering group of $\mathrm{SO}_0(2,1)$.)

For instance, substituting the SU(1,1) matrices (1.90) into (1.98), making use of (1.92), we find that the corresponding $\mathrm{SO}_0(2,1)$ matrices are given by

$$a^i{}_j = \begin{cases} \cos\alpha\,\delta^i_j + (\cos\alpha - 1)\,n^i n_j + \sin\alpha\,\varepsilon^{ikl} n_k g_{lj}, & \text{if } n^k n_k = -1, \\ \delta^i_j - \frac{1}{2}\alpha^2 n^i n_j + \alpha\,\varepsilon^{ikl} n_k g_{lj}, & \text{if } n^k n_k = 0, \\ \cosh\alpha\,\delta^i_j + (1 - \cosh\alpha)\,n^i n_j + \sinh\alpha\,\varepsilon^{ikl} n_k g_{lj}, & \text{if } n^k n_k = 1. \end{cases}$$

(Recall that $\varepsilon^{123} = -1$.) By analogy with (1.48), the expression (1.96) can also be written in the form

$$x^i = \mathrm{i}\psi^\dagger \eta \varepsilon^{\mathrm{t}} \varepsilon \tilde{\sigma}^i \psi = \mathrm{i}(\varepsilon\eta\overline{\psi})^{\mathrm{t}} \varepsilon\tilde{\sigma}^i \psi. \tag{1.101}$$

The product $\varepsilon\eta\overline{\psi}$ transforms in the same manner as ψ; making use of (1.87) and (1.45) we find that if $\psi \mapsto Q\psi$, then

$$\begin{aligned} \varepsilon\eta\overline{\psi} \mapsto \varepsilon\eta\overline{Q\psi} &= \varepsilon\eta(Q^\dagger)^{\mathrm{t}}\overline{\psi} = -(Q^\dagger\eta\varepsilon)^{\mathrm{t}}\overline{\psi} = -(\eta Q^{-1}\varepsilon)^{\mathrm{t}}\overline{\psi} \\ &= (\eta\varepsilon Q^{\mathrm{t}}\varepsilon\varepsilon)^{\mathrm{t}}\overline{\psi} = -(\eta\varepsilon Q^{\mathrm{t}})^{\mathrm{t}}\overline{\psi} = Q(\varepsilon\eta\overline{\psi}). \end{aligned}$$

In the present case the mate of a spinor ψ will be defined by

$$\widehat{\psi} \equiv -\mathrm{i}\varepsilon\eta\overline{\psi}, \tag{1.102}$$

then, (1.101) amounts to

$$x^i = -\widehat{\psi}^{\mathrm{t}}\varepsilon\tilde{\sigma}^i\psi. \tag{1.103}$$

We also have, $\widehat{\psi}^1 = \mathrm{i}\overline{\psi^2}$, $\widehat{\psi}^2 = \mathrm{i}\overline{\psi^1}$, and $\widehat{\widehat{\psi}} = \psi$.

In the same way as the spinors ψ and $\widehat{\psi}$ yield the components of a vector in (1.103), we can form the vector

$$M^i \equiv \psi^{\mathrm{t}}\varepsilon\tilde{\sigma}^i\psi \tag{1.104}$$

[*cf.* (1.43)]. (The components M^i are given explicitly by $M^1 = \mathrm{i}(\psi^1)^2 + \mathrm{i}(\psi^2)^2$, $M^2 = (\psi^1)^2 - (\psi^2)^2$, $M^3 = -2\mathrm{i}\psi^1\psi^2$.) By virtue of (1.45) and (1.98), under the transformation (1.95), the components M^i transform according to $M'^i = a^i{}_j M^j$. The vector M^i is null (*i.e.*, $g_{ij}M^iM^j = 0$) and orthogonal to x^i. It can be shown that $\overline{M^i} = \widehat{\psi}^{\mathrm{t}}\varepsilon\tilde{\sigma}^i\widehat{\psi}$ and that M^i is real if and only if $\widehat{\psi} = \pm\psi$.

The products $\varepsilon\tilde{\sigma}_i$ appearing in (1.101) are given by

$$\varepsilon\tilde{\sigma}_1 = \begin{pmatrix} \mathrm{i} & 0 \\ 0 & \mathrm{i} \end{pmatrix}, \quad \varepsilon\tilde{\sigma}_2 = \begin{pmatrix} 1 & 0 \\ 0 & -1 \end{pmatrix}, \quad \varepsilon\tilde{\sigma}_3 = \begin{pmatrix} 0 & -\mathrm{i} \\ -\mathrm{i} & 0 \end{pmatrix}, \tag{1.105}$$

and denoting by $\tilde{\sigma}_{iAB}$ the components of these matrices [consistent with (1.57)] we find that

$$\tilde{\sigma}_{iAB} = \tilde{\sigma}_{iBA} \tag{1.106}$$

and

$$\overline{\tilde{\sigma}_{iAB}} = -\eta_{AC}\eta_{BD}\tilde{\sigma}_i{}^{CD}, \tag{1.107}$$

which is equivalent to (1.91).

Making use of the connection symbols (1.105), the components (1.103) and (1.104) can be written as

$$x^i = -\tilde{\sigma}^i{}_{AB}\widehat{\psi}^A\psi^B, \qquad M^i = \tilde{\sigma}^i{}_{AB}\psi^A\psi^B, \tag{1.108}$$

where, according to (1.57), (1.58), and (1.102),

$$\widehat{\psi}_A = \mathrm{i}\eta_{AB}\overline{\psi^B} \qquad \text{or} \qquad \widehat{\psi}^A = -\mathrm{i}\eta^{AB}\overline{\psi_B}.$$

Another property of the connection symbols (1.105), which will be useful later, follows from (1.92)

$$\tilde{\sigma}_i\tilde{\sigma}_j + \tilde{\sigma}_j\tilde{\sigma}_i = 2g_{ij}I, \tag{1.109}$$

hence

$$\tilde{\sigma}_{iAB}\tilde{\sigma}_j{}^B{}_C + \tilde{\sigma}_{jAB}\tilde{\sigma}_i{}^B{}_C = 2g_{ij}\varepsilon_{AC}$$

and

$$\tilde{\sigma}_{iAB}\tilde{\sigma}_j{}^{AB} = -2g_{ij} \tag{1.110}$$

[*cf.* (1.64)–(1.66)]. According to (1.109), the matrices $\tilde{\sigma}_i$ form a representation of the generators of the Clifford algebra corresponding to the indefinite metric $(g_{ij}) = \mathrm{diag}(1, 1, -1)$.

As an application of the formalism developed in this section, we shall consider a particle in a repulsive central potential with Hamiltonian

$$H = \frac{1}{2m}(p_x^2 + p_y^2) - \frac{m\omega^2}{2}(x^2 + y^2), \tag{1.111}$$

where m and ω are real constants. As we shall show, in the present case it is convenient to combine the canonical coordinates x, y, p_x, p_y, to form the two-component spinor

$$\psi = \begin{pmatrix} \psi^1 \\ \psi^2 \end{pmatrix} = \begin{pmatrix} p_y + \mathrm{i}p_x \\ m\omega(x + \mathrm{i}y) \end{pmatrix}.$$

Then the Hamiltonian can be written as

$$H = \frac{1}{2m}\psi^\dagger\eta\psi = \frac{\mathrm{i}}{2m}\widehat{\psi}^A\psi_A$$

and the nonvanishing Poisson brackets between the components ψ^A and their conjugates are given by $\{\psi^A, \psi^B\} = -2im\omega\varepsilon^{AB}$. Since $\widehat{\psi}^A\psi_A$ and the Poisson brackets between the components ψ^A and $\widehat{\psi}^A$ are invariant under the SU(1,1) transformations (1.95), these are canonical transformations that leave the Hamiltonian invariant or dynamical symmetries.

In order to find the generating functions of these symmetries we find that the rate of change of ψ^A under the SU(1,1) transformation $\psi \mapsto Q\psi$, with Q given by (1.90), is $\mathrm{d}\psi/\mathrm{d}\alpha = \frac{1}{2}n^k\tilde{\sigma}_k\psi$, thus

$$\begin{aligned}\frac{\mathrm{d}\psi^A}{\mathrm{d}\alpha} &= \frac{1}{2}n^i\tilde{\sigma}_i{}^A{}_B\psi^B = -\frac{1}{2}n^i\tilde{\sigma}_{iCB}\varepsilon^{AC}\psi^B = \frac{1}{4im\omega}n^i\tilde{\sigma}_{iCB}\{\psi^A, \psi^C\}\psi^B \\ &= \frac{1}{8im\omega}n^i\tilde{\sigma}_{iCB}\{\psi^A, \psi^C\psi^B\} \\ &= \frac{1}{4m\omega}n^i\{\psi^A, \operatorname{Im}(\tilde{\sigma}_{iCB}\psi^C\psi^B)\},\end{aligned}$$

which implies that the functions $K_i \equiv (4m\omega)^{-1}\operatorname{Im}(\tilde{\sigma}_{iAB}\psi^A\psi^B)$ are constants of the motion that generate the action of SU(1,1) on the phase space. From

$$\begin{aligned}\{\tilde{\sigma}_{iAB}\psi^A\psi^B, \tilde{\sigma}_{jCD}\psi^C\psi^D\} &= 4\tilde{\sigma}_{iAB}\tilde{\sigma}_{jCD}\psi^A\psi^C\{\psi^B, \psi^D\} \\ &= 8im\omega\tilde{\sigma}_{iAB}\tilde{\sigma}_j{}^B{}_C\psi^A\psi^C \\ &= 8im\omega\varepsilon_{ijk}\tilde{\sigma}^k{}_{AC}\psi^A\psi^C,\end{aligned}$$

it follows that $\{K_i, K_j\} = \varepsilon_{ijm}K^m$. According to our previous results, under the SU(1,1) transformation, $\psi \mapsto Q\psi$, the functions K_i transform linearly by means of the $\mathrm{SO}_0(2,1)$ matrix $(a^i{}_j)$ defined by (1.98).

The Hamiltonian of the Kepler problem in two dimensions written in terms of the parabolic coordinates, u, v, is

$$H = \frac{1}{u^2+v^2}\left[\frac{1}{2m}(p_u^2+p_v^2) - 2k\right]$$

[see (1.78)]. Hence, the hypersurface in phase space $H = E$ corresponds to $h_{\mathrm{E}} = 2k$, where

$$h_{\mathrm{E}} \equiv \frac{1}{2m}(p_u^2+p_v^2) - E(u^2+v^2),$$

which is of the form (1.111) with $E = \frac{1}{2}m\omega^2 (> 0)$. Since the hypersurface $h_{\mathrm{E}} = 2k$ is invariant under the canonical transformations $\psi \mapsto Q\psi$, with $Q \in \mathrm{SU}(1,1)$ and

$$\psi = \begin{pmatrix} p_v + \mathrm{i}p_u \\ m\omega(u+\mathrm{i}v) \end{pmatrix},$$

so is the hypersurface $H = E$. Taking into account that (u, v, p_u, p_v) and $(-u, -v, -p_u, -p_v)$ correspond to the same point (x, y, p_x, p_y), it follows that

$SO_0(2,1)$ acts on the phase space as a dynamical symmetry group of the two-dimensional Kepler problem with positive energy (note that k may be positive or negative).

Alternative definition

Another procedure for defining spinors in a three-dimensional space with indefinite metric, which shows the existence of a homomorphism of $SO_0(2,1)$ with $SL(2,\mathbb{R})$, is obtained by considering the stereographic projection of the circle onto the extended real line. Considering again the space $\mathbb{R}^3$ with the indefinite metric (1.79), we have a "null cone" at the origin given by $x^2 + y^2 - z^2 = 0$. The intersection of this null cone with the plane $z = 1$ is a circle that can be identified with $S^1 = \{(x, y) \in \mathbb{R}^2 \mid x^2 + y^2 = 1\}$.

The points of S^1 can be put into a one-to-one correspondence with the points of the extended real line in the following manner. Any point $(x, y) \in S^1$, different from (1,0), can be joined with (1,0) by means of a straight line that intersects the y axis at some point $(0, \zeta)$. The points of the line through (1,0) and (x, y) are of the form $(1, 0) + t[(x, y) - (1, 0)] = (1 + t(x - 1), ty)$ and, for $t = 1/(1 - x)$, this line intersects the y axis at $(0, y/(1 - x))$; therefore, under this projection the point $(x, y) \in S^1$ corresponds to the *real* number

$$\zeta = \frac{y}{1 - x}.$$

Thus,

$$x = \frac{\zeta^2 - 1}{\zeta^2 + 1}, \qquad y = \frac{2\zeta}{\zeta^2 + 1}. \tag{1.112}$$

Under a rotation through an angle α about the z axis the condition $z = 1$ is preserved and ζ is transformed into

$$\begin{aligned} \zeta' &= \frac{y \cos\alpha + x \sin\alpha}{1 - (x \cos\alpha - y \sin\alpha)} = \frac{2\zeta \cos\alpha + (\zeta^2 - 1) \sin\alpha}{\zeta^2 + 1 - (\zeta^2 - 1) \cos\alpha + 2\zeta \sin\alpha} \\ &= \frac{\zeta \cos\frac{1}{2}\alpha - \sin\frac{1}{2}\alpha}{\zeta \sin\frac{1}{2}\alpha + \cos\frac{1}{2}\alpha}. \end{aligned}$$

This linear fractional transformation can be represented by the (real) 2×2 matrix

$$\begin{pmatrix} \cos\frac{1}{2}\alpha & -\sin\frac{1}{2}\alpha \\ \sin\frac{1}{2}\alpha & \cos\frac{1}{2}\alpha \end{pmatrix}, \tag{1.113}$$

whose determinant is equal to 1.

If ζ is written as u/v, with u and v real, then the linear fractional transformation $\zeta' = (a\zeta + b)/(c\zeta + d)$, where a, b, c, and d are real, follows from the linear transformation

$$\begin{pmatrix} u' \\ v' \end{pmatrix} = \begin{pmatrix} a & b \\ c & d \end{pmatrix} \begin{pmatrix} u \\ v \end{pmatrix} \tag{1.114}$$

and we can assume that the determinant $ad - bc$ is equal to 1. Substituting $\zeta = u/v$ into (1.112) we obtain the expressions

$$x = \frac{u^2 - v^2}{u^2 + v^2}, \qquad y = \frac{2uv}{u^2 + v^2}, \tag{1.115}$$

which duly satisfy the condition $x^2 + y^2 = 1$ since

$$(u^2 - v^2)^2 + (2uv)^2 = (u^2 + v^2)^2. \tag{1.116}$$

Thus, the expressions

$$M_1 = u^2 - v^2, \qquad M_2 = 2uv, \qquad M_3 = u^2 + v^2 \tag{1.117}$$

[*cf.* (1.115)], give the components of a null, real vector: $(M_1)^2 + (M_2)^2 - (M_3)^2 = 0$ with $M_3 \geqslant 0$. Conversely, as in the case of (1.52), given a null vector M_i, such that $M_3 \geqslant 0$, there exists a spinor $\psi = \begin{pmatrix} u \\ v \end{pmatrix}$, defined up to sign, such that (1.117) hold.

It will be shown that all the transformations (1.114), which contain as special cases the rotations represented by (1.113), give rise to isometries of the indefinite metric (1.79).

By analogy with (1.59) and (1.108), the components of the null vector (1.117) can be expressed in the form

$$M_i = s_{iAB}\psi^A\psi^B \tag{1.118}$$

provided that $\psi = \begin{pmatrix} \psi^1 \\ \psi^2 \end{pmatrix}$ and we let

$$(s_{1AB}) = \begin{pmatrix} 1 & 0 \\ 0 & -1 \end{pmatrix}, \qquad (s_{2AB}) = \begin{pmatrix} 0 & 1 \\ 1 & 0 \end{pmatrix}, \qquad (s_{3AB}) = \begin{pmatrix} 1 & 0 \\ 0 & 1 \end{pmatrix}. \tag{1.119}$$

The connection symbols s_{iAB} are real and have the properties

$$s_{iAB} = s_{iBA}$$

and

$$s_{iAB}s_j{}^{AB} = -2g_{ij},$$

with the spinor indices being raised or lowered following the conventions (1.57) and (1.58) [*cf.* (1.63), (1.66), (1.106), and (1.110)].

Raising the first spinor index of s_{iAB}, *i.e.*, $s_i{}^A{}_B = -\varepsilon^{AC} s_{iCB}$, and denoting by s_i the matrix $(s_i{}^A{}_B)$ we obtain

$$s_1 = \begin{pmatrix} 0 & 1 \\ 1 & 0 \end{pmatrix}, \qquad s_2 = \begin{pmatrix} -1 & 0 \\ 0 & 1 \end{pmatrix}, \qquad s_3 = \begin{pmatrix} 0 & -1 \\ 1 & 0 \end{pmatrix}. \tag{1.120}$$

Then (1.118) is equivalent to

$$M_i = \psi^{\mathrm{t}} \varepsilon s_i \psi. \tag{1.121}$$

The matrices (1.120) are real, have vanishing trace and form a basis for the real, traceless 2×2 matrices (this means that the matrices s_i form a basis for the Lie algebra of the group SL(2,$\mathbb{R}$), which consists of the 2×2 real matrices with determinant equal to 1). Hence, if $Q \in \mathrm{SL}(2, \mathbb{R})$ [as the matrix (1.113)], then $Q^{-1} s^i Q$ is real and traceless, therefore

$$Q^{-1} s^i Q = a^i{}_j s^j, \tag{1.122}$$

where $(a^i{}_j)$ is some real 3×3 matrix [*cf.* (1.35) and (1.98)].

The products of the matrices s^i are given by

$$s_i s_j = g_{ij} I + \varepsilon_{ijk} s^k \tag{1.123}$$

and by combining (1.122) and (1.123) one can show directly that the matrix $(a^i{}_j)$ appearing in (1.122) belongs to the group $\mathrm{SO}_0(2,1)$. It is more convenient, however, to notice that (1.123) is identical to (1.92) and, therefore, there exists a matrix, U, with determinant equal to 1, defined up to a sign, such that

$$s_i = U \tilde{\sigma}_i U^{-1}. \tag{1.124}$$

It can be verified that the matrix U can be taken as

$$U = -\tfrac{1}{2} \begin{pmatrix} 1 + \mathrm{i} & -1 - \mathrm{i} \\ 1 - \mathrm{i} & 1 - \mathrm{i} \end{pmatrix}$$

and a direct computation gives

$$U \eta U^{\dagger} = \mathrm{i}\varepsilon. \tag{1.125}$$

If $Q \in \mathrm{SL}(2, \mathbb{R})$, then $\tilde{Q} \equiv U^{-1} Q U \in \mathrm{SU}(1, 1)$. In effect, since Q is real, $Q^{\dagger} = Q^{\mathrm{t}}$, and using (1.125), (1.45), and (1.125) again, we find that

$$\tilde{Q} \eta (\tilde{Q})^{\dagger} = U^{-1} Q U \eta U^{\dagger} Q^{\mathrm{t}} (U^{-1})^{\dagger} = \mathrm{i} U^{-1} Q \varepsilon Q^{\mathrm{t}} (U^{-1})^{\dagger} = \mathrm{i} U^{-1} \varepsilon (U^{-1})^{\dagger} = \eta,$$

i.e., $\tilde{Q}$ satisfies the condition (1.87). The mapping $Q \mapsto U^{-1}QU$ is an isomorphism of SL(2,$\mathbb{R}$) onto SU(1,1) and, substituting (1.124) into (1.122), we obtain

$$(\tilde{Q})^{-1}\tilde{\sigma}^i\tilde{Q} = a^i{}_j\tilde{\sigma}^j.$$

Hence, comparing this last equation with (1.98), we conclude that the matrix $(a^i{}_j)$ appearing in (1.122) belongs to the group $SO_0(2,1)$ and that the matrices $Q \in SL(2, \mathbb{R})$ and $\tilde{Q} \in SU(1, 1)$ give rise to the same $SO_0(2,1)$ matrix, by means of (1.122) and (1.98), respectively.

With any SL(2,$\mathbb{R}$) spinor ψ [*i.e.*, a two-component spinor associated with the connection symbols (1.119)] it is natural to associate the SU(1,1) spinor

$$\tilde{\psi} \equiv U^{-1}\psi, \tag{1.126}$$

so that the transformation $\psi \mapsto Q\psi$ is equivalent to $\tilde{\psi} \mapsto \tilde{Q}\tilde{\psi}$. Then, requiring that the mate of a SL(2,$\mathbb{R}$) spinor, ψ, be equal to the SL(2,$\mathbb{R}$) spinor corresponding to the mate of $\tilde{\psi}$, *i.e.*, $\hat{\tilde{\psi}} = \tilde{\hat{\psi}}$, from (1.102) and (1.126) we have $\hat{\tilde{\psi}} = -\mathrm{i}\varepsilon\eta\overline{(U^{-1}\psi)} = U^{-1}\widehat{\psi}$, hence, $\widehat{\psi} = -\mathrm{i}U\varepsilon\eta\overline{U^{-1}}\,\overline{\psi} = \overline{\psi}$, *i.e.*,

$$\widehat{\psi}^A \equiv \overline{\psi^A}, \tag{1.127}$$

which clearly shows that, in the case of an indefinite metric, $\widehat{\widehat{\psi}} = \psi$. Equation (1.127) makes sense, since the matrix appearing in the spinor transformation (1.114) is real and, therefore, the components ψ^A and their conjugates transform in the same way. If the spinor ψ in (1.118) is complex, then M_i is still null but complex. With a complex spinor ψ, we can also form the vector [analogous to (1.60)]

$$R_i = -s_{iAB}\widehat{\psi}^A\psi^B,$$

which is real and orthogonal to M_i (according to the metric g_{ij}). When ψ is real, *i.e.*, $\widehat{\psi} = \psi$, the vectors $-R_i$ and M_i coincide.

Isometries of the hyperbolic plane

Equations (1.81) allow us to use ζ and $\overline{\zeta}$ as coordinates on the hyperboloid $M = \{(x, y, z) \in \mathbb{R}^3 \mid x^2 + y^2 - z^2 = -1, z \geqslant 1\}$; then the (positive definite) metric induced by (1.79) on M takes the form $4\mathrm{d}\zeta\,\mathrm{d}\overline{\zeta}/(1-\zeta\overline{\zeta})^2$. Since M is mapped onto itself by the $SO_0(2,1)$ transformations, the metric $4\mathrm{d}\zeta\,\mathrm{d}\overline{\zeta}/(1-\zeta\overline{\zeta})^2$ is invariant under the linear fractional transformations (1.85). Taking $4\mathrm{d}\zeta\,\mathrm{d}\overline{\zeta}/(1-\zeta\overline{\zeta})^2$ as the metric of the open disc $D \equiv \{\zeta \in \mathbb{C} \mid |\zeta| < 1\}$, it becomes a space with constant Gaussian curvature equal to -1 and the stereographic projection (1.80) is an isometry. Therefore, the group $SO_0(2,1)$ acts as an isometry group of D

by means of the transformations (1.85), where $\begin{pmatrix} \beta & \gamma \\ \delta & \epsilon \end{pmatrix}$ is one of the SU(1,1) corresponding to a given element of $SO_0(2,1)$.

In the same manner as the complex number ζ is expressed as the quotient of the two components of a spinor $\psi = \begin{pmatrix} u \\ v \end{pmatrix}$, with $\zeta = u/v$, one can consider the quotient, ξ, of the components of the SL(2,$\mathbb{R}$) spinor $U\psi$ [see (1.126)]. Then, the relationship between ζ and ξ is

$$\xi = \mathrm{i}\frac{\zeta - 1}{\zeta + 1} = \mathrm{i}\frac{|\zeta|^2 - 1 + \zeta - \bar{\zeta}}{|\zeta + 1|^2}, \qquad (1.128)$$

which shows that the points of D, where $|\zeta| < 1$, correspond to the points of the lower half-plane $\mathrm{Im}\,\xi < 0$ (similarly, $|\zeta| > 1$, corresponds to $\mathrm{Im}\,\xi > 0$).

From (1.128) one obtains $\zeta = (\mathrm{i} + \xi)/(\mathrm{i} - \xi)$ and it follows that $4\mathrm{d}\zeta\mathrm{d}\bar{\zeta}/(1 - \zeta\bar{\zeta})^2 = \mathrm{d}\xi\mathrm{d}\bar{\xi}/(\mathrm{Im}\,\xi)^2$. Taking $\mathrm{d}\xi\mathrm{d}\bar{\xi}/(\mathrm{Im}\,\xi)^2$ as the metric of the half-plane $\mathrm{Im}\,\xi < 0$, the mapping (1.128) is an isometry and, according to the preceding results, the linear fractional transformations

$$\xi \mapsto \frac{a\xi + b}{c\xi + d},$$

with $a, b, c, d \in \mathbb{R}$ such that $ad - bc = 1$, are isometries of $\mathrm{d}\xi\mathrm{d}\bar{\xi}/(\mathrm{Im}\,\xi)^2$. The half-plane $\mathrm{Im}\,\xi < 0$ with the metric $\mathrm{d}\xi\mathrm{d}\bar{\xi}/(\mathrm{Im}\,\xi)^2$ is also isometric to the upper half-plane $\{(x, y) \in \mathbb{R}^2 \mid y > 0\}$ with the metric $(\mathrm{d}x^2 + \mathrm{d}y^2)/y^2$, known as the hyperbolic plane (see, *e.g.*, Stillwell 1992) or the Poincaré half-plane, which models the Lobachevsky geometry.

2
Spin-Weighted Spherical Harmonics

2.1 Spherical harmonics

The spherical harmonics can be defined in various ways; they are eigenfunctions of the Laplace–Beltrami operator of the sphere and they are the angular part of the separable solutions in spherical coordinates of the Laplace equation in Euclidean space. Another useful characterization is given by the following result.

Proposition. Let x_i be Cartesian coordinates in the n-dimensional Euclidean space, the homogeneous polynomial of degree l,

$$f(x_1, x_2, \ldots, x_n) \equiv d_{ij\ldots k} x_i x_j \cdots x_k, \tag{2.1}$$

where the constant (real or complex) coefficients $d_{ij\ldots k}$ are symmetric in their l indices ($i, j, \ldots = 1, \ldots, n$), is a solution of the Laplace equation, $\partial_i \partial_i f = 0$, if and only if the trace of $d_{ij\ldots k}$ vanishes,

$$d_{iik\ldots m} = 0. \tag{2.2}$$

(Since the coefficients $d_{ij\ldots k}$ are totally symmetric, the trace of $d_{ij\ldots k}$ can be calculated by contracting any pair of indices.)

Proof. Considering the polynomial defined by (2.1) we have

$$\begin{aligned}
\partial_i \partial_i f &= d_{jkm\ldots p} \partial_i \partial_i (x_j x_k x_m \cdots x_p) \\
&= d_{jkm\ldots p} \partial_i (\delta_{ij} x_k x_m \cdots x_p + \cdots + x_j x_k x_m \cdots \delta_{ip}) \\
&= l d_{ikm\ldots p} \partial_i (x_k x_m \cdots x_p) \\
&= l d_{ikm\ldots p} (\delta_{ik} x_m \cdots x_p + \cdots + x_k x_m \cdots \delta_{ip}) \\
&= l(l-1) d_{iim\ldots p} x_m \cdots x_p,
\end{aligned}$$

which shows the validity of the proposition. (Note that for $l = 1$, the coefficients in (2.1) have only one index and therefore the trace is not defined; however, any homogeneous polynomial of degree 1 satisfies the Laplace equation.)

Thus, writing $N_i \equiv x_i/r$, where $r \equiv \sqrt{x_1^2 + \cdots + x_n^2}$, we have

$$d_{ij\ldots k} x_i x_j \cdots x_k = r^l d_{ij\ldots k} N_i N_j \cdots N_k$$

and therefore $d_{ij\ldots k} N_i N_j \cdots N_k$, being the angular part of a solution of the Laplace equation, is a spherical harmonic (of order l) provided that the trace of $d_{ij\ldots k}$ vanishes.

In the specific case of three-dimensional Euclidean space, the components N_i, which correspond to a point of the sphere S^2, can be expressed in the form (1.60)

$$N_i = -\sigma_{iAB} \widehat{o}^A o^B,$$

where o^A are the components of a spinor normalized in such a way that $o^A \widehat{o}_A = 1$. Hence, a spherical harmonic of order l can be written as

$$\begin{aligned} d_{ij\ldots k} N_i N_j \cdots N_k & \\ &= (-1)^l d_{ij\ldots k} \sigma_{iAB} \sigma_{jCD} \cdots \sigma_{kEF} \widehat{o}^A \widehat{o}^C \cdots \widehat{o}^E o^B o^D \cdots o^F \\ &= d_{ABCD\ldots EF} \widehat{o}^A \widehat{o}^C \cdots \widehat{o}^E o^B o^D \cdots o^F, \end{aligned}$$

where we have defined

$$d_{ABCD\ldots EF} \equiv (-1)^l d_{ij\ldots k} \sigma_{iAB} \sigma_{jCD} \cdots \sigma_{kEF}. \tag{2.3}$$

(This definition differs by a constant factor from the definition of the spinor equivalent of a tensor given in Section 5.1.) Owing to the symmetry of the connection symbols, $\sigma_{iAB} = \sigma_{iBA}$ [see (1.63)], the coefficients $d_{ABCD\ldots EF}$ are symmetric in each pair of indices AB, CD, ..., EF, *e.g.*, $d_{ABCD\ldots EF} = d_{BACD\ldots EF}$, and the symmetry of $d_{ij\ldots k}$ implies that $d_{ABCD\ldots EF}$ is symmetric under the interchange of a pair of indices AB, CD, ..., EF, with another of these pairs, *e.g.*, $d_{ABCD\ldots EF} = d_{CDAB\ldots EF}$. It will be shown that the condition (2.2) is equivalent to the symmetry of $d_{ABCD\ldots EF}$ under the interchange of indices belonging to different pairs. First, we note that any difference of the form $M_{AB} - M_{BA}$ vanishes if $A = B$ and changes sign when A and B are interchanged; therefore $M_{AB} - M_{BA}$ is proportional to ε_{AB}. Specifically, $M_{AB} - M_{BA} = (M_{12} - M_{21})\varepsilon_{AB}$, that is,

$$M_{AB} - M_{BA} = \varepsilon^{RS} M_{RS} \varepsilon_{AB}. \tag{2.4}$$

Then, using (2.4), (2.3), the symmetry of $d_{ij\ldots k}$ and (1.65), we have, for instance,

$$\begin{aligned} d_{ABCD\ldots EF} - d_{ACBD\ldots EF} & \\ &= \varepsilon^{RS} d_{ARSD\ldots EF} \varepsilon_{BC} \end{aligned}$$

$$
\begin{aligned}
&= (-1)^l d_{ij\ldots k}\varepsilon^{RS}\sigma_{iAR}\sigma_{jSD}\cdots\sigma_{kEF}\varepsilon_{BC} \\
&= (-1)^{l+1} d_{ij\ldots k}\sigma_{iAS}\sigma_j{}^S{}_D\cdots\sigma_{kEF}\varepsilon_{BC} \\
&= (-1)^{l+1} d_{ij\ldots k}\tfrac{1}{2}(\sigma_{iAS}\sigma_j{}^S{}_D + \sigma_{jAS}\sigma_i{}^S{}_D)\cdots\sigma_{kEF}\varepsilon_{BC} \\
&= (-1)^{l+1} d_{ij\ldots k}\delta_{ij}\varepsilon_{AD}\cdots\sigma_{kEF}\varepsilon_{BC} \\
&= (-1)^{l+1} d_{ii\ldots k}\varepsilon_{AD}\cdots\sigma_{kEF}\varepsilon_{BC}
\end{aligned}
$$

and, therefore, $d_{ABCD\ldots EF} = d_{ACBD\ldots EF}$ if and only if $d_{ii\ldots k} = 0$.

Thus, any spherical harmonic of order l has the expression

$$
d_{AB\ldots CDE\ldots F}\underbrace{o^A o^B \cdots o^C}_{l}\underbrace{\widehat{o}^D\widehat{o}^E\cdots\widehat{o}^F}_{l}, \tag{2.5}
$$

where $d_{AB\ldots F}$ are real or complex constants *totally symmetric* in their $2l$ indices. All the components $d_{AB\ldots F}$ can be expressed in terms of $d_{11\ldots 11}, d_{11\ldots 12}, d_{11\ldots 22}, \ldots, d_{22\ldots 22}$, where the number of indices with the value 2 is $0, 1, 2, \ldots, 2l$, respectively. This shows that there are $2l+1$ linearly independent spherical harmonics of order l (*cf.* Hochstadt 1971). By virtue of the symmetry of the coefficients in (2.5), this expression is also equivalent to $d_{AB\ldots CDE\ldots F}o^{(A}o^B\cdots o^C\widehat{o}^D\widehat{o}^E\cdots\widehat{o}^{F)}$, where the parentheses denote symmetrization on the indices enclosed, *e.g.*, $M^{(AB)} = \frac{1}{2}(M^{AB} + M^{BA})$, $M^{(ABC)} = \frac{1}{6}(M^{ABC} + M^{BCA} + M^{CAB} + M^{ACB} + M^{CBA} + M^{BAC})$.

Writing the components of o in terms of the spherical coordinates, from (1.27), with $r = 1$ and $\chi = 0$, we have

$$
\begin{pmatrix} o^1 \\ o^2 \end{pmatrix} = \begin{pmatrix} \mathrm{e}^{-\mathrm{i}\phi/2}\cos\frac{1}{2}\theta \\ \mathrm{e}^{\mathrm{i}\phi/2}\sin\frac{1}{2}\theta \end{pmatrix}, \qquad \begin{pmatrix} \widehat{o}^1 \\ \widehat{o}^2 \end{pmatrix} = \begin{pmatrix} -\mathrm{e}^{-\mathrm{i}\phi/2}\sin\frac{1}{2}\theta \\ \mathrm{e}^{\mathrm{i}\phi/2}\cos\frac{1}{2}\theta \end{pmatrix}. \tag{2.6}
$$

(Note that these are the columns of the matrix (1.55) with $\chi = 0$.) Then, for instance, any spherical harmonic of order 1 is of the form

$$
\begin{aligned}
&d_{AB}o^A\widehat{o}^B \\
&\quad = d_{11}o^1\widehat{o}^1 + 2d_{12}o^{(1}\widehat{o}^{2)} + d_{22}o^2\widehat{o}^2 \\
&\quad = d_{11}(-\tfrac{1}{2}\mathrm{e}^{-\mathrm{i}\phi}\sin\theta) + d_{12}\cos\theta + d_{22}(\tfrac{1}{2}\mathrm{e}^{\mathrm{i}\phi}\sin\theta) \\
&\quad = d_{11}\left(-\sqrt{\frac{2\pi}{3}}\,Y_{1,-1}\right) + d_{12}\left(\sqrt{\frac{4\pi}{3}}\,Y_{1,0}\right) + d_{22}\left(-\sqrt{\frac{2\pi}{3}}\,Y_{1,1}\right),
\end{aligned}
$$

where we have made use of the standard notation for the spherical harmonics.

Since each component o^1 or $\widehat{o}^1$ contains a factor $\mathrm{e}^{-\mathrm{i}\phi/2}$ and each component o^2 or $\widehat{o}^2$ contains a factor $\mathrm{e}^{\mathrm{i}\phi/2}$, the spherical harmonic of order l, $o^{(A}o^B\cdots o^C\widehat{o}^D\widehat{o}^E\cdots\widehat{o}^{F)}$, is an eigenfunction of the operator $L_z \equiv -\mathrm{i}\partial/\partial\phi$ with eigenvalue $\frac{1}{2}(n_2 - n_1)$, where n_1 [resp. n_2] is the number of indices $AB \ldots F$ taking the value 1 [resp. 2]. The integral numbers n_1 and n_2 satisfy the condition $n_1 + n_2 = 2l$ and therefore n_1 and n_2 must be both even or odd, which implies that $m \equiv \frac{1}{2}(n_2 - n_1)$ is an integer that can take the $2l + 1$ values $-l, -l+1, \ldots, -1, 0, 1, \ldots, l-1, l$.

The integral

$$(f, g) \equiv \int_{S^2} \overline{f} g \, d\Omega = \int_0^{2\pi} \int_0^{\pi} \overline{f(\theta, \phi)} g(\theta, \phi) \sin\theta \, d\theta \, d\phi, \tag{2.7}$$

gives an inner product for the complex-valued functions defined on the sphere. Since $\overline{o^A} = \widehat{o}_A$, the inner product of two spherical harmonics (2.5) leads to integrals of the form

$$\int_{S^2} \underbrace{o^A o^B \cdots o^C}_{n} \underbrace{\widehat{o}_P \widehat{o}_R \cdots \widehat{o}_S}_{n} \, d\Omega. \tag{2.8}$$

By virtue of the invariance under rotations of the solid angle element $d\Omega$ and of the symmetry of the integrand in the indices $A, B, \ldots, C$ and $P, R, \ldots, S$, the integral (2.8) must be of the form

$$\int_{S^2} \underbrace{o^A o^B \cdots o^C}_{n} \underbrace{\widehat{o}_P \widehat{o}_R \cdots \widehat{o}_S}_{n} \, d\Omega = \lambda_{(n)} \delta_P^{(A} \delta_R^B \cdots \delta_S^{C)},$$

where $\lambda_{(n)}$ is some constant. By contracting on a pair of indices, *e.g.*, A and P, we obtain

$$\int_{S^2} o^B \cdots o^C \widehat{o}_R \cdots \widehat{o}_S \, d\Omega = \lambda_{(n)} \frac{n+1}{n} \delta_R^{(B} \cdots \delta_S^{C)},$$

hence, $\lambda_{n-1} = \lambda_{(n)}(n+1)/n$, which means that the product $(n+1)\lambda_{(n)}$ is independent of n; therefore, $(n+1)\lambda_{(n)} = 1\lambda_{(0)} = 4\pi$ and

$$\int_{S^2} \underbrace{o^A o^B \cdots o^C}_{n} \underbrace{\widehat{o}_P \widehat{o}_R \cdots \widehat{o}_S}_{n} \, d\Omega = \frac{4\pi}{n+1} \delta_P^{(A} \delta_R^B \cdots \delta_S^{C)}.$$

Let $d_{A \ldots BC \ldots D} o^A \cdots o^B \widehat{o}^C \cdots \widehat{o}^D$ and $h_{P \ldots RS \ldots T} o^P \cdots o^R \widehat{o}^S \cdots \widehat{o}^T$ be two spherical harmonics of order l and l', respectively, with $d_{A \ldots D}$ and $h_{P \ldots T}$ being completely symmetric. Then, making use of the definition

$$\widehat{d}_{AB \ldots D} \equiv \overline{d^{AB \ldots D}},$$

[*cf.* (1.61)] the inner product of these functions is

$$\begin{aligned}
&(d_{A \ldots BC \ldots D} o^A \cdots o^B \widehat{o}^C \cdots \widehat{o}^D, h_{P \ldots RS \ldots T} o^P \cdots o^R \widehat{o}^S \cdots \widehat{o}^T) \\
&= (-1)^{l'} \widehat{d}^{A \ldots B}{}_{C \ldots D} h_{P \ldots R}{}^{S \ldots T} \int_{S^2} o^C \cdots o^D o^P \cdots o^R \widehat{o}_A \cdots \widehat{o}_B \widehat{o}_S \cdots \widehat{o}_T \, d\Omega \\
&= (-1)^{l'} \frac{4\pi}{l + l' + 1} \widehat{d}^{A \ldots B}{}_{C \ldots D} h_{P \ldots R}{}^{S \ldots T} \delta_A^{(C} \cdots \delta_B^D \delta_S^P \cdots \delta_T^{R)}.
\end{aligned} \tag{2.9}$$

Since $\widehat{d}_{AB...D}$ and $h_{PR...T}$ are completely symmetric, their contractions vanish, $\widehat{d}^B{}_{B...D} = 0$, $h^R{}_{R...T} = 0$ [see (2.4)], therefore, if $l' \neq l$, the last expression in (2.9) is equal to zero, showing the well-known fact that two spherical harmonics of different orders are orthogonal to each other. Thus, assuming that $l' = l$, the right-hand side of (2.9) amounts to

$$\begin{aligned}
&(-1)^l \frac{4\pi}{2l+1} \frac{l!\,l!}{(2l)!} \widehat{d}^{A...B}{}_{C...D}\, h_{A...B}{}^{C...D} \\
&= \frac{4\pi}{2l+1} \frac{l!\,l!}{(2l)!} \widehat{d}_{A...BC...D}\, h^{A...BC...D} \\
&= \frac{4\pi}{2l+1} \frac{l!\,l!}{(2l)!} \sum_{m=-l}^{l} (-1)^{l+m} \frac{(2l)!}{(l+m)!(l-m)!} \widehat{d}_{\underbrace{1...1}_{l+m}\underbrace{2...2}_{l-m}}\, h_{\underbrace{2...2}_{l+m}\underbrace{1...1}_{l-m}} \\
&= \frac{4\pi}{2l+1} (l!)^2 \sum_{m=-l}^{l} \frac{1}{(l+m)!(l-m)!} \overline{d_{\underbrace{1...1}_{l-m}\underbrace{2...2}_{l+m}}}\, h_{\underbrace{1...1}_{l-m}\underbrace{2...2}_{l+m}}.
\end{aligned}$$

This last expression shows that

$$\sqrt{\frac{4\pi}{2l+1} \frac{1}{(l+m)!(l-m)!}}\, l!\, d_{\underbrace{1...1}_{l-m}\underbrace{2...2}_{l+m}} \tag{2.10}$$

are components of the spherical harmonics $d_{A...BC...D}o^A \cdots o^B \widehat{o}^C \cdots \widehat{o}^D$ with respect to an orthonormal basis. Thus, expressing the spherical harmonic $d_{A...BC...D}o^A \cdots o^B \widehat{o}^C \cdots \widehat{o}^D$ in terms of the components (2.10) we have

$$\begin{aligned}
&d_{A...BC...D}o^A \cdots o^B \widehat{o}^C \cdots \widehat{o}^D \\
&= \sum_{m=-l}^{l} (-1)^m \sqrt{\frac{4\pi}{2l+1} \frac{1}{(l+m)!)l-m)!}}\, l!\, d_{\underbrace{1...1}_{l-m}\underbrace{2...2}_{l+m}} \\
&\quad \times (-1)^m \frac{(2l)!}{l!} \sqrt{\frac{2l+1}{4\pi} \frac{1}{(l+m)!(l-m)!}}\, \overbrace{o^{(1} \cdots o^1 \widehat{o}^1 \cdots \widehat{o}^{2)}}^{(l-m)\ 1\text{'s},\ (l+m)\ 2\text{'s}},
\end{aligned}$$

where the factors $(-1)^m$ have been introduced in order to get agreement with the convention employed in quantum mechanics. Thus, the symmetrized products $o^{(A}o^B \cdots o^C \widehat{o}^D \widehat{o}^E \cdots \widehat{o}^{F)}$ are related to the (normalized) spherical harmonics, Y_{lm}, by

$$Y_{lm} = (-1)^m \frac{(2l)!}{l!} \sqrt{\frac{2l+1}{4\pi} \frac{1}{(l+m)!(l-m)!}}\, \overbrace{o^{(1} \cdots o^1 \widehat{o}^1 \cdots \widehat{o}^{2)}}^{(l-m)\ 1\text{'s},\ (l+m)\ 2\text{'s}}. \tag{2.11}$$

This last expression is equivalent to

$$\begin{aligned} Y_{lm} &= (-1)^m \frac{(2l)!}{l!}\sqrt{\frac{2l+1}{4\pi}\frac{1}{(l+m)!(l-m)!}} \\ &\quad \times \frac{1}{\binom{2l}{l}} \sum_{k=0}^{l} \binom{l-m}{k}\binom{l+m}{l-k} (\widehat{o}^1)^k (\widehat{o}^2)^{l-k} (o^1)^{l-m-k} (o^2)^{m+k} \\ &= (-1)^m l! \sqrt{\frac{2l+1}{4\pi}(l+m)!(l-m)!} \\ &\quad \times \sum_{k=0}^{l} \frac{(-1)^k (\sin\frac{1}{2}\theta)^{m+2k} (\cos\frac{1}{2}\theta)^{2l-m-2k}}{k!(l-m-k)!(l-k)!(m+k)!} e^{im\phi}, \end{aligned} \tag{2.12}$$

where we have made use of (2.6).

2.2 Spin weight

The transformation

$$o \mapsto o' = e^{i\alpha/2} o, \tag{2.13}$$

with α real, leaves the point of S^2 with coordinates $N_i = -\sigma_{iAB}\widehat{o}^A o^B$ invariant, but produces a rotation through α of the vectors $\operatorname{Re}\mathbf{M}$ and $\operatorname{Im}\mathbf{M}$, which form an orthonormal basis of the tangent plane to the sphere at the point N_i, with $M_i = \sigma_{iAB} o^A o^B$. A quantity η has spin weight s if under the transformation (2.13), it transforms as (Newman and Penrose 1966)

$$\eta' = e^{is\alpha}\eta. \tag{2.14}$$

Thus, by definition, the components o^A have spin weight 1/2. If η has spin weight s, its complex conjugate, $\overline{\eta}$, has spin weight $-s$. The product of two quantities with spin weights s and s' has spin weight $s+s'$. From (1.61) it follows that the components $\widehat{o}^A$ have spin weight $-1/2$.

The expression (2.5) is invariant under the transformation (2.13) and therefore the spherical harmonics have spin weight 0. The spherical harmonics can be generalized by considering functions of the form (2.5) where the number of factors $o^A o^B \cdots o^C$ does not coincide with the number of factors $\widehat{o}^D \widehat{o}^E \cdots \widehat{o}^F$. An expression of the form

$$d_{AB\ldots CDE\ldots F}\, \underbrace{o^A o^B \cdots o^C}_{j+s} \underbrace{\widehat{o}^D \widehat{o}^E \cdots \widehat{o}^F}_{j-s}, \tag{2.15}$$

where the constant coefficients $d_{AB...F}$ are totally symmetric in their $2j$ indices ($j = 0, 1/2, 1, 3/2, \ldots$), has spin weight s and will be called spin-weighted spherical harmonic of order j and spin weight s (see also Penrose and Rindler 1984). Since $j + s$ and $j - s$ must be nonnegative integers, it follows that

$$|s| \leqslant j \tag{2.16}$$

and that j and s are both integers or "half-integers." A spin-weighted spherical harmonic of spin weight 0 is an ordinary spherical harmonic (and, necessarily, its order is integral). Making use of (2.15), given j and s, the spin-weighted spherical harmonics can be easily constructed; for instance, the spin-weighted spherical harmonics of order 1 and spin weight 1 are of the form

$$\begin{aligned} d_{AB}o^A o^B &= d_{11}(\mathrm{e}^{-\mathrm{i}\phi/2}\cos\tfrac{1}{2}\theta)^2 + 2d_{12}\sin\tfrac{1}{2}\theta\cos\tfrac{1}{2}\theta + d_{22}(\mathrm{e}^{\mathrm{i}\phi/2}\sin\tfrac{1}{2}\theta)^2 \\ &= \tfrac{1}{2}d_{11}\mathrm{e}^{-\mathrm{i}\phi}(1+\cos\theta) + d_{12}\sin\theta + \tfrac{1}{2}d_{22}\mathrm{e}^{\mathrm{i}\phi}(1-\cos\theta), \end{aligned}$$

where d_{11}, d_{12}, and d_{22} are arbitrary constants. As in the case of the spherical harmonics (2.5), the spin-weighted spherical harmonic (2.15) is an eigenfunction of $-\mathrm{i}\partial_\phi$ with eigenvalue $m = \frac{1}{2}(n_2 - n_1)$, where n_1 [resp. n_2] is the number of superscripts $A, B, \ldots, F$ taking the value 1 [resp. 2]. Then, m can take the $2j+1$ values $-j, -j+1, \ldots, j$, and both j and m take integral or half-integral values.

The derivatives of a quantity with a given spin weight may not have a well-defined spin weight. However, the operators $\eth$ ("eth") and $\bar{\eth}$ ("eth bar") defined below produce quantities with a well-defined spin weight when applied to a quantity with a definite spin weight. If η has spin weight s, $\eth\eta$ and $\bar{\eth}\eta$ are defined by (Newman and Penrose 1966)

$$\begin{aligned} \eth\eta &= -\left(\partial_\theta + \frac{\mathrm{i}}{\sin\theta}\partial_\phi - s\cot\theta\right)\eta = -\sin^s\theta\left(\partial_\theta + \frac{\mathrm{i}}{\sin\theta}\partial_\phi\right)(\eta\sin^{-s}\theta), \\ \bar{\eth}\eta &= -\left(\partial_\theta - \frac{\mathrm{i}}{\sin\theta}\partial_\phi + s\cot\theta\right)\eta = -\sin^{-s}\theta\left(\partial_\theta - \frac{\mathrm{i}}{\sin\theta}\partial_\phi\right)(\eta\sin^s\theta), \end{aligned} \tag{2.17}$$

then $\eth\eta$ has spin weight $s+1$ and $\bar{\eth}\eta$ has spin weight $s-1$. Furthermore, $\overline{\eth\eta} = \bar{\eth}\,\bar{\eta}$ and $\eth(\eta\kappa) = \eta\eth\kappa + \kappa\eth\eta$, $\bar{\eth}(\eta\kappa) = \eta\bar{\eth}\kappa + \kappa\bar{\eth}\eta$. It will be shown that the operators $\eth$ and $\bar{\eth}$ arise in a natural way when an expression involving derivatives of vector or spinor fields is written in terms of spin-weighted combinations of the field components [see, *e.g.*, (3.5)–(3.8)].

By means of a direct computation, using (2.6) and (2.17), taking into account that o^A and $\widehat{o}^A$ have spin weight 1/2 and $-1/2$, respectively, we obtain

$$\eth o^A = 0, \qquad \eth\widehat{o}^A = o^A, \tag{2.18}$$

and

$$\bar{\eth} o^A = -\hat{o}^A, \qquad \bar{\eth}\hat{o}^A = 0. \tag{2.19}$$

These relations imply that $\eth$ or $\bar{\eth}$ applied to a spin-weighted spherical harmonic yields another spin-weighted spherical harmonic. In effect, using the fact that $d_{AB\ldots F}$ are totally symmetric,

$$\begin{aligned}
&\eth(d_{AB\ldots CDE\ldots F}\underbrace{o^A o^B \cdots o^C}_{j+s}\underbrace{\hat{o}^D\hat{o}^E\cdots\hat{o}^F}_{j-s}) \\
&\quad = d_{AB\ldots CDE\ldots F}o^A o^B \cdots o^C \eth(\hat{o}^D\hat{o}^E\cdots\hat{o}^F) \\
&\quad = d_{AB\ldots CDE\ldots F}o^A o^B \cdots o^C (o^D\hat{o}^E\cdots\hat{o}^F + \hat{o}^D o^E \cdots \hat{o}^F + \cdots \\
&\qquad\qquad + \hat{o}^D\hat{o}^E\cdots o^F) \\
&\quad = (j-s)d_{AB\ldots CDE\ldots F}\underbrace{o^A o^B \cdots o^C o^D}_{j+s+1}\underbrace{\hat{o}^E\cdots\hat{o}^F}_{j-s-1}
\end{aligned} \tag{2.20}$$

and

$$\begin{aligned}
&\bar{\eth}(d_{AB\ldots CDE\ldots F}\underbrace{o^A o^B \cdots o^C}_{j+s}\underbrace{\hat{o}^D\hat{o}^E\cdots\hat{o}^F}_{j-s}) \\
&\quad = d_{AB\ldots CDE\ldots F}\,\bar{\eth}(o^A o^B \cdots o^C)\hat{o}^D\hat{o}^E\cdots\hat{o}^F \\
&\quad = d_{AB\ldots CDE\ldots F}(-\hat{o}^A o^B \cdots o^C - o^A\hat{o}^B\cdots\hat{o}^C - \cdots - o^A o^B \cdots \hat{o}^C) \\
&\qquad \times \hat{o}^D\hat{o}^E\cdots\hat{o}^F \\
&\quad = -(j+s)d_{AB\ldots CDE\ldots F}\underbrace{o^A o^B \cdots}_{j+s-1}\underbrace{\hat{o}^C\hat{o}^D\hat{o}^E\cdots\hat{o}^F}_{j-s+1},
\end{aligned} \tag{2.21}$$

i.e., apart from a constant factor, the effect of $\eth$ or $\bar{\eth}$ on a spin-weighted spherical harmonic is to replace a factor $\hat{o}$ by a factor o or vice versa. Thus, if $j \neq \pm s$, after applying $\eth$ and $\bar{\eth}$, in any order, to a spin-weighted spherical harmonic, the result is a multiple of the same spin-weighted spherical harmonic. If ${}_s\mathcal{P}_j$ denotes a spin-weighted spherical harmonic of order j and spin weight s, from (2.20) and (2.21) it follows that

$$\begin{aligned}
\bar{\eth}\eth\,{}_s\mathcal{P}_j &= -(j+s+1)(j-s)\,{}_s\mathcal{P}_j = [s(s+1) - j(j+1)]\,{}_s\mathcal{P}_j, \\
\eth\bar{\eth}\,{}_s\mathcal{P}_j &= -(j-s+1)(j+s)\,{}_s\mathcal{P}_j = [s(s-1) - j(j+1)]\,{}_s\mathcal{P}_j,
\end{aligned} \tag{2.22}$$

showing that the spin-weighted spherical harmonics are eigenfunctions of $\bar{\eth}\eth$ and of $\eth\bar{\eth}$.

A direct computation, using (2.17), shows that if η has spin weight s,

$$\bar{\eth}\eth\eta = \left(\frac{1}{\sin\theta}\partial_\theta \sin\theta\, \partial_\theta + \frac{1}{\sin^2\theta}\partial_\phi^2 + \frac{2is\cos\theta}{\sin^2\theta}\partial_\phi - \frac{s^2}{\sin^2\theta} + s(s+1)\right)\eta,$$
$$\eth\bar{\eth}\eta = \left(\frac{1}{\sin\theta}\partial_\theta \sin\theta\, \partial_\theta + \frac{1}{\sin^2\theta}\partial_\phi^2 + \frac{2is\cos\theta}{\sin^2\theta}\partial_\phi - \frac{s^2}{\sin^2\theta} + s(s-1)\right)\eta, \tag{2.23}$$

therefore,

$$(\bar{\eth}\eth - \eth\bar{\eth})\eta = 2s\eta \tag{2.24}$$

and, if f has spin weight 0,

$$\bar{\eth}\eth f = \eth\bar{\eth} f = -L^2 f, \tag{2.25}$$

where $L^2 \equiv (-i\mathbf{r} \times \nabla)^2$. (Note that by combining (2.22) and (2.25) it follows that the ordinary spherical harmonics of order j are eigenfunctions of L^2 with eigenvalue $j(j+1)$.)

By analogy with the ordinary spherical harmonics of order l, Y_{lm}, which are eigenfunctions of $-i\partial_\phi$ with eigenvalue m, normalized with respect to the inner product (2.7), ${}_sY_{jm}$ will denote a normalized spin-weighted spherical harmonic of order j and spin weight s that is an eigenfunction of $-i\partial_\phi$ with eigenvalue m. Since ∂_ϕ commutes with $\eth$ and $\bar{\eth}$, $\eth\, {}_sY_{jm}$ and $\bar{\eth}\, {}_sY_{jm}$ must be proportional to ${}_{s+1}Y_{jm}$ and ${}_{s-1}Y_{jm}$, respectively. Writing $\eth\, {}_sY_{jm} = C(j,s)\, {}_{s+1}Y_{jm}$ and $\bar{\eth}\, {}_sY_{jm} = D(j,s)\, {}_{s-1}Y_{jm}$, where $C(j,s)$ and $D(j,s)$ are some constants to be determined, and making use of the fact that if f and g are functions with spin weight s and $s-1$, respectively (so that the spin weight of $\overline{f}\eth g$ is equal to 0),

$$(f, \eth g) = -(\bar{\eth} f, g), \tag{2.26}$$

and (2.22) we have

$$|C(j,s)|^2 = (\eth\, {}_sY_{jm}, \eth\, {}_sY_{jm}) = -(\bar{\eth}\eth\, {}_sY_{jm}, {}_sY_{jm}) = j(j+1) - s(s+1)$$

and

$$\bar{\eth}\eth\, {}_sY_{jm} = C(j,s)\bar{\eth}\, {}_{s+1}Y_{jm} = C(j,s)D(j,s+1)\, {}_sY_{jm},$$

which must coincide with $[s(s+1) - j(j+1)]\, {}_sY_{jm}$. Therefore, choosing the phase of $C(j,s)$ in such a way that $C(j,s) = [j(j+1) - s(s+1)]^{1/2}$, we obtain, $D(j,s) = -[j(j+1) - s(s-1)]^{1/2}$, *i.e.*,

$$\eth\, {}_sY_{jm} = [j(j+1) - s(s+1)]^{1/2}\, {}_{s+1}Y_{jm},$$
$$\bar{\eth}\, {}_sY_{jm} = -[j(j+1) - s(s-1)]^{1/2}\, {}_{s-1}Y_{jm}. \tag{2.27}$$

In particular, if s is an integer (and, hence, j is also an integer), taking ${}_0Y_{jm} = Y_{jm}$, it follows that

$$
{}_sY_{jm} = \begin{cases} \left[\dfrac{(j-s)!}{(j+s)!}\right]^{1/2} \eth^s \, Y_{jm}, & \text{if } 0 \leqslant s \leqslant j, \\[2ex] (-1)^s \left[\dfrac{(j+s)!}{(j-s)!}\right]^{1/2} \bar{\eth}^{-s} \, Y_{jm}, & \text{if } -j \leqslant s \leqslant 0. \end{cases} \tag{2.28}
$$

Thus, making use of (2.18), (2.20), and (2.21) we find that, for j and s integral,

$$
\begin{aligned}
{}_sY_{jm} &= (-1)^m (2j)! \sqrt{\frac{2j+1}{4\pi} \frac{1}{(j+m)!(j-m)!} \frac{1}{(j+s)!(j-s)!}} \\
&\quad \times \overbrace{\underbrace{o^{(1} o^1 \cdots o^1}_{j+s} \underbrace{\widehat{o}^1 \widehat{o}^2 \cdots \widehat{o}^{2)}}_{j-s}}^{(j-m)\ 1\text{'s},\ (j+m)\ 2\text{'s}} .
\end{aligned} \tag{2.29}
$$

With one slight modification in the derivation given in the previous section, it can be shown directly that, for each value of s (integral or half-integral), the functions (2.29) form an orthonormal set. (However, two spin-weighted spherical harmonics of different spin weight need not be orthogonal to each other.)

Expression (2.29) can also be written in the form

$$
\begin{aligned}
{}_sY_{jm} &= (-1)^m (2j)! \sqrt{\frac{2j+1}{4\pi} \frac{1}{(j+m)!(j-m)!} \frac{1}{(j+s)!(j-s)!}} \\
&\quad \times \frac{1}{\binom{2j}{j-s}} \sum_{k=0}^{j-s} \binom{j-m}{k} \binom{j+m}{j-s-k} (\widehat{o}^1)^k (\widehat{o}^2)^{j-s-k} (o^1)^{j-m-k} (o^2)^{m+s+k} \\
&= (-1)^m \sqrt{\frac{2j+1}{4\pi} (j+m)!(j-m)!(j+s)!(j-s)!} \\
&\quad \times \sum_{k=0}^{j-s} \frac{(-1)^k (\sin \frac{1}{2}\theta)^{m+s+2k} (\cos \frac{1}{2}\theta)^{2j-m-s-2k}}{k!(j-m-k)!(j-s-k)!(m+s+k)!} \mathrm{e}^{im\phi}
\end{aligned} \tag{2.30}
$$

(note that in the nonvanishing terms contained in these sums, k ranges from $\max\{0, -m-s\}$ to $\min\{j-s, j-m\}$). Hence,

$$
\overline{{}_sY_{jm}} = (-1)^{m+s} \, {}_{-s}Y_{j,-m} \tag{2.31}
$$

and

$$
{}_sY_{jm}(0, \phi) = \begin{cases} 0, & \text{if } m \neq -s, \\[1ex] (-1)^{-s} \sqrt{\dfrac{2j+1}{4\pi}} \, \mathrm{e}^{-is\phi}, & \text{if } m = -s. \end{cases} \tag{2.32}
$$

Completeness

For each value of s, the spin-weighted spherical harmonics, ${}_sY_{jm}$, form a complete (orthonormal) set (Newman and Penrose 1966, Penrose and Rindler 1984) in the sense that any function, f, defined on the sphere S^2 with spin weight s can be expanded in a series of the ${}_sY_{jm}$,

$$f = \sum_{j=|s|}^{\infty} \sum_{m=-j}^{j} c_{jm}\, {}_sY_{jm}. \tag{2.33}$$

In effect, if f is a function with spin weight $s > 0$, the product

$$f \underbrace{\widehat{o}^A \widehat{o}^B \cdots \widehat{o}^C}_{2s}$$

has spin weight 0 (when f has spin weight $s < 0$, we consider $\overline{f}$ in place of f). Then, assuming the completeness of the ordinary spherical harmonics, we have

$$f \widehat{o}^A \widehat{o}^B \cdots \widehat{o}^C = \sum_{j=0}^{\infty} \sum_{m=-j}^{j} b^{AB\ldots C}(j, m)\, Y_{jm}, \tag{2.34}$$

where $b^{AB\ldots C}(j, m)$ are some constants totally symmetric in the $2s$ indices A, B, $\ldots$, C. Then, contracting both sides of (2.34) with $o_A o_B \cdots o_C$ and using the fact that $o^A \widehat{o}_A = 1$, we have [see (2.29)]

$$\begin{aligned} f &= \sum_{j=0}^{\infty} \sum_{m=-j}^{j} b_{AB\ldots C}(j, m) o^A o^B \cdots o^C\, Y_{jm} \\ &= \sum_{j=0}^{\infty} \sum_{m=-j}^{j} \left(\sum_{m'=-s}^{s} b_{jmm'}\, {}_sY_{sm'} \right) Y_{jm}. \end{aligned} \tag{2.35}$$

Each product ${}_sY_{sm'}\, Y_{jm}$ can be expressed as a linear combination of spin-weighted spherical harmonics of spin weight s and orders $j + s$, $j + s - 1$, ..., s,

$${}_sY_{sm'}\, Y_{jm} = \sum_{j'=s}^{j+s} B_{sm'jmj'}\, {}_sY_{j',m+m'}. \tag{2.36}$$

In fact, apart from constant factors, a product of the form ${}_sY_{sm'}\, Y_{jm}$ is given by

$$\underbrace{o^A o^B \cdots o^C}_{2s}\, \underbrace{o^{(D} \cdots o^E}_{j}\, \underbrace{\widehat{o}^F \cdots \widehat{o}^{G)}}_{j}, \tag{2.37}$$

which is not necessarily totally symmetric and, therefore, is not necessarily a spin-weighted spherical harmonic. However, the difference between the product (2.37) and the symmetrized product $o^{(A}o^{B}\cdots o^{C}o^{D}\cdots o^{E}\widehat{o}^{F}\cdots\widehat{o}^{G)}$ can be reduced making use repeatedly of the fact that $o^1\widehat{o}^2 - o^2\widehat{o}^1 = o^A\widehat{o}_A = 1$; this process eliminates pairs of factors $o^A\widehat{o}^B$, leaving the spin weight unchanged and reducing the order by one unit in each step. When all the factors $\widehat{o}^A$ have been eliminated, the resulting expression is a product of $2s$ components o^A, which is necessarily symmetric and therefore is a spin-weighted spherical harmonic of order s. For instance, the difference between the product $o^1o^2o^{(1}o^1\widehat{o}^1\widehat{o}^{2)}$ and $o^{(1}o^2o^1o^1\widehat{o}^1\widehat{o}^{2)}$ is given by

$$\begin{aligned}
o^1o^2o^{(1}o^1\widehat{o}^1\widehat{o}^{2)} &- o^{(1}o^2o^1o^1\widehat{o}^1\widehat{o}^{2)}\\
&= o^1o^2\tfrac{1}{2}(o^1o^1\widehat{o}^1\widehat{o}^2 + o^1o^2\widehat{o}^1\widehat{o}^1)\\
&\quad - \tfrac{1}{15}(o^1o^1o^1o^1\widehat{o}^2\widehat{o}^2 + 8o^1o^1o^1o^2\widehat{o}^1\widehat{o}^2 + 6o^1o^1o^2o^2\widehat{o}^1\widehat{o}^1)\\
&= \tfrac{1}{15}o^1o^1(o^2o^2\widehat{o}^1\widehat{o}^1 - o^1o^1\widehat{o}^2\widehat{o}^2) + \tfrac{1}{30}o^1o^1o^2(o^2\widehat{o}^1 - o^1\widehat{o}^2)\widehat{o}^1\\
&= \tfrac{1}{15}o^1o^1(o^2o^2\widehat{o}^1\widehat{o}^1 - o^1(o^2\widehat{o}^1 + 1)\widehat{o}^2) - \tfrac{1}{30}o^1o^1o^2\widehat{o}^1\\
&= \tfrac{1}{15}o^1o^1o^2(o^2\widehat{o}^1 - o^1\widehat{o}^2)\widehat{o}^1 - \tfrac{1}{15}o^1o^1o^1\widehat{o}^2 - \tfrac{1}{30}o^1o^1o^2\widehat{o}^1\\
&= -\tfrac{1}{10}o^1o^1o^2\widehat{o}^1 - \tfrac{1}{15}o^1o^1o^1\widehat{o}^2\\
&= -\tfrac{1}{24}(o^1o^1o^1\widehat{o}^2 + 3o^1o^1o^2\widehat{o}^1) - \tfrac{1}{40}o^1o^1(o^1\widehat{o}^2 - o^2\widehat{o}^1)\\
&= -\tfrac{1}{6}o^{(1}o^1o^1\widehat{o}^{2)} - \tfrac{1}{40}o^1o^1,
\end{aligned}$$

thus,

$$o^1o^2o^{(1}o^1\widehat{o}^1\widehat{o}^{2)} = o^{(1}o^1o^1o^1\widehat{o}^2\widehat{o}^{2)} - \tfrac{1}{6}o^{(1}o^1o^1\widehat{o}^{2)} - \tfrac{1}{40}o^1o^1$$

or, equivalently, according to (2.11) and (2.29),

$${}_1Y_{1,0}\,Y_{2,-1} = \frac{2}{5}\sqrt{\frac{5}{7\pi}}\,{}_1Y_{3,-1} - \frac{1}{4\sqrt{\pi}}\,{}_1Y_{2,-1} - \frac{3}{20}\sqrt{\frac{5}{3\pi}}\,{}_1Y_{1,-1}$$

[see also (3.33), (3.152) and (3.153)]. As we shall see in the next section, the coefficients $B_{sm'jmj'}$ appearing in (2.36) are, apart from constant factors, products of Clebsch–Gordan coefficients. Finally, substituting (2.36) into (2.35) we obtain (2.33).

Alternative conventions

In place of (2.4), the spinors o and $\widehat{o}$ can be taken as

$$\begin{pmatrix} o^1 & \widehat{o}^1 \\ o^2 & \widehat{o}^2 \end{pmatrix} = \begin{pmatrix} e^{-i(\phi+\chi)/2}\cos\tfrac{1}{2}\theta & -e^{-i(\phi-\chi)/2}\sin\tfrac{1}{2}\theta \\ e^{i(\phi-\chi)/2}\sin\tfrac{1}{2}\theta & e^{i(\phi+\chi)/2}\cos\tfrac{1}{2}\theta \end{pmatrix} \tag{2.38}$$

[*cf.* (1.55)], where now χ is some function of θ and ϕ. The basic relations (2.18) and (2.19) hold provided that the operators $\eth$ and $\bar{\eth}$ are defined by

$$\begin{aligned} \eth\eta &= -\mathrm{e}^{-\mathrm{i}(s+1)\chi/2}\left(\partial_\theta + \frac{\mathrm{i}}{\sin\theta}\partial_\phi - s\cot\theta\right)(\mathrm{e}^{\mathrm{i}s\chi/2}\eta), \\ \bar{\eth}\eta &= -\mathrm{e}^{-\mathrm{i}(s-1)\chi/2}\left(\partial_\theta - \frac{\mathrm{i}}{\sin\theta}\partial_\phi + s\cot\theta\right)(\mathrm{e}^{\mathrm{i}s\chi/2}\eta) \end{aligned} \tag{2.39}$$

[*cf.* (2.17)]. The expression for the ordinary spherical harmonics is not altered by this change [see (2.12)] and many of the preceding formulas hold with the definitions (2.38) and (2.39) [*e.g.*, (2.20)–(2.22), (2.24), and (2.26)–(2.31)] but now the final expression (2.30) contains an additional factor $\mathrm{e}^{-\mathrm{i}s\chi}$.

The case where $\chi = \phi$ is distinguished by the fact that, in terms of the complex coordinates $\zeta = \mathrm{e}^{\mathrm{i}\phi}\cot\frac{1}{2}\theta$ and its complex conjugate, $\bar{\zeta}$, from (2.39) we have

$$\begin{aligned} \eth\eta &= (1+\zeta\bar{\zeta})^{1-s}\partial_\zeta\big((1+\zeta\bar{\zeta})^s\eta\big), \\ \bar{\eth}\eta &= (1+\zeta\bar{\zeta})^{1+s}\partial_{\bar{\zeta}}\big((1+\zeta\bar{\zeta})^{-s}\eta\big) \end{aligned} \tag{2.40}$$

(*cf.* Eastwood and Tod 1982, Penrose and Rindler 1984, Stewart 1990). In what follows, we will use the definitions (2.6) and (2.17) (which correspond to $\chi = 0$).

Relationship with other special functions

The spin-weighted spherical harmonics ${}_sY_{jm}(\theta,\phi)$ satisfy the differential equations

$$\bar{\eth}\eth\,{}_sY_{jm} = [s(s+1) - j(j+1)]\,{}_sY_{jm}, \qquad -\mathrm{i}\partial_\phi({}_sY_{jm}) = m\,{}_sY_{jm}$$

[see (2.22)]; these equations together with (2.23) imply that ${}_sY_{jm}(\theta,\phi) = {}_s\mathcal{Y}_{jm}(\theta)\mathrm{e}^{im\phi}$, where ${}_s\mathcal{Y}_{jm}(\theta)$ is a function of θ only that satisfies the ordinary differential equation

$$\left[\frac{1}{\sin\theta}\frac{\mathrm{d}}{\mathrm{d}\theta}\sin\theta\frac{\mathrm{d}}{\mathrm{d}\theta} - \frac{m^2 + 2ms\cos\theta + s^2}{\sin^2\theta} + j(j+1)\right]{}_s\mathcal{Y}_{jm}(\theta) = 0, \tag{2.41}$$

which reduces to the associated Legendre equation when m or s is equal to 0. In terms of the variable $x = \cos\theta$, the differential equation (2.41) takes the form

$$\left[(1-x^2)\frac{\mathrm{d}^2}{\mathrm{d}x^2} - 2x\frac{\mathrm{d}}{\mathrm{d}x} - \frac{m^2 + 2msx + s^2}{1-x^2} + j(j+1)\right]{}_s\mathcal{Y}_{jm} = 0 \tag{2.42}$$

or, equivalently,

$$\left[(1-x^2)\frac{\mathrm{d}^2}{\mathrm{d}x^2} - 2x\frac{\mathrm{d}}{\mathrm{d}x} - \frac{(m+s)^2}{2(1-x)} - \frac{(m-s)^2}{2(1+x)} + j(j+1)\right]{}_s\mathcal{Y}_{jm} = 0,$$

which is the generalized associated Legendre equation (see, *e.g.*, Virchenko and Fedotova 2001, and the references cited therein), therefore ${}_s\mathcal{Y}_{jm}$ is proportional to the generalized associated Legendre function of the first kind $P_j^{m+s,m-s}$.

Substituting ${}_s\mathcal{Y}_{jm}(x) = (1-x)^{\alpha/2}(1+x)^{\beta/2} f(x)$, where α and β are some constants, into (2.42) we find that f obeys the Jacobi equation

$$\left[(1-x^2)\frac{\mathrm{d}^2}{\mathrm{d}x^2} + [(\beta-\alpha) - (\alpha+\beta+2)x]\frac{\mathrm{d}}{\mathrm{d}x} + n(n+\alpha+\beta+1)\right] f(x) = 0,$$

provided that

$$\alpha = |m+s|, \qquad \beta = |m-s|, \qquad n = j - \tfrac{1}{2}(\alpha+\beta) = j - \max\{|m|, |s|\}.$$

Thus,

$${}_sY_{jm}(\theta,\phi) = A(1-\cos\theta)^{\alpha/2}(1+\cos\theta)^{\beta/2} P_n^{(\alpha,\beta)}(\cos\theta)\, \mathrm{e}^{\mathrm{i}m\phi},$$

where A is a normalization constant and $P_n^{(\alpha,\beta)}(x)$ is a Jacobi polynomial (the subscript n is equal to the degree of the polynomial).

Similarly, letting ${}_s\mathcal{Y}_{jm}(x) = (1-x)^{(m+s)/2}(1+x)^{j-(m+s)/2} g\left(\frac{x-1}{x+1}\right)$, we find that g obeys the hypergeometric equation

$$x(1-x)\frac{\mathrm{d}^2 g(x)}{\mathrm{d}x^2} + [m+s+1-(1+m+s-2j)x]\frac{\mathrm{d}g(x)}{\mathrm{d}x} - (m-j)(s-j)g(x) = 0,$$

hence, for $m+s \geqslant 0$,

$$\begin{aligned} {}_sY_{jm}(\theta,\phi) &= B\left(\sin\tfrac{1}{2}\theta\right)^{m+s}\left(\cos\tfrac{1}{2}\theta\right)^{2j-(m+s)} \\ &\quad \times {}_2F_1(m-j, s-j, m+s+1; -\tan^2\tfrac{1}{2}\theta)\, \mathrm{e}^{\mathrm{i}m\phi}, \end{aligned}$$

where B is some constant and ${}_2F_1$ denotes the hypergeometric function. Making use of the explicit expression of the hypergeometric series, from (2.30) one finds that, for $m+s \geqslant 0$,

$$\begin{aligned} {}_sY_{jm}(\theta,\phi) &= \frac{(-1)^m}{(m+s)!}\sqrt{\frac{2j+1}{4\pi}\frac{(j+s)!(j+m)!}{(j-s)!(j-m)!}}\left(\sin\tfrac{1}{2}\theta\right)^{m+s} \\ &\quad \times \left(\cos\tfrac{1}{2}\theta\right)^{2j-(m+s)} {}_2F_1(m-j, s-j, m+s+1; -\tan^2\tfrac{1}{2}\theta)\, \mathrm{e}^{\mathrm{i}m\phi}. \end{aligned}$$

It will be shown in the next section that the spin-weighted spherical harmonics are also related to the Wigner functions, which arise in the study of representations of the rotation group SO(3) when the representation space is that of the ordinary spherical harmonics. Further properties of the operators $\eth$ and $\bar{\eth}$ and of the spin-weighted spherical harmonics can be found in Penrose and Rindler (1984) and Goldberg *et al.* (1967).

2.3 Wigner functions

Under any rotation, $\mathcal{R}$, of three-dimensional space, a point with Cartesian coordinates x_i is transformed into another point with coordinates $x'_i = a_{ij}x_j$, with $(a_{ij}) \in SO(3)$ [see (1.37)] and, under this rotation, any scalar function f transforms into another function, f' or $\mathcal{R}f$, defined by

$$f'(x_i) \equiv f(\tilde{a}_{ij}x_j), \tag{2.43}$$

where $(\tilde{a}_{ij})$ is the inverse of (a_{ij}) (so that, $f'(x'_i) = f(x_i)$). Employing the usual (pointwise) operations between scalar functions, one finds that the map $f \mapsto \mathcal{R}f$ is linear: $\mathcal{R}(af+bg) = a\mathcal{R}f+b\mathcal{R}g$, for any pair of complex constants a, b. If $\mathcal{R}_1$ and $\mathcal{R}_2$ are the rotations given by the SO(3) matrices (a_{ij}) and (b_{ij}), respectively, then the composition $\mathcal{R}_1\mathcal{R}_2$ corresponds to the matrix product $c_{ij} = a_{ik}b_{kj}$ and using repeatedly the definition (2.43) we have

$$\begin{aligned}(\mathcal{R}_1\mathcal{R}_2)f(x_i) &= f(\tilde{c}_{ij}x_j) = f(\tilde{b}_{ik}\tilde{a}_{kj}x_j) \\ &= (\mathcal{R}_2 f)(\tilde{a}_{kj}x_j) = [\mathcal{R}_1(\mathcal{R}_2 f)](x_k),\end{aligned}$$

i.e.,

$$(\mathcal{R}_1\mathcal{R}_2)f = \mathcal{R}_1(\mathcal{R}_2 f). \tag{2.44}$$

If f is a homogeneous polynomial of degree l, $f(x_i) = d_{ij\ldots k}x_i x_j \cdots x_k$, then $(\mathcal{R}f)(x_i) = d_{ij\ldots k}\tilde{a}_{ip}\tilde{a}_{jq}\cdots\tilde{a}_{ks}x_p x_q \cdots x_s$ is also a homogeneous polynomial of degree l with coefficients $d'_{pq\cdots s} \equiv d_{ij\ldots k}\tilde{a}_{ip}\tilde{a}_{jq}\cdots\tilde{a}_{ks}$ which are symmetric if and only if $d_{ij\ldots k}$ are. Furthermore, making use of (1.39) it follows that, $d'_{ppq\ldots s} = d_{ijk\ldots m}\tilde{a}_{ip}\tilde{a}_{jp}\tilde{a}_{kq}\cdots\tilde{a}_{ms} = d_{iik\ldots m}\tilde{a}_{kq}\cdots\tilde{a}_{ms}$, which means that the trace of $d'_{ij\ldots k}$ vanishes if and only if the trace of $d_{ij\ldots k}$ vanishes, hence, the image of a spherical harmonic of order l under any rotation about the origin is another spherical harmonic of the same order.

Since the spherical harmonics of order l form a (complex) vector space of dimension $2l+1$, for each (integral) value of l, from (2.43) it follows that the spherical harmonics of order l form a basis for a linear representation of SO(3). In terms of the basis for the spherical harmonics of order l given by the functions Y_{lm} ($m = 0, \pm 1, \ldots, \pm l$), any rotation $\mathcal{R}$ is represented by a $(2l+1)\times(2l+1)$ matrix $D^l_{m'm}(\mathcal{R})$, $(m', m = 0, \pm 1, \ldots, \pm l)$, defined by

$$\mathcal{R}\,Y_{lm} = \sum_{m'=-l}^{l} D^l_{m'm}(\mathcal{R})\,Y_{lm'}. \tag{2.45}$$

For l fixed, the matrices $D^l_{m'm}(\mathcal{R})$ form a representation of SO(3) since, according to (2.44) and (2.45), for any two rotations,

$$
\begin{aligned}
(\mathcal{R}_1\mathcal{R}_2)\, Y_{lm} &= \sum_{m''=-l}^{l} D^l_{m''m}(\mathcal{R}_1\mathcal{R}_2)\, Y_{lm''} \\
&= \mathcal{R}_1(\mathcal{R}_2\, Y_{lm}) = \mathcal{R}_1\left(\sum_{m'=-l}^{l} D^l_{m'm}(\mathcal{R}_2)\, Y_{lm'}\right) \\
&= \sum_{m'=-l}^{l} D^l_{m'm}(\mathcal{R}_2)\, \mathcal{R}_1\, Y_{lm'} \\
&= \sum_{m'=-l}^{l} D^l_{m'm}(\mathcal{R}_2) \sum_{m''=-l}^{l} D^l_{m''m'}(\mathcal{R}_1)\, Y_{lm''}
\end{aligned}
$$

which implies that

$$
D^l_{m''m}(\mathcal{R}_1\mathcal{R}_2) = \sum_{m'=-l}^{l} D^l_{m''m'}(\mathcal{R}_1)\, D^l_{m'm}(\mathcal{R}_2). \tag{2.46}
$$

The functions $D^l_{m'm} : \mathrm{SO}(3) \to \mathbb{C}$ are known as *Wigner functions* (see, *e.g.*, Messiah 1962, Goldberg *et al.* 1967, Tung 1985, Sakurai 1994 and the references cited therein).

The Wigner functions can be easily obtained making use of the expression (2.11) for the spherical harmonics and the fact that, under any rotation, the components of the spinors o and $\hat{o}$ transform by means of the same SU(2) matrix; thus, from (2.43) and (2.11), we have

$$
\mathcal{R}Y_{lm} = (-1)^m \frac{(2l)!}{l!}\sqrt{\frac{2l+1}{4\pi}\frac{1}{(l+m)!(l-m)!}}\; \overbrace{\tilde{Q}^{(1}_{A} \quad \cdots \quad \tilde{Q}^{2)}_{B}}^{(l-m)\text{ 1's, }(l+m)\text{ 2's}}\, o^A \cdots \hat{o}^B, \tag{2.47}
$$

where $(\tilde{Q}^A_B)$ is the inverse of SU(2) matrix corresponding to the rotation $\mathcal{R}$. (Recall that there are two SU(2) matrices representing a given rotation, which differ by a sign; however, since (2.47) contains an even number of factors $\tilde{Q}^A_B$, the same result is obtained using (Q^A_B) or $-(Q^A_B)$.) Collecting terms we have

$$
\begin{aligned}
\mathcal{R}Y_{lm} &= (-1)^m \frac{(2l)!}{l!}\sqrt{\frac{2l+1}{4\pi}\frac{1}{(l+m)!(l-m)!}} \sum_{m'=-l}^{l} \frac{(2l)!}{(l+m')!(l-m')!} \\
&\quad \times \underbrace{\overbrace{\tilde{Q}^{(1}_{1} \quad \cdots \quad \tilde{Q}^{2)}_{2}}^{(l-m)\text{ 1's, }(l+m)\text{ 2's}}}_{(l-m')\text{ 1's, }(l+m')\text{ 2's}}\, \overbrace{o^1 \quad \cdots \quad \hat{o}^2}^{(l-m')\text{ 1's, }(l+m')\text{ 2's}},
\end{aligned}
$$

then, comparing with (2.45) and using (2.11) again, it follows that

$$D^l_{m'm}(\mathcal{R}) = \frac{(2l)!}{\sqrt{(l+m)!(l-m)!(l+m')!(l-m')!}} \overbrace{\underbrace{\tilde{Q}_1^{(1} \quad \cdots \quad \tilde{Q}_2^{2)}}_{(l-m')\ 1\text{'s},\ (l+m')\ 2\text{'s}}}^{(l-m)\ 1\text{'s},\ (l+m)\ 2\text{'s}} \tag{2.48}$$

or, using the fact that (Q^A_B) is unitary, $\tilde{Q}^A_B = \overline{Q^B_A}$, therefore,

$$\overline{D^l_{m'm}(\mathcal{R})} = \frac{(2l)!}{\sqrt{(l+m)!(l-m)!(l+m')!(l-m')!}} \overbrace{\underbrace{Q_1^{(1} \quad \cdots \quad Q_2^{2)}}_{(l-m)\ 1\text{'s},\ (l+m)\ 2\text{'s}}}^{(l-m')\ 1\text{'s},\ (l+m')\ 2\text{'s}}. \tag{2.49}$$

From (2.48) and (2.49) it follows that

$$D^l_{m'm}(\mathcal{R}^{-1}) = \overline{D^l_{mm'}(\mathcal{R})}, \tag{2.50}$$

which means that the representation of SO(3) given by the matrices $D^l_{m'm}$ is unitary. (This conclusion also follows from the fact that, for any rotation, $(\mathcal{R}f, \mathcal{R}g) = (f, g)$.)

Recalling that if the rotation $\mathcal{R}$ is parametrized by the Euler angles ϕ, θ, χ, then

$$Q_1^A = \mathrm{e}^{-\mathrm{i}\chi/2} o^A, \qquad Q_2^A = \mathrm{e}^{\mathrm{i}\chi/2} \widehat{o}^A,$$

with o^A and $\widehat{o}^A$ defined by (2.6) [see (1.55)], writing $D^l_{m'm}(\mathcal{R}(\phi, \theta, \chi)) \equiv D^l_{m'm}(\phi, \theta, \chi)$, from (2.49) we have

$$\overline{D^l_{m'm}(\phi, \theta, \chi)} = \frac{(2l)!}{\sqrt{(l+m)!(l-m)!(l+m')!(l-m')!}} \mathrm{e}^{im\chi} \times \overbrace{\underbrace{o^{(1} o^1 \cdots o^1}_{l-m} \underbrace{\widehat{o}^1 \widehat{o}^2 \cdots \widehat{o}^{2)}}_{l+m}}^{(l-m')\ 1\text{'s},\ (l+m')\ 2\text{'s}},$$

thus, comparing with (2.29) we find that

$$\overline{D^l_{m'm}(\phi, \theta, \chi)} = (-1)^{-m} \sqrt{\frac{4\pi}{2l+1}}\, {}_{-m}Y_{lm'}(\theta, \phi) \mathrm{e}^{\mathrm{i}m\chi}, \tag{2.51}$$

or, owing to (2.31),

$$D^l_{m'm}(\phi, \theta, \chi) = (-1)^{m'} \sqrt{\frac{4\pi}{2l+1}}\, {}_mY_{l,-m'}(\theta, \phi) \mathrm{e}^{-\mathrm{i}m\chi} \tag{2.52}$$

(*cf.* Goldberg *et al.* 1967, Torres del Castillo and Hernández-Guevara 1995). This last equation shows that $D^l_{m'm}(\phi, \theta, \chi)$ is the product of three one-variable functions,

$$D^l_{m'm}(\phi, \theta, \chi) = \mathrm{e}^{-\mathrm{i}m'\phi} d^l_{m'm}(\theta) \mathrm{e}^{-\mathrm{i}m\chi}. \tag{2.53}$$

Then, in terms of the functions $d^l_{m'm}$ defined in (2.53), the spin-weighted spherical harmonics are given by

$$_sY_{jm}(\theta, \phi) = (-1)^m \sqrt{\frac{2j+1}{4\pi}}\, d^j_{-m,s}(\theta) \mathrm{e}^{\mathrm{i}m\phi}$$

and, comparing with (2.30), we have

$$\begin{aligned} d^l_{m'm}(\theta) &= \sqrt{(l+m)!(l-m)!(l+m')!(l-m')!} \\ &\quad \times \sum_k \frac{(-1)^k (\sin \frac{1}{2}\theta)^{m-m'+2k} (\cos \frac{1}{2}\theta)^{2l-m+m'-2k}}{k!(l+m'-k)!(l-m-k)!(m-m'+k)!}. \end{aligned}$$

The product of two spin-weighted spherical harmonics with the same argument can be expressed as a linear combination of spin-weighted spherical harmonics, taking advantage of the relationship of these functions with the Wigner functions. The decomposition of the direct product of representations of SO(3) given by

$$D^j_{ms}(\mathcal{R}) D^{j'}_{m's'}(\mathcal{R}) = \sum_{J,M,S} \langle jj'; mm' | jj'; JM \rangle \, \langle jj'; ss' | jj'; JS \rangle \, D^J_{MS}(\mathcal{R}),$$

where the $\langle jj'; mm' | jj'; JM \rangle$ denote the Clebsch–Gordan coefficients (see, *e.g.*, Messiah 1962, Brink and Satcher 1993, Sakurai 1994 and the references cited therein), amounts to

$$\begin{aligned} {}_sY_{jm}\, {}_{s'}Y_{j'm'} &= \sum_{J,M,S} (-1)^{j+j'-J} \sqrt{\frac{(2j+1)(2j'+1)}{4\pi(2J+1)}} \langle jj'; mm' | jj'; JM \rangle \\ &\quad \times \langle jj'; ss' | jj'; JS \rangle \; {}_SY_{JM}, \end{aligned} \tag{2.54}$$

while the formula

$$D^J_{MS}(\mathcal{R}) = \sum_{m,m',s,s'} \langle jj'; mm' | jj'; JM \rangle \, D^j_{ms}(\mathcal{R}) \, D^{j'}_{m's'}(\mathcal{R}) \, \langle jj'; ss' | jj'; JS \rangle$$

is equivalent to

$$\begin{aligned} {}_SY_{JM} &= \sum_{m,m',s,s'} (-1)^{j+j'-J} \sqrt{\frac{4\pi(2J+1)}{(2j+1)(2j'+1)}} \langle jj'; mm' | jj'; JM \rangle \\ &\quad \times \langle jj'; ss' | jj'; JS \rangle \; {}_sY_{jm}\, {}_{s'}Y_{j'm'}. \end{aligned} \tag{2.55}$$

Addition theorems

By virtue of (2.51) and (2.52), the relations (2.46) can be translated into identities satisfied by the spin-weighted spherical harmonics. For instance, using (2.51), (2.50), and the fact that the inverse of the rotation with Euler angles (ϕ, θ, χ) is that with Euler angles $(-\chi, -\theta, -\phi)$ [see (1.55)], we have

$$
\begin{aligned}
\sum_{m=-j}^{j} & \overline{{}_{s'}Y_{jm}(\theta_2, \phi_2)}\, {}_sY_{jm}(\theta_1, \phi_1) \\
&= \frac{2j+1}{4\pi}(-1)^{-s-s'} \sum_{m=-j}^{j} D^j_{m,-s'}(\phi_2, \theta_2, 0)\, \overline{D^j_{m,-s}(\phi_1, \theta_1, 0)} \\
&= \frac{2j+1}{4\pi}(-1)^{-s-s'} \sum_{m=-j}^{j} D^j_{-s,m}(0, -\theta_1, -\phi_1)\, D^j_{m,-s'}(\phi_2, \theta_2, 0). \quad (2.56)
\end{aligned}
$$

On the other hand, according to (2.46) and (2.52),

$$
\begin{aligned}
\sum_{m=-j}^{j} & D^j_{-s,m}(0, -\theta_1, -\phi_1)\, D^j_{m,-s'}(\phi_2, \theta_2, 0) \\
&= D^j_{-s,-s'}(\phi_3, \theta_3, \chi_3) \\
&= (-1)^{-s}\sqrt{\frac{4\pi}{2j+1}}\, {}_{-s'}Y_{j,s}(\theta_3, \phi_3) e^{\mathrm{i}s'\chi_3}, \quad (2.57)
\end{aligned}
$$

where $(\phi_3, \theta_3, \chi_3)$ are the Euler angles of the composition of the rotations with Euler angles $(0, -\theta_1, -\phi_1)$ and $(\phi_2, \theta_2, 0)$; hence, making use of (1.55),

$$
Q(\phi_3, \theta_3, \chi_3) = Q(0, -\theta_1, -\phi_1)\, Q(\phi_2, \theta_2, 0),
$$

which gives

$$
\cos\theta_3 = \cos\theta_1 \cos\theta_2 + \sin\theta_1 \sin\theta_2 \cos(\phi_2 - \phi_1)
$$

and

$$
\begin{aligned}
& e^{-\mathrm{i}(\phi_3+\chi_3)/2} \\
&= \frac{\cos\frac{1}{2}(\phi_2-\phi_1)\cos\frac{1}{2}(\theta_2-\theta_1) - \mathrm{i}\sin\frac{1}{2}(\phi_2-\phi_1)\cos\frac{1}{2}(\theta_1+\theta_2)}{\sqrt{\cos^2\frac{1}{2}(\phi_2-\phi_1)\cos^2\frac{1}{2}(\theta_2-\theta_1) + \sin^2\frac{1}{2}(\phi_2-\phi_1)\cos^2\frac{1}{2}(\theta_1+\theta_2)}}, \\
& e^{\mathrm{i}(\phi_3-\chi_3)/2} \\
&= \frac{\cos\frac{1}{2}(\phi_2-\phi_1)\sin\frac{1}{2}(\theta_2-\theta_1) + \mathrm{i}\sin\frac{1}{2}(\phi_2-\phi_1)\sin\frac{1}{2}(\theta_1+\theta_2)}{\sqrt{\cos^2\frac{1}{2}(\phi_2-\phi_1)\sin^2\frac{1}{2}(\theta_2-\theta_1) + \sin^2\frac{1}{2}(\phi_2-\phi_1)\sin^2\frac{1}{2}(\theta_1+\theta_2)}}.
\end{aligned}
$$

Thus, substituting (2.57) into (2.56) we obtain

$$\sum_{m=-j}^{j} \overline{{}_{s'}Y_{jm}(\theta_2, \phi_2)}\, {}_sY_{jm}(\theta_1, \phi_1) = (-1)^{-2s-s'} \sqrt{\frac{2j+1}{4\pi}}\, {}_{-s'}Y_{js}(\theta_3, \phi_3) e^{is'\chi_3} \tag{2.58}$$

or, equivalently, according to (2.31),

$$\sum_{m=-j}^{j} \overline{{}_sY_{jm}(\theta_1, \phi_1)}\, {}_{s'}Y_{jm}(\theta_2, \phi_2) = (-1)^{-s} \sqrt{\frac{2j+1}{4\pi}}\, {}_{s'}Y_{j,-s}(\theta_3, \phi_3) e^{-is'\chi_3}. \tag{2.59}$$

This last identity takes simpler forms in some special cases. Letting, for example, $s = s' = 0$ and using the fact that ${}_0Y_{j,0}(\theta_3, \phi_3) = Y_{j,0}(\theta_3, \phi_3) = \sqrt{(2j+1)/(4\pi)}\, P_j(\cos\theta_3)$, where P_j is the Legendre polynomial of order j, (2.58) yields the addition theorem for the spherical harmonics

$$\sum_{m=-j}^{j} \overline{Y_{jm}(\theta_1, \phi_1)}\, Y_{jm}(\theta_2, \phi_2) = \frac{2j+1}{4\pi} P_j(\cos\theta_3). \tag{2.60}$$

Similarly, taking $(\theta_1, \phi_1) = (\theta_2, \phi_2) \equiv (\theta, \phi)$, we obtain $\phi_3 = \theta_3 = \chi_3 = 0$ and making use of (2.32) it follows that

$$\sum_{m=-j}^{j} \overline{{}_{s'}Y_{jm}(\theta, \phi)}\, {}_sY_{jm}(\theta, \phi) = \frac{2j+1}{4\pi} \delta_{ss'}. \tag{2.61}$$

The foregoing equations apply for integral or half-integral values of the spin weights. Some additional properties of the spin-weighted spherical harmonics are derived in Sect. 3.1; other properties can be obtained from those of the special functions related to them (generalized associated Legendre functions, Jacobi polynomials, hypergeometric functions, Wigner functions).

Spherical harmonics in four dimensions

The Wigner functions themselves are spherical harmonics in four dimensions. The explicit expression (2.48) shows that the functions $D^l_{m'm}$ are homogeneous polynomials of degree $2l$ in the Cartesian coordinates of the points of the sphere S^3. In effect, the Cartesian coordinates of any point of the four-dimensional Euclidean space, x_μ ($\mu, \nu, \ldots = 1, 2, 3, 4$), can be expressed in the spinor form

$$x_\mu = \sigma_{\mu A\dot{B}} x^{A\dot{B}} \tag{2.62}$$

($A, B, \ldots = 1, 2$; $\dot{A}, \dot{B}, \ldots = \dot{1}, \dot{2}$), where the Infeld–van der Waerden symbols $\sigma_{\mu A\dot{B}}$ satisfy the conditions

$$\sigma_{\mu A\dot{B}}\sigma_{\mu C\dot{D}} = -2\varepsilon_{AC}\varepsilon_{\dot{B}\dot{D}}$$

and

$$\overline{\sigma_{\mu A\dot{B}}} = -\sigma_{\mu}{}^{A\dot{B}}, \tag{2.63}$$

where

$$(\varepsilon_{\dot{A}\dot{B}}) = \begin{pmatrix} 0 & 1 \\ -1 & 0 \end{pmatrix} = (\varepsilon^{\dot{A}\dot{B}})$$

and the dotted indices are raised or lowered in the same manner as the undotted ones, *e.g.*, $\psi_{\dot{A}} = \varepsilon_{\dot{A}\dot{B}}\psi^{\dot{B}}$. For instance, we can choose,

$$\begin{aligned} (\sigma_{1A\dot{B}}) &= \begin{pmatrix} 1 & 0 \\ 0 & -1 \end{pmatrix}, & (\sigma_{2A\dot{B}}) &= \begin{pmatrix} \mathrm{i} & 0 \\ 0 & \mathrm{i} \end{pmatrix}, \\ (\sigma_{3A\dot{B}}) &= \begin{pmatrix} 0 & -1 \\ -1 & 0 \end{pmatrix}, & (\sigma_{4A\dot{B}}) &= \begin{pmatrix} 0 & \mathrm{i} \\ -\mathrm{i} & 0 \end{pmatrix}. \end{aligned} \tag{2.64}$$

Then, the coordinates x_μ in (2.62) are real if and only if $\overline{x^{A\dot{B}}} = -x_{A\dot{B}}$. Furthermore, the effect of any rotation about the origin in four-dimensional Euclidean space is equivalent to a transformation of the form $x'^{A\dot{B}} = L^A{}_C M^{\dot{B}}{}_{\dot{D}} x^{C\dot{D}}$, where $(L^A{}_C)$ and $(M^{\dot{B}}{}_{\dot{D}})$ are SU(2) matrices. If $(R^A{}_{\dot{B}})$ is a 2×2 matrix belonging to SU(2) (with the superscript labeling rows and the subscript labeling columns), then $\overline{R_{A\dot{B}}} = R^{A\dot{B}}$ and therefore

$$N_\mu = \frac{\mathrm{i}}{2}\sigma_{\mu A}{}^{\dot{B}} R^A{}_{\dot{B}} \tag{2.65}$$

are the Cartesian components of a (real) unit vector in four-dimensional Euclidean space. With the $\sigma_{\mu A\dot{B}}$ taken as in (2.64), the matrix $(R^A{}_{\dot{B}})$ is given explicitly by

$$(R^A{}_{\dot{B}}) = \begin{pmatrix} N_4 - \mathrm{i}N_3 & -N_2 - \mathrm{i}N_1 \\ N_2 - \mathrm{i}N_1 & N_4 + \mathrm{i}N_3 \end{pmatrix}. \tag{2.66}$$

Expression (2.65) [or (2.66)] gives a one-to-one correspondence between the points of the sphere S^3 and the elements of SU(2). Thus, any spherical harmonic of order l in four dimensions can be written as

$$\begin{aligned} d_{\mu\nu\ldots\rho}N_\mu N_\nu \cdots N_\rho &= \left(\frac{\mathrm{i}}{2}\right)^l d_{\mu\nu\ldots\rho}\sigma_{\mu A}{}^{\dot{B}}\sigma_{\nu C}{}^{\dot{D}} \cdots \sigma_{\rho E}{}^{\dot{F}} R^A{}_{\dot{B}} R^C{}_{\dot{D}} \cdots R^E{}_{\dot{F}} \\ &\equiv \left(\frac{\mathrm{i}}{2}\right)^l d_{AC\ldots E}^{\dot{B}\dot{D}\ldots\dot{F}} R^A{}_{\dot{B}} R^C{}_{\dot{D}} \cdots R^E{}_{\dot{F}}. \end{aligned} \tag{2.67}$$

The complete symmetry of the coefficients $d_{\mu\nu\ldots\rho}$ in their l indices and the vanishing of their traces are equivalent to the symmetry of the coefficients $d_{AC\ldots E}^{\dot{B}\dot{D}\ldots\dot{F}}$ in the dotted indices and in the undotted ones, separately; this implies that there are $(l+1)^2$ linearly independent spherical harmonics of order l in four dimensions. By means of the expression (2.49), we can consider the Wigner functions as functions defined on SU(2) or, equivalently, as functions defined on S^3 and, by comparing (2.49) and (2.67), it follows that the Wigner functions are spherical harmonics in four dimensions (*cf.* Bander and Itzykson 1966).

3
Spin-Weighted Spherical Harmonics. Applications

3.1 3.1 Solution of the vector Helmholtz equation

The orthonormal basis, $\{\mathbf{e}_r, \mathbf{e}_\theta, \mathbf{e}_\phi\}$, induced by the spherical coordinates (r, θ, ϕ) is related to the spinor o given in (2.6) by means of

$$\mathbf{e}_r = o^\dagger \sigma o, \qquad \mathbf{e}_\theta + \mathrm{i}\mathbf{e}_\phi = o^{\mathrm{t}} \varepsilon \sigma o$$

[*cf.* (1.26) and (1.43)] and the transformation (2.13) produces the rotation through α about $\mathbf{e}_r$ given by

$$\mathbf{e}_r \mapsto \mathbf{e}_r, \qquad \mathbf{e}_\theta + \mathrm{i}\mathbf{e}_\phi \mapsto \mathrm{e}^{\mathrm{i}\alpha}(\mathbf{e}_\theta + \mathrm{i}\mathbf{e}_\phi). \tag{3.1}$$

Any vector field, $\mathbf{F}$, can be expressed as

$$\mathbf{F} = F_r \mathbf{e}_r + F_\theta \mathbf{e}_\theta + F_\phi \mathbf{e}_\phi$$

or, equivalently,

$$\mathbf{F} = -\sqrt{2}\, F_0\, \mathbf{e}_r - \frac{1}{\sqrt{2}} F_{-1}(\mathbf{e}_\theta + \mathrm{i}\mathbf{e}_\phi) + \frac{1}{\sqrt{2}} F_{+1}(\mathbf{e}_\theta - \mathrm{i}\mathbf{e}_\phi), \tag{3.2}$$

where

$$F_0 \equiv -\frac{1}{\sqrt{2}} \mathbf{F} \cdot \mathbf{e}_r, \qquad F_{\pm 1} \equiv \pm \frac{1}{\sqrt{2}} \mathbf{F} \cdot (\mathbf{e}_\theta \pm \mathrm{i}\mathbf{e}_\phi). \tag{3.3}$$

By virtue of (3.1), F_s $(s = 0, \pm 1)$ has spin weight s. (These combinations come from the spinor formalism, see (6.61)). The vector field $\mathbf{F}$ is real if and only if its spin-weighted components F_s satisfy

$$\overline{F_s} = (-1)^s F_{-s}.$$

The standard vector operators,

$$
\begin{aligned}
\nabla f &= (\partial_r f)\,\mathbf{e}_r + \frac{1}{r}(\partial_\theta f)\,\mathbf{e}_\theta + \frac{1}{r\sin\theta}(\partial_\phi f)\,\mathbf{e}_\phi, \qquad (3.4)\\
\nabla\cdot\mathbf{F} &= \frac{1}{r^2}\partial_r(r^2 F_r) + \frac{1}{r\sin\theta}\partial_\theta(F_\theta\sin\theta) + \frac{1}{r\sin\theta}\partial_\phi F_\phi,\\
\nabla\times\mathbf{F} &= \frac{1}{r\sin\theta}\Big[\partial_\theta(F_\phi\sin\theta) - \partial_\phi F_\theta\Big]\mathbf{e}_r + \frac{1}{r}\left[\frac{1}{\sin\theta}\partial_\phi F_r - \partial_r(rF_\phi)\right]\mathbf{e}_\theta\\
&\quad + \frac{1}{r}\Big[\partial_r(rF_\theta) - \partial_\theta F_r\Big]\mathbf{e}_\phi,
\end{aligned}
$$

can be written in terms of the spin-weighted combinations $\mathbf{e}_r$, $\mathbf{e}_\theta + \mathrm{i}\mathbf{e}_\phi$, $\mathbf{e}_\theta - \mathrm{i}\mathbf{e}_\phi$ and (3.3). Assuming that the function f has spin weight 0, taking into account that F_0 and $F_{\pm 1}$ have spin weight 0 and ± 1, respectively, making use of the definitions (2.17) we obtain

$$
\nabla f = (\partial_r f)\,\mathbf{e}_r - \frac{1}{2r}\,\bar{\eth} f\,(\mathbf{e}_\theta + \mathrm{i}\mathbf{e}_\phi) - \frac{1}{2r}\,\eth f\,(\mathbf{e}_\theta - \mathrm{i}\mathbf{e}_\phi), \qquad (3.5)
$$

$$
\nabla\cdot\mathbf{F} = -\frac{\sqrt{2}}{r^2}\partial_r(r^2F_0) + \frac{1}{\sqrt{2}\,r}(\eth F_{-1} - \bar{\eth} F_{+1}), \qquad (3.6)
$$

$$
\begin{aligned}
\nabla\times\mathbf{F} &= \frac{\mathrm{i}}{\sqrt{2}\,r}(\eth F_{-1} + \bar{\eth} F_{+1})\,\mathbf{e}_r + \frac{\mathrm{i}}{\sqrt{2}\,r}\Big[\partial_r(rF_{-1}) + \bar{\eth} F_0\Big](\mathbf{e}_\theta + \mathrm{i}\mathbf{e}_\phi)\\
&\quad + \frac{\mathrm{i}}{\sqrt{2}\,r}\Big[\partial_r(rF_{+1}) - \eth F_0\Big](\mathbf{e}_\theta - \mathrm{i}\mathbf{e}_\phi). \qquad (3.7)
\end{aligned}
$$

Hence,

$$
\nabla^2 f = \frac{1}{r^2}\partial_r(r^2\partial_r f) + \frac{1}{r^2}\bar{\eth}\eth f
$$

and, by virtue of the identity $\nabla\times(\nabla\times\mathbf{F}) = \nabla(\nabla\cdot\mathbf{F}) - \nabla^2\mathbf{F}$, we obtain

$$
\begin{aligned}
\nabla^2\mathbf{F} &= -\sqrt{2}\left[\partial_r\frac{1}{r^2}\partial_r(r^2F_0) + \frac{1}{r^2}\bar{\eth}\eth F_0 + \frac{1}{r^2}\eth F_{-1} - \frac{1}{r^2}\bar{\eth} F_{+1}\right]\mathbf{e}_r\\
&\quad - \frac{1}{\sqrt{2}}\left[\frac{1}{r}\partial_r^2(rF_{-1}) + \frac{1}{r^2}\bar{\eth}\eth F_{-1} - \frac{2}{r^2}\bar{\eth} F_0\right](\mathbf{e}_\theta + \mathrm{i}\mathbf{e}_\phi)\\
&\quad + \frac{1}{\sqrt{2}}\left[\frac{1}{r}\partial_r^2(rF_{+1}) + \frac{1}{r^2}\eth\bar{\eth} F_{+1} + \frac{2}{r^2}\eth F_0\right](\mathbf{e}_\theta - \mathrm{i}\mathbf{e}_\phi). \qquad (3.8)
\end{aligned}
$$

The vector Helmholtz equation,

$$
\nabla^2\mathbf{F} + k^2\mathbf{F} = 0, \qquad (3.9)
$$

where k is a constant, can be solved by separation of variables, looking for solutions of the form

$$
F_s = g_s(r)\,{}_sY_{jm}(\theta,\phi), \qquad (s = 0, \pm 1), \qquad (3.10)
$$

where the F_s are the spin-weighted components of $\mathbf{F}$, the $g_s(r)$ are functions of r only, and j is an integer greater than, or equal to, 1 [see (2.16)] (the case where $j = 0$ is considered below). Substituting (3.10) into (3.9), making use of (2.22), (2.27), (3.2), and (3.8), we obtain the system of ordinary differential equations

$$\begin{aligned} \frac{\mathrm{d}}{\mathrm{d}r}\frac{1}{r^2}\frac{\mathrm{d}}{\mathrm{d}r}(r^2 g_0) - \frac{j(j+1)}{r^2}g_0 + \frac{\sqrt{j(j+1)}}{r^2}(g_1 + g_{-1}) + k^2 g_0 &= 0, \\ \frac{1}{r}\frac{\mathrm{d}^2}{\mathrm{d}r^2}(r g_{\pm 1}) - \frac{j(j+1)}{r^2}g_{\pm 1} + \frac{2\sqrt{j(j+1)}}{r^2}g_0 + k^2 g_{\pm 1} &= 0, \end{aligned} \tag{3.11}$$

or, making use of the combinations

$$G \equiv \frac{\sqrt{j(j+1)}}{2}(g_1 - g_{-1}), \qquad H \equiv \frac{\sqrt{j(j+1)}}{2}(g_1 + g_{-1}), \tag{3.12}$$

we have

$$\left[\frac{\mathrm{d}^2}{\mathrm{d}r^2} + \frac{2}{r}\frac{\mathrm{d}}{\mathrm{d}r} + k^2 - \frac{j(j+1)}{r^2}\right] G = 0, \tag{3.13}$$

$$\frac{\mathrm{d}^2 g_0}{\mathrm{d}r^2} + \frac{2}{r}\frac{\mathrm{d}g_0}{\mathrm{d}r} - \frac{2g_0}{r^2} - \frac{j(j+1)}{r^2}g_0 + \frac{2H}{r^2} + k^2 g_0 = 0, \tag{3.14}$$

$$\frac{\mathrm{d}^2 H}{\mathrm{d}r^2} + \frac{2}{r}\frac{\mathrm{d}H}{\mathrm{d}r} - \frac{j(j+1)}{r^2}H + \frac{2j(j+1)}{r^2}g_0 + k^2 H = 0. \tag{3.15}$$

If $k \neq 0$, the solution of (3.13) is a linear combination of spherical Bessel functions, *e.g.*,

$$G = A j_j(kr) + B n_j(kr), \tag{3.16}$$

where A and B are arbitrary constants. On the other hand, (3.14) and (3.15) are equivalent to

$$\left[\frac{\mathrm{d}^2}{\mathrm{d}r^2} + \frac{2}{r}\frac{\mathrm{d}}{\mathrm{d}r} + k^2 - \frac{(j-1)j}{r^2}\right](H + j g_0) = 0,$$

$$\left[\frac{\mathrm{d}^2}{\mathrm{d}r^2} + \frac{2}{r}\frac{\mathrm{d}}{\mathrm{d}r} + k^2 - \frac{(j+1)(j+2)}{r^2}\right]\big(H - (j-1)g_0\big) = 0,$$

therefore,

$$\begin{aligned} H + j g_0 &= C j_{j-1}(kr) + D n_{j-1}(kr), \\ H - (j+1) g_0 &= E j_{j+1}(kr) + F n_{j+1}(kr), \end{aligned} \tag{3.17}$$

where C, D, E, and F are arbitrary constants. Thus, from (3.10), (3.12), (3.16), and (3.17) we find, for $j > 0$,

$$F_0 = \frac{1}{2j+1}\big[C j_{j-1}(kr) + D n_{j-1}(kr) - E j_{j+1}(kr) - F n_{j+1}(kr)\big] Y_{jm},$$

$$F_{\pm 1} = \frac{1}{\sqrt{j(j+1)}}\left[\pm A j_j(kr) \pm B n_j(kr) + \frac{j+1}{2j+1}\big(C j_{j-1}(kr) + D n_{j-1}(kr)\big) + \frac{j}{2j+1}\big(E j_{j+1}(kr) + F n_{j+1}(kr)\big)\right] {}_{\pm 1}Y_{jm}. \tag{3.18}$$

It may be remarked that whereas the spin-weighted components $F_{\pm 1}$ are separable, the components F_θ and F_ϕ are not.

In the case where $j = 0$, only F_0 can be different from zero, then (since ${}_0Y_{00}$ is a constant), looking for a solution of (3.9) of the form

$$F_0 = g_0(r), \qquad F_{\pm 1} = 0, \tag{3.19}$$

making use of (3.8), we obtain

$$\left[\frac{\mathrm{d}^2}{\mathrm{d}r^2} + \frac{2}{r}\frac{\mathrm{d}}{\mathrm{d}r} - \frac{2}{r^2} + k^2\right] g_0 = 0,$$

hence, if $k \neq 0$,

$$g_0(r) = A j_1(kr) + B n_1(kr), \tag{3.20}$$

where A and B are arbitrary constants.

Divergenceless solutions of the Helmholtz vector equation

Expressions (3.18) reduce considerably if the divergence of **F** vanishes. Indeed, substituting (3.18) into (3.6), making use of (2.27) and the recurrence relations

$$\begin{aligned} \frac{1}{x} z_l(x) &= \frac{1}{2l+1}[z_{l-1}(x) + z_{l+1}(x)], \\ \frac{\mathrm{d}z_l}{\mathrm{d}x}(x) &= \frac{1}{2l+1}[l z_{l-1}(x) - (l+1) z_{l+1}(x)], \end{aligned} \tag{3.21}$$

where the z_l are any of the spherical Bessel functions j_l, n_l, $h_l^{(1)}$, or $h_l^{(2)}$, it follows that the divergence of the vector field (3.18) is equal to zero if and only if

$$E = -C, \qquad F = -D. \tag{3.22}$$

When $j = 0$, from (3.6), (3.19), and (3.20) it follows that the divergence of **F** does not vanish for $k \neq 0$.

Assuming that the relations (3.22) hold, using (3.21) we find that

$$\begin{aligned} F_0 &= \frac{1}{kr}\big(C j_j(kr) + D n_j(kr)\big) Y_{jm}, \\ F_{\pm 1} &= \frac{1}{\sqrt{j(j+1)}}\left[\pm \big(A j_j(kr) + B n_j(kr)\big) + \frac{1}{kr}\frac{\mathrm{d}}{\mathrm{d}r} r\big(C j_j(kr) + D n_j(kr)\big)\right] {}_{\pm 1}Y_{jm} \end{aligned} \tag{3.23}$$

or, equivalently, making use of (2.27),

$$\begin{aligned} F_0 &= -\frac{1}{\sqrt{2}\,kr}\bar{\eth}\eth\psi_2, \\ F_{+1} &= -\frac{\mathrm{i}}{\sqrt{2}}\eth\psi_1 + \frac{1}{\sqrt{2}\,kr}\partial_r r\eth\psi_2, \\ F_{-1} &= -\frac{\mathrm{i}}{\sqrt{2}}\bar{\eth}\psi_1 - \frac{1}{\sqrt{2}\,kr}\partial_r r\bar{\eth}\psi_2, \end{aligned} \tag{3.24}$$

where

$$\begin{aligned} \psi_1 &\equiv \frac{\mathrm{i}\sqrt{2}}{j(j+1)}\big(Aj_j(kr) + Bn_j(kr)\big)\,Y_{jm}, \\ \psi_2 &\equiv \frac{\sqrt{2}}{j(j+1)}\big(Cj_j(kr) + Dn_j(kr)\big)\,Y_{jm}, \end{aligned} \tag{3.25}$$

are solutions of the scalar Helmholtz equation $\nabla^2\psi + k^2\psi = 0$.

According to (3.5) and (3.7), equations (3.24) are equivalent to

$$\mathbf{F} = \mathbf{r} \times \nabla\psi_1 + \frac{1}{k}\nabla \times (\mathbf{r} \times \nabla\psi_2), \tag{3.26}$$

or

$$\mathbf{F} = \mathrm{i}\mathbf{L}\psi_1 + \frac{\mathrm{i}}{k}\nabla \times \mathbf{L}\psi_2, \tag{3.27}$$

where

$$\mathbf{L} \equiv -\mathrm{i}\mathbf{r} \times \nabla. \tag{3.28}$$

From the completeness of the spin-weighted spherical harmonics and the linearity of the operators appearing in (3.24) and of the scalar Helmholtz equation, it follows that any divergenceless solution of the vector Helmholtz equation (3.9) can be expressed in the form (3.26), where the scalar potentials ψ_i are solutions of the scalar Helmholtz equation (see also Campbell and Morgan 1971). If the potentials ψ_i are real, then $\mathbf{F}$ is also real. The scalar potentials ψ_i are known as Debye potentials.

Relationship with the vector spherical harmonics

The expression (3.27) is useful in the study of electromagnetic radiation (see, *e.g.*, Eyges 1972, Jackson 1975). If the Debye potentials are expressed as series in the separable functions (3.25), then, using (3.27), the vector field $\mathbf{F}$ is given in terms of the vector spherical harmonics

$$\mathbf{X}_{lm} \equiv [l(l+1)]^{-1/2}\,\mathbf{L}Y_{lm} = [l(l+1)]^{-1/2}\,(-\mathrm{i}\mathbf{r} \times \nabla)Y_{lm}. \tag{3.29}$$

From (3.2) and (3.5) it follows that the components of $\mathbf{X}_{lm}$ are

$$(\mathbf{X}_{lm})_0 = 0, \qquad (\mathbf{X}_{lm})_{\pm 1} = \mp\frac{1}{\sqrt{2}}\,{}_{\pm 1}Y_{lm}. \tag{3.30}$$

These components can also be obtained making use of the relations

$$\mathbf{e}_\theta \pm \mathrm{i}\mathbf{e}_\phi = \tfrac{1}{2}\mathrm{e}^{-\mathrm{i}\phi}(\cos\theta \pm 1)(\mathbf{e}_x + \mathrm{i}\mathbf{e}_y) + \tfrac{1}{2}\mathrm{e}^{\mathrm{i}\phi}(\cos\theta \mp 1)(\mathbf{e}_x - \mathrm{i}\mathbf{e}_y) - \sin\theta\,\mathbf{e}_z$$

and

$$L_\pm Y_{lm} = \sqrt{l(l+1) - m(m\pm 1)}\,Y_{l,m\pm 1}, \qquad L_z Y_{lm} = m\,Y_{lm}, \tag{3.31}$$

where $L_\pm \equiv L_x \pm \mathrm{i}L_y$, which give

$$\begin{aligned}(\mathbf{X}_{lm})_{\pm 1} &= \pm\frac{1}{\sqrt{2l(l+1)}}\left\{\tfrac{1}{2}\mathrm{e}^{-\mathrm{i}\phi}(\cos\theta \pm 1)\sqrt{l(l+1) - m(m+1)}\,Y_{l,m+1}\right.\\ &\left. + \tfrac{1}{2}\mathrm{e}^{\mathrm{i}\phi}(\cos\theta \mp 1)\sqrt{l(l+1) - m(m-1)}\,Y_{l,m-1} - \sin\theta\,mY_{lm}\right\}.\end{aligned} \tag{3.32}$$

A comparison of (3.30) and (3.32) yields

$$\begin{aligned}{}_1Y_{lm} &= -[l(l+1)]^{-1/2}\left\{\tfrac{1}{2}\mathrm{e}^{-\mathrm{i}\phi}(\cos\theta + 1)\sqrt{l(l+1) - m(m+1)}\,Y_{l,m+1}\right.\\ &\quad\left. + \tfrac{1}{2}\mathrm{e}^{\mathrm{i}\phi}(\cos\theta - 1)\sqrt{l(l+1) - m(m-1)}\,Y_{l,m-1} - \sin\theta\,mY_{lm}\right\}\\ &= \sqrt{\frac{4\pi}{3}\frac{1}{l(l+1)}}\left\{\sqrt{l(l+1) - m(m+1)}\,{}_1Y_{1,-1}\,Y_{l,m+1}\right.\\ &\quad\left. - \sqrt{l(l+1) - m(m-1)}\,{}_1Y_{1,1}\,Y_{l,m-1} + \sqrt{2}\,m\,{}_1Y_{1,0}\,Y_{lm}\right\}\end{aligned} \tag{3.33}$$

and its complex conjugate.

From (3.30) and (2.61) it follows that

$$\sum_{m=-l}^{l} \overline{Y_{lm}}\,\mathbf{X}_{lm} = 0, \qquad \sum_{m=-l}^{l} \overline{\mathbf{X}_{lm}}\cdot\mathbf{X}_{lm} = \frac{2l+1}{4\pi}.$$

EXAMPLE. Superconducting sphere in a uniform magnetic field.

We shall consider a superconducting sphere of radius a placed in an originally uniform magnetic induction $B_b\mathbf{e}_z$. Outside the sphere, the magnetic induction and the magnetic field satisfy the equations $\nabla\cdot\mathbf{B} = 0$ and $\nabla\times\mathbf{H} = 0$, with $\mathbf{B} = \mathbf{H}$ (in Gaussian units). Hence, $\mathbf{B}$ is the gradient of some function, $\mathbf{B} = -\nabla\varphi_{\mathrm{M}}$, and from $\nabla \cdot \mathbf{B} = 0$ it follows that $\nabla^2 \varphi_M = 0$. If the origin is at the center of the sphere, the axial symmetry and the fact that as $r \to \infty$, $\mathbf{B} \to B_b \mathbf{e}_z$, imply that

$$\begin{aligned}\varphi_M &= -B_b r \cos\theta + \sum_{j=0}^{\infty} \frac{b_j}{r^{j+1}} P_j(\cos\theta) \\ &= -\sqrt{\frac{4\pi}{3}}\, B_b r Y_{1,0} + \sum_{j=0}^{\infty} \sqrt{\frac{4\pi}{2j+1}}\, \frac{b_j}{r^{j+1}} Y_{j,0},\end{aligned}$$

where the b_j are real constants. Then, from $\mathbf{B} = -\nabla\varphi_M$, (3.5), and (2.27) we find that the spin-weighted components of $\mathbf{B}$ are given by

$$\begin{aligned}B_0 &= -\sqrt{\frac{2\pi}{3}}\, B_b Y_{1,0} - \sum_{j=0}^{\infty} \sqrt{\frac{2\pi}{2j+1}}\, (j+1) \frac{b_j}{r^{j+2}} Y_{j,0}, \\ B_{\pm 1} &= -\sqrt{\frac{4\pi}{3}}\, B_b\, {}_{\pm 1}Y_{1,0} + \sum_{j=0}^{\infty} \sqrt{\frac{2\pi j(j+1)}{2j+1}}\, \frac{b_j}{r^{j+2}}\, {}_{\pm 1}Y_{j,0}.\end{aligned} \tag{3.34}$$

It will be assumed that, inside the sphere, the magnetic induction $\mathbf{B}$ obeys the equation $\nabla^2 \mathbf{B} = \lambda^{-2}\mathbf{B}$, where λ is the penetration depth (see, *e.g.*, Reitz, Milford, and Christy 1993); this is the vector Helmholtz equation (3.9) if we take $k = (\mathrm{i}\lambda)^{-1}$. Since the divergence of $\mathbf{B}$ vanishes, $\mathbf{B}$ is a superposition of fields of the form (3.23) with $k = (\mathrm{i}\lambda)^{-1}$ and $m = 0$, owing to the axial symmetry, which only contain the spherical Bessel functions j_j, since the functions n_j diverge at the origin,

$$\begin{aligned}B_0 &= \sum_{j=1}^{\infty} \frac{\mathrm{i}\lambda}{r} C_j j_j\!\left(\frac{r}{\mathrm{i}\lambda}\right) Y_{j,0}, \\ B_{\pm 1} &= \sum_{j=1}^{\infty} \frac{1}{\sqrt{j(j+1)}} \left[\pm A_j j_j\!\left(\frac{r}{\mathrm{i}\lambda}\right) + \frac{\mathrm{i}\lambda}{r} \frac{\mathrm{d}}{\mathrm{d}r} r\, C_j j_j\!\left(\frac{r}{\mathrm{i}\lambda}\right)\right] {}_{\pm 1}Y_{j,0}.\end{aligned} \tag{3.35}$$

The continuity of $\mathbf{B}$ at the boundary of the sphere implies that, at $r = a$, each component in (3.34) must be equal to the corresponding component in (3.35); hence, making use of the linear independence of the spin-weighted spherical harmonics for each value of the spin weight, it follows that the only nonvanishing coefficients are b_1 and C_1, which are related by

$$\begin{aligned}-\sqrt{\frac{2\pi}{3}}\, B_b - \sqrt{\frac{2\pi}{3}}\, \frac{2b_1}{a^3} &= \frac{\mathrm{i}\lambda}{a} C_1 j_1\!\left(\frac{a}{\mathrm{i}\lambda}\right), \\ -\sqrt{\frac{4\pi}{3}}\, B_b + \sqrt{\frac{4\pi}{3}}\, \frac{b_1}{a^3} &= \frac{\mathrm{i}\lambda}{\sqrt{2}\,a} \frac{\mathrm{d}}{\mathrm{d}r}\left[r\, C_1 j_1\!\left(\frac{r}{\mathrm{i}\lambda}\right)\right]\bigg|_{r=a}.\end{aligned}$$

Then, using the explicit expression

$$j_1\left(\frac{r}{\mathrm{i}\lambda}\right) = \mathrm{i}\left(\frac{\lambda^2}{r^2}\sinh\frac{r}{\lambda} - \frac{\lambda}{r}\cosh\frac{r}{\lambda}\right),$$

we obtain

$$b_1 = \frac{3}{2}B_b a^3\left(\frac{\lambda}{a}\coth\frac{a}{\lambda} - \frac{\lambda^2}{a^2} - \frac{1}{3}\right), \qquad C_1 = -\frac{\sqrt{6\pi}\, B_b\,(a/\lambda)}{\sinh(a/\lambda)}.$$

Characterization of the separable solutions. Angular momentum

The angular momentum operators are obtained by considering the rate of change of a scalar function, or another kind of field, under rotations. If $(a_{ij}) \in \mathrm{SO}(3)$ corresponds to the rotation about $\mathbf{n}$ through the angle α, then the inverse of (a_{ij}) is equal to its transpose, $\tilde{a}_{ij} = a_{ji}$ [see (1.38)] and from (1.36) we have $\mathrm{d}\tilde{a}_{ij}/\mathrm{d}\alpha\big|_{\alpha=0} = \varepsilon_{ijk}n_k$; hence, making use of the chain rule and (2.43) we find that the rate of change of a scalar function, f, under rotations about $\mathbf{n}$ is given by

$$\frac{\mathrm{d}}{\mathrm{d}\alpha}f(\tilde{a}_{ij}x_j)\bigg|_{\alpha=0} = \varepsilon_{ijk}n_k x_j \partial_i f(x_m) = n_k(-\mathrm{i}L_k f)(x_m)$$

where

$$L_k \equiv -\mathrm{i}\varepsilon_{kji}x_j\partial_i$$

are the Cartesian components of the angular momentum operator $\mathbf{L} = -\mathrm{i}\mathbf{r}\times\nabla$. Writing the operators L_k in terms of the spherical coordinates one finds the standard expressions

$$\begin{aligned} L_1 &= \mathrm{i}(\sin\phi\,\partial_\theta + \cot\theta\cos\phi\,\partial_\phi), \\ L_2 &= \mathrm{i}(-\cos\phi\,\partial_\theta + \cot\theta\sin\phi\,\partial_\phi), \\ L_3 &= -\mathrm{i}\partial_\phi. \end{aligned} \tag{3.36}$$

In order to find the rate of change of a vector, tensor or spinor field under rotations, it is necessary to take into account the fact that a rotation transforms the space points as well as the value of the field at each point. For example, in the case of a vector field, $\mathbf{F}$, the components of the image of $\mathbf{F}$ under the rotation $\mathcal{R}$ corresponding to (a_{ij}) are given by

$$[\mathcal{R}\mathbf{F}(x_i)]_j \equiv a_{jk}F_k(\tilde{a}_{im}x_m),$$

thus,

$$\begin{aligned} \frac{\mathrm{d}}{\mathrm{d}\alpha}[a_{jk}F_k(\tilde{a}_{im}x_m)]\bigg|_{\alpha=0} &= -\varepsilon_{jkp}n_p F_k(x_m) + \varepsilon_{imp}n_p x_m \partial_i F_j(x_k) \\ &= n_k\big(-\mathrm{i}J_k\mathbf{F}(x_m)\big)_j, \end{aligned}$$

where

$$J_k\mathbf{F} = L_k\mathbf{F} + \mathrm{i}\mathbf{e}_k \times \mathbf{F} \tag{3.37}$$

and $\{\mathbf{e}_1, \mathbf{e}_2, \mathbf{e}_3\}$ is the ordered basis $\{\mathbf{e}_x, \mathbf{e}_y, \mathbf{e}_z\}$. In the language of quantum mechanics, the operators J_k are the components of the total angular momentum operator, L_k corresponds to the orbital angular momentum and the term $\mathrm{i}\mathbf{e}_k \times \mathbf{F}$ corresponds to the intrinsic angular momentum of the vector field $\mathbf{F}$.

Expressing the vector field $\mathbf{F}$ in the form (3.2), we have

$$\begin{aligned}
J_k & \left[-\sqrt{2}\, F_0\, \mathbf{e}_r - \frac{1}{\sqrt{2}} F_{-1}(\mathbf{e}_\theta + \mathrm{i}\mathbf{e}_\phi) + \frac{1}{\sqrt{2}} F_{+1}(\mathbf{e}_\theta - \mathrm{i}\mathbf{e}_\phi)\right] \\
&= -\sqrt{2}(L_k F_0)\, \mathbf{e}_r - \frac{1}{\sqrt{2}}(L_k F_{-1})(\mathbf{e}_\theta + \mathrm{i}\mathbf{e}_\phi) + \frac{1}{\sqrt{2}}(L_k F_{+1})(\mathbf{e}_\theta - \mathrm{i}\mathbf{e}_\phi) \\
&\quad - \sqrt{2}\, F_0(L_k\mathbf{e}_r + \mathrm{i}\mathbf{e}_k \times \mathbf{e}_r) - \frac{1}{\sqrt{2}} F_{-1}\big(L_k(\mathbf{e}_\theta + \mathrm{i}\mathbf{e}_\phi) + \mathrm{i}\mathbf{e}_k \times (\mathbf{e}_\theta + \mathrm{i}\mathbf{e}_\phi)\big) \\
&\quad + \frac{1}{\sqrt{2}} F_{+1}\big(L_k(\mathbf{e}_\theta - \mathrm{i}\mathbf{e}_\phi) + \mathrm{i}\mathbf{e}_k \times (\mathbf{e}_\theta - \mathrm{i}\mathbf{e}_\phi)\big).
\end{aligned} \tag{3.38}$$

Making use of the relations

$$\begin{aligned}
\mathbf{e}_1 &= \sin\theta\cos\phi\, \mathbf{e}_r + \cos\theta\cos\phi\, \mathbf{e}_\theta - \sin\phi\, \mathbf{e}_\phi, \\
\mathbf{e}_2 &= \sin\theta\sin\phi\, \mathbf{e}_r + \cos\theta\sin\phi\, \mathbf{e}_\theta + \cos\phi\, \mathbf{e}_\phi, \\
\mathbf{e}_3 &= \cos\theta\, \mathbf{e}_r - \sin\theta\, \mathbf{e}_\theta,
\end{aligned}$$

together with

$$\begin{aligned}
\frac{\partial \mathbf{e}_r}{\partial\theta} &= \mathbf{e}_\theta, & \frac{\partial(\mathbf{e}_\theta + \mathrm{i}\mathbf{e}_\phi)}{\partial\theta} &= -\mathbf{e}_r, \\
\frac{\partial \mathbf{e}_r}{\partial\phi} &= \sin\theta\, \mathbf{e}_\phi, & \frac{\partial(\mathbf{e}_\theta + \mathrm{i}\mathbf{e}_\phi)}{\partial\phi} &= -\mathrm{i}\sin\theta\, \mathbf{e}_r - \mathrm{i}\cos\theta\, (\mathbf{e}_\theta + \mathrm{i}\mathbf{e}_\phi)
\end{aligned}$$

and (3.36), we find that

$$L_k\mathbf{e}_r + \mathrm{i}\mathbf{e}_k \times \mathbf{e}_r = 0, \qquad (k = 1, 2, 3)$$

which corresponds to the fact that $\mathbf{e}_r$ is invariant under any rotation about the origin, and

$$\begin{aligned}
L_1(\mathbf{e}_\theta \pm \mathrm{i}\mathbf{e}_\phi) + \mathrm{i}\mathbf{e}_1 \times (\mathbf{e}_\theta \pm \mathrm{i}\mathbf{e}_\phi) &= \pm\frac{\cos\phi}{\sin\theta}(\mathbf{e}_\theta \pm \mathrm{i}\mathbf{e}_\phi), \\
L_2(\mathbf{e}_\theta \pm \mathrm{i}\mathbf{e}_\phi) + \mathrm{i}\mathbf{e}_2 \times (\mathbf{e}_\theta \pm \mathrm{i}\mathbf{e}_\phi) &= \pm\frac{\sin\phi}{\sin\theta}(\mathbf{e}_\theta \pm \mathrm{i}\mathbf{e}_\phi), \\
L_3(\mathbf{e}_\theta \pm \mathrm{i}\mathbf{e}_\phi) + \mathrm{i}\mathbf{e}_3 \times (\mathbf{e}_\theta \pm \mathrm{i}\mathbf{e}_\phi) &= 0.
\end{aligned} \tag{3.39}$$

Thus, from (3.38) and (3.39), it follows that the spin-weighted components of $J_k\mathbf{F}$ can be expressed as

$$(J_k\mathbf{F})_s = J_k^{(s)} F_s, \tag{3.40}$$

where we have introduced the operators

$$\begin{aligned} J_1^{(s)} &\equiv L_1 - s\frac{\cos\phi}{\sin\theta} = \mathrm{i}(\sin\phi\,\partial_\theta + \cot\theta\cos\phi\,\partial_\phi + \mathrm{i}s\csc\theta\cos\phi), \\ J_2^{(s)} &\equiv L_2 - s\frac{\sin\phi}{\sin\theta} = \mathrm{i}(-\cos\phi\,\partial_\theta + \cot\theta\sin\phi\,\partial_\phi + \mathrm{i}s\csc\theta\sin\phi), \\ J_3^{(s)} &\equiv L_3 = -\mathrm{i}\partial_\phi. \end{aligned} \tag{3.41}$$

Note that $J_k^{(0)} = L_k$. According to (3.40), the total angular momentum operators $J_k^{(s)}$ do not change the spin weight.

A straightforward computation [using (3.41) or (3.37)] gives

$$[J_i^{(s)}, J_j^{(s)}] = \mathrm{i}\varepsilon_{ijk} J_k^{(s)} \tag{3.42}$$

and, using (2.23),

$$J^{(s)2} \equiv J_1^{(s)2} + J_2^{(s)2} + J_3^{(s)2} = -\bar{\eth}\eth + s(s+1),$$

therefore, the spin-weighted spherical harmonics ${}_sY_{jm}$ are eigenfunctions of $J^{(s)2}$ and $J_3^{(s)}$,

$$J^{(s)2}\,{}_sY_{jm} = j(j+1)\,{}_sY_{jm}, \qquad J_3^{(s)}\,{}_sY_{jm} = m\,{}_sY_{jm}. \tag{3.43}$$

Furthermore, from the definitions (2.17) and (3.41) it follows that, for s integral or half-integral,

$$J_k^{(s+1)}\eth = \eth J_k^{(s)}, \qquad J_k^{(s-1)}\bar{\eth} = \bar{\eth} J_k^{(s)}. \tag{3.44}$$

Following the standard procedure, from (3.42) and (3.43) one concludes that $(J_1^{(s)} \pm \mathrm{i}J_2^{(s)})\,{}_sY_{jm}$ is proportional to ${}_sY_{j,m\pm1}$. In particular, for s integral, using (2.28), (3.31), and (3.44),

$$(J_1^{(s)} \pm \mathrm{i}J_2^{(s)})\,{}_sY_{jm} = \sqrt{j(j+1) - m(m\pm1)}\,{}_sY_{j,m\pm1}. \tag{3.45}$$

According to (3.41), the raising and lowering operators $J_\pm^{(s)} \equiv J_1^{(s)} \pm \mathrm{i}J_2^{(s)}$ are given in terms of the spherical coordinates by

$$J_\pm^{(s)} = \pm e^{\pm\mathrm{i}\phi}\left(\partial_\theta \pm \mathrm{i}\cot\theta\,\partial_\phi \mp \frac{s}{\sin\theta}\right)$$

[*cf.* (2.17)].

Alternatively, from (2.6) and (3.41) we obtain

$$J_-^{(1/2)} o^1 = 0, \qquad J_+^{(1/2)} o^1 = -o^2,$$
$$J_+^{(1/2)} o^2 = 0, \qquad J_-^{(1/2)} o^2 = -o^1. \tag{3.46}$$

Then, noting that

$$\overline{J_\pm^{(s)} \eta} = -J_\mp^{(-s)} \overline{\eta},$$

from (3.46) we have

$$J_-^{(-1/2)} \widehat{o^1} = 0, \qquad J_+^{(-1/2)} \widehat{o^1} = -\widehat{o^2},$$
$$J_+^{(-1/2)} \widehat{o^2} = 0, \qquad J_-^{(-1/2)} \widehat{o^2} = -\widehat{o^1}. \tag{3.47}$$

Furthermore, if η and κ have spin weight s and s', respectively,

$$J_k^{(s+s')}(\eta\kappa) = \eta J_k^{(s')} \kappa + \kappa J_k^{(s)} \eta, \tag{3.48}$$

hence, making use of the first equality in (2.30), and (3.46)–(3.48) one finds that the relation (3.45) holds for all values of s.

It may be remarked that the spin-weighted spherical harmonics can be constructed using the fact that $J_+^{(s)}\, {}_sY_{jj} = 0$ and applying $J_-^{(s)}$ repeatedly to ${}_sY_{jj}$, in a form analogous to that employed in some textbooks to obtain the ordinary spherical harmonics.

From (3.40) and (3.43) it follows that a vector field $\mathbf{F}$ is an eigenfunction of J_3 and $J^2 \equiv J_1{}^2 + J_2{}^2 + J_3{}^2$, with eigenvalues m and $j(j+1)$, if and only if its spin-weighted components are of the form $F_s = g_s(r)\, {}_sY_{jm}$ [see (3.10)].

The fact that the radial equations (3.11) can be reduced to three independent second-order differential equations is related to the existence of an operator that commutes with J^2, J_3 and ∇^2. We start by noticing that the separable solutions (3.18) can be written in the form

$$\begin{pmatrix} F_{-1} \\ F_0 \\ F_{+1} \end{pmatrix} = \frac{f_j(kr)}{\sqrt{2}} \begin{pmatrix} {}_{-1}Y_{jm} \\ 0 \\ -({}_1Y_{jm}) \end{pmatrix} + \frac{f_{j-1}(kr)}{\sqrt{2(2j+1)}} \begin{pmatrix} -\sqrt{j+1}\, {}_{-1}Y_{jm} \\ -\sqrt{j}\, Y_{jm} \\ -\sqrt{j+1}\, {}_1Y_{jm} \end{pmatrix}$$
$$+ \frac{f_{j+1}(kr)}{\sqrt{2(2j+1)}} \begin{pmatrix} -\sqrt{j}\, {}_{-1}Y_{jm} \\ \sqrt{j+1}\, Y_{jm} \\ -\sqrt{j}\, {}_1Y_{jm} \end{pmatrix}, \tag{3.49}$$

where f_l is a spherical Bessel function of order l. The vector fields

$$\mathbf{X}_{jm} = \frac{1}{\sqrt{2}} \begin{pmatrix} {}_{-1}Y_{jm} \\ 0 \\ -({}_1Y_{jm}) \end{pmatrix},$$

$$\mathbf{W}_{jm} = \frac{1}{\sqrt{2(2j+1)}} \begin{pmatrix} -\sqrt{j+1}\,{}_{-1}Y_{jm} \\ -\sqrt{j}\,Y_{jm} \\ -\sqrt{j+1}\,{}_{1}Y_{jm} \end{pmatrix}, \tag{3.50}$$

$$\mathbf{V}_{jm} = \frac{1}{\sqrt{2(2j+1)}} \begin{pmatrix} -\sqrt{j}\,{}_{-1}Y_{jm} \\ \sqrt{j+1}\,Y_{jm} \\ -\sqrt{j}\,{}_{1}Y_{jm} \end{pmatrix},$$

appearing in (3.49), are eigenvectors of the differential operator

$$K \equiv \begin{pmatrix} 0 & -\bar{\eth} & 0 \\ \frac{1}{2}\bar{\eth} & -1 & -\frac{1}{2}\bar{\eth} \\ 0 & \eth & 0 \end{pmatrix},$$

with eigenvalues 0, j, and $-j-1$, respectively (the entries of the columns in (3.49) and (3.50) give the spin-weighted components of the vector fields). According to (3.30), the vector field $\mathbf{X}_{jm}$ given by (3.50) is the usual vector spherical harmonic (3.29); the vector fields $\mathbf{W}_{jm}$ and $\mathbf{V}_{jm}$ are the vector spherical harmonics defined in Hill (1954) and Arfken (1985).

A straightforward computation shows that, with respect to the Cartesian basis $\{\mathbf{e}_x, \mathbf{e}_y, \mathbf{e}_z\}$, the operator K is given by

$$K = I + \mathbf{L} \cdot \mathbf{S},$$

where $\mathbf{L}$ and $\mathbf{S}$ are the orbital and spin angular momentum operators with components $(L_k)_{jl} = -\mathrm{i}\delta_{jl}\varepsilon_{krs}x_r\partial/\partial x_s$, $(S_k)_{jl} = \mathrm{i}\varepsilon_{jkl}$ [see (3.37)], and I is the 3×3 unit matrix. Since $S^2 = S_k S_k = 2I$, the square of the total angular momentum, J^2, can be expressed as $J^2 = L^2 + 2\mathbf{L}\cdot\mathbf{S} + S^2 = L^2 + 2K$; hence, $L^2 = J^2 - 2K$, which shows that the vector fields $\mathbf{X}_{jm}$, $\mathbf{W}_{jm}$, and $\mathbf{V}_{jm}$, being eigenvectors of J^2 and K, are eigenvectors of L^2 with eigenvalue $l(l+1)$, and $l = j, j-1, j+1$, respectively. (Note that, according to the rules for the addition of angular momenta, for a spin-1 field, the only possible values of l are $j-1$, j, and $j+1$, provided that j is different from zero.) It may be noticed that the value of l of each vector field on the right-hand side of (3.49) coincides with the order of the accompanying spherical Bessel function.

The separable solution of the vector Helmholtz equation with $j = 0$, given by (3.19) and (3.20), is an eigenfunction of K with eigenvalue -1 (in this case, $\mathbf{F}$ is proportional to $\mathbf{V}_{00}$; $\mathbf{X}_{jm}$ and $\mathbf{W}_{jm}$ are different from zero only if $j \geqslant 1$); therefore, it is an eigenfunction of L^2 with $l = 1$.

Under the inversion $\mathbf{r} \mapsto -\mathbf{r}$, a spin-weighted spherical harmonic of integral order, ${}_sY_{jm}$, is mapped into $(-1)^j\,{}_{-s}Y_{jm}$. Hence, assuming that under the inversion $\mathbf{e}_r$ and $\mathbf{e}_\phi$ are left unchanged while $\mathbf{e}_\theta$ changes sign (as in Davydov 1988), it follows that the parity of the vector fields $\mathbf{X}_{jm}$, $\mathbf{W}_{jm}$ and $\mathbf{V}_{jm}$ is $(-1)^{j+1}$, $(-1)^j$ and

$(-1)^j$, respectively. (According to the convention followed in Jackson 1975 and Arfken 1985, the parity of $\mathbf{X}_{jm}$, $\mathbf{W}_{jm}$ and $\mathbf{V}_{jm}$ is $(-1)^j$, $(-1)^{j+1}$ and $(-1)^{j+1}$, respectively.)

Eigenfunctions of the curl operator

The vector field $\mathbf{u}$ is an eigenfunction of the curl operator with eigenvalue λ if

$$\nabla \times \mathbf{u} = \lambda \mathbf{u}. \tag{3.51}$$

Taking the divergence on both sides of (3.51) we obtain $0 = \lambda \nabla \cdot \mathbf{u}$; therefore, if $\lambda \neq 0$,

$$\nabla \cdot \mathbf{u} = 0. \tag{3.52}$$

Equations (3.51) and (3.52) imply that $\mathbf{u}$ obeys the Helmholtz equation $\nabla^2 \mathbf{u} + \lambda^2 \mathbf{u} = 0$; thus, the eigenfunctions of the curl operator with eigenvalue $\lambda \neq 0$ are of the form $\mathbf{u} = \mathbf{r} \times \nabla \psi_1 + \lambda^{-1} \nabla \times (\mathbf{r} \times \nabla \psi_2)$ [see (3.26)], where ψ_1 and ψ_2 satisfy the scalar Helmholtz equation. Then, $\nabla \times \mathbf{u} = \nabla \times (\mathbf{r} \times \nabla \psi_1) + \lambda \mathbf{r} \times \nabla \psi_2$, which coincides with $\lambda \mathbf{u}$ if $\psi_1 = \psi_2$. Hence, making $\psi \equiv \lambda^{-1} \psi_1$, we find that the eigenfunctions of the curl operator with nonzero eigenvalue can be expressed in the form

$$\mathbf{u} = \lambda \, \mathbf{r} \times \nabla \psi + \nabla \times (\mathbf{r} \times \nabla \psi), \tag{3.53}$$

where ψ is a solution of the Helmholtz equation

$$\nabla^2 \psi + \lambda^2 \psi = 0. \tag{3.54}$$

3.2 The source-free electromagnetic field

The electric and magnetic fields in vacuum, in a source-free region, are divergenceless and, if it is assumed that they have a harmonic time dependence with frequency ω, satisfy the vector Helmholtz equation

$$\nabla^2 \mathbf{E} + k^2 \mathbf{E} = 0, \qquad \nabla^2 \mathbf{B} + k^2 \mathbf{B} = 0,$$

with $k = \omega / c$. Thus, according to (3.26), the electric field of a monochromatic wave can be expressed as

$$\begin{aligned} \mathbf{E} &= \mathrm{Re}\left[\left(\mathbf{r} \times \nabla \psi_1 + \frac{1}{k} \nabla \times (\mathbf{r} \times \nabla \psi_2)\right) \mathrm{e}^{-\mathrm{i}\omega t}\right] \\ &= \mathrm{Re}\left[\frac{\mathrm{i}}{\omega} \partial_t (\mathbf{r} \times \nabla \psi_1 \mathrm{e}^{-\mathrm{i}\omega t}) + \frac{1}{k} \nabla \times (\mathbf{r} \times \nabla \psi_2 \mathrm{e}^{-\mathrm{i}\omega t})\right] \\ &= \frac{1}{c} \partial_t (\mathbf{r} \times \nabla \chi_{\mathrm{M}}) - \nabla \times (\mathbf{r} \times \nabla \chi_{\mathrm{E}}), \end{aligned} \tag{3.55}$$

where $\chi_{\rm M} = {\rm Re}\,({\rm i}/k)\psi_1 {\rm e}^{-{\rm i}\omega t}$ and $\chi_{\rm E} = -{\rm Re}\,(1/k)\psi_2 {\rm e}^{-{\rm i}\omega t}$ obey the wave equation, $\nabla^2\chi - (1/c^2)\,\partial_t^2\chi = 0$. Then, making use of the equation $\nabla \times \mathbf{E} = -(1/c)\,\partial_t \mathbf{B}$ one finds that

$$\mathbf{B} = -\frac{1}{c}\partial_t(\mathbf{r} \times \nabla\chi_{\rm E}) - \nabla \times (\mathbf{r} \times \nabla\chi_{\rm M}). \tag{3.56}$$

Owing to the linearity of the wave equation and the fact that any electromagnetic field can be expressed as a superposition of monochromatic waves, any solution of the source-free Maxwell equations can be written in the form (3.55) and (3.56). If the potentials $\chi_{\rm E}$ and $\chi_{\rm M}$ are real, then the fields $\mathbf{E}$ and $\mathbf{B}$ are also real. Given the fields $\mathbf{E}$ and $\mathbf{B}$, the potentials $\chi_{\rm E}$ and $\chi_{\rm M}$ can be obtained, noting that (3.55) and (3.56) lead to

$$\mathbf{r} \cdot \mathbf{E} = L^2\chi_{\rm E}, \qquad \mathbf{r} \cdot \mathbf{B} = L^2\chi_{\rm M}. \tag{3.57}$$

The usual electromagnetic potentials, φ and $\mathbf{A}$, can also be expressed in terms of the scalar potentials $\chi_{\rm E}$ and $\chi_{\rm M}$; using the fact that the latter satisfy the wave equation one can verify that the potentials

$$\varphi = -\frac{1}{r}\mathbf{r} \cdot \nabla(r\chi_{\rm E}), \qquad \mathbf{A} = \frac{1}{c}(\partial_t\chi_{\rm E})\,\mathbf{r} - \mathbf{r} \times \nabla\chi_{\rm M}$$

correspond to the electromagnetic field (3.55) and (3.56).

As shown below, it is convenient to make use, in place of $\mathbf{E}$ and $\mathbf{B}$, of the *complex* vector field

$$\mathbf{F} \equiv \mathbf{E} + {\rm i}\mathbf{B}. \tag{3.58}$$

According to (3.55) and (3.56), the vector field $\mathbf{F}$ can be written as

$$\mathbf{F} = -\frac{\rm i}{c}\partial_t(\mathbf{r} \times \nabla\chi) - \nabla \times (\mathbf{r} \times \nabla\chi), \tag{3.59}$$

where $\chi \equiv \chi_{\rm E} + {\rm i}\chi_{\rm M}$ is a (possibly complex) solution of the wave equation, thus showing that any solution of the source-free Maxwell equations in vacuum can be written in terms of a single complex scalar potential. From the definitions (3.3) and (3.58) it follows that

$$\begin{aligned} F_0 &= -\frac{1}{\sqrt{2}}(E_r + {\rm i}B_r), \\ F_{\pm1} &= -\frac{1}{\sqrt{2}}[\pm E_\theta - B_\phi + {\rm i}(E_\phi \pm B_\theta)], \end{aligned} \tag{3.60}$$

therefore, the radial component of the Poynting vector, given by $S_r = (c/4\pi)(E_\theta B_\phi - E_\phi B_\theta)$, amounts to

$$S_r = \frac{c}{8\pi}(|F_{-1}|^2 - |F_{+1}|^2). \tag{3.61}$$

On the other hand, the spin-weighted components of (3.59) are

$$\begin{aligned} F_{+1} &= -\frac{1}{\sqrt{2}\,r}\left(\frac{1}{c}\partial_t + \partial_r\right) r\eth\chi, \\ F_0 &= \frac{1}{\sqrt{2}\,r}\bar{\eth}\eth\chi, \\ F_{-1} &= -\frac{1}{\sqrt{2}\,r}\left(\frac{1}{c}\partial_t - \partial_r\right) r\bar{\eth}\chi. \end{aligned} \tag{3.62}$$

Asymptotic behavior of the solutions

The scalar wave equation admits separable solutions in spherical coordinates of the form

$$\chi = \left(Ah_j^{(1)}(kr) + Bh_j^{(2)}(kr)\right)\mathrm{i}^j Y_{jm}(\theta,\phi)\mathrm{e}^{-\mathrm{i}\omega t}, \tag{3.63}$$

where A, B are arbitrary constants, $h_j^{(1)}$ and $h_j^{(2)}$ are spherical Hankel functions, and the factor i^j is introduced for convenience. Making use of the asymptotic form of the spherical Hankel functions

$$h_j^{(1)}(kr) \sim (-\mathrm{i})^{j+1}\frac{\mathrm{e}^{\mathrm{i}kr}}{kr}\left(1 + \frac{\mathrm{i}}{2}\frac{(j+1)!}{(j-1)!}\frac{1}{kr} - \frac{1}{2^2 2!}\frac{(j+2)!}{(j-2)!}\frac{1}{(kr)^2} + \cdots\right), \tag{3.64}$$

we find that

$$\begin{aligned} \left(\mathrm{i} + \frac{1}{kr}\partial_r r\right) h_j^{(1)}(kr) &\sim 2\mathrm{i}(-\mathrm{i})^{j+1}\frac{\mathrm{e}^{\mathrm{i}kr}}{kr}\left(1 + O\left(\frac{1}{kr}\right)\right), \\ \left(-\mathrm{i} + \frac{1}{kr}\partial_r r\right) h_j^{(1)}(kr) &\sim -\frac{\mathrm{i}}{2}(-\mathrm{i})^{j+1} j(j+1)\frac{\mathrm{e}^{\mathrm{i}kr}}{(kr)^3}\left(1 + O\left(\frac{1}{kr}\right)\right), \end{aligned} \tag{3.65}$$

therefore, if the potential χ in (3.63) only contains outgoing waves, *i.e.*, $B = 0$, substituting (3.63) into (3.62), with the aid of (2.27), (3.64), and (3.65), we obtain the asymptotic expressions

$$\begin{aligned} F_{+1} &\sim \frac{Ak}{2\sqrt{2}}[j(j+1)]^{3/2}\frac{\mathrm{e}^{\mathrm{i}(kr-\omega t)}}{(kr)^3}\,{}_1Y_{jm}, \\ F_0 &\sim \mathrm{i}\frac{Ak}{\sqrt{2}}j(j+1)\frac{\mathrm{e}^{\mathrm{i}(kr-\omega t)}}{(kr)^2}\,Y_{jm}, \\ F_{-1} &\sim -\sqrt{2}\,Ak[j(j+1)]^{1/2}\frac{\mathrm{e}^{\mathrm{i}(kr-\omega t)}}{kr}\,{}_{-1}Y_{jm}. \end{aligned}$$

Therefore, for outgoing waves, F_{-1} is the dominant component at large distances and

$$F_s = O\left(\frac{1}{r^{2+s}}\right). \tag{3.66}$$

(Relations similar to (3.66) apply to the massless fields of any spin in an asymptotically simple space-time and this result is known as the "peeling theorem" (see, *e.g.*, Penrose and Rindler 1986, Stewart 1990).) From (3.61) and (3.66) it follows that the outgoing energy flux per unit time and unit solid angle is

$$\frac{d^2 E_{\text{out}}}{dt d\Omega} = \lim_{r\to\infty} \frac{c}{8\pi} r^2 |F_{-1}|^2. \tag{3.67}$$

Similarly, making use of the complex conjugates of (3.65) and the fact that $h_j^{(2)}(kr) = \overline{h_j^{(1)}(kr)}$, one finds that the potential

$$\chi = B i^j h_j^{(2)}(kr) Y_{jm}(\theta, \phi) e^{-i\omega t}$$

produces an electromagnetic field such that

$$\begin{aligned} F_{+1} &\sim (-1)^j [j(j+1)]^{1/2} \frac{e^{-i(kr+\omega t)}}{kr} \, {}_1Y_{jm}, \\ F_0 &\sim (-1)^j i j(j+1) \frac{e^{-i(kr+\omega t)}}{(kr)^2} \, Y_{jm}, \\ F_{-1} &\sim (-1)^{j+1} [j(j+1)]^{3/2} \frac{e^{-i(kr+\omega t)}}{(kr)^3} \, {}_{-1}Y_{jm}. \end{aligned}$$

Hence, for ingoing waves, F_{+1} is the dominant component at large distances and

$$F_s = O\left(\frac{1}{r^{2-s}}\right) \tag{3.68}$$

[*cf.* (3.66)]. Equations (3.61) and (3.68) imply that the energy flux per unit time and unit solid angle of the ingoing waves is

$$\frac{d^2 E_{\text{in}}}{dt d\Omega} = \lim_{r\to\infty} \frac{c}{8\pi} r^2 |F_{+1}|^2.$$

Thus, in the radiation zone, the component F_{-1} represents the outgoing field, while F_{+1} represents the ingoing field.

It should be remarked that in order to study the asymptotic behavior of the solutions of the Maxwell equations, it is not necessary to assume that the electromagnetic field has a time dependence of the form $\exp(-i\omega t)$. Such a dependence appeared in (3.63) because we considered separable solutions of the wave equation in the variables t, r, θ, ϕ. In this context, it is more convenient to employ the null coordinate $u \equiv ct - r$, or $v \equiv ct + r$, together with r, θ, ϕ. For instance, in terms of the coordinates u, r, θ, ϕ the spin-weighted components (3.62) take the form

$$F_{+1} = -\frac{1}{\sqrt{2}r} \partial_r (r \eth \chi),$$

$$F_0 = \frac{1}{\sqrt{2}r}\bar{\eth}\eth\chi, \tag{3.69}$$

$$F_{-1} = -\frac{1}{\sqrt{2}r}(2\partial_u - \partial_r)r\bar{\eth}\chi$$

and the wave equation is

$$2\partial_u\partial_r\chi + \frac{2}{r}\partial_u\chi - \frac{1}{r^2}\partial_r(r^2\partial_r\chi) - \frac{1}{r^2}\bar{\eth}\eth\chi = 0. \tag{3.70}$$

Looking for solutions of the wave equation such that [*cf.* (3.63) and (3.64)]

$$\chi = \sum_{n=1}^{N}\frac{f_n(u,\theta,\phi)}{r^n} + O\left(\frac{1}{r^{N+1}}\right), \tag{3.71}$$

from (3.70) we obtain the relations

$$2n\partial_u f_{n+1} + n(n-1)f_n + \bar{\eth}\eth f_n = 0. \tag{3.72}$$

Then, from (3.69) and (3.71) we obtain, *e.g.*,

$$F_{+1} = \frac{1}{\sqrt{2}}\sum_{n=3}^{N+1}\frac{F_{+1}^{(n)}(u,\theta,\phi)}{r^n} + O\left(\frac{1}{r^{N+2}}\right), \tag{3.73}$$

with

$$F_{+1}^{(n)} = (n-2)\eth f_{n-1}. \tag{3.74}$$

The Newman–Penrose conserved quantities

By virtue of the completeness of the spherical harmonics, the function f_n can be expanded as

$$f_n(u,\theta,\phi) = \sum_{l=0}^{\infty}\sum_{m=-l}^{l} a_{nlm}(u)\, Y_{lm}(\theta,\phi). \tag{3.75}$$

Using (2.22) and the linear independence of the spherical harmonics we find that (3.72) is equivalent to

$$2n\frac{\mathrm{d}}{\mathrm{d}u}a_{n+1,lm} + [n(n-1) - l(l+1)]a_{nlm} = 0.$$

Hence,

$$a_{l+2,lm} = \text{const.} \tag{3.76}$$

These constants can be expressed directly in terms of the components of the electromagnetic field. Indeed, from (3.75) and (2.27) one finds

$$F_{+1}^{(n)} = (n-2)\sum_{l,m}\sqrt{l(l+1)}\,a_{n-1,lm}(u)\,{}_1Y_{lm}(\theta,\phi)$$

and, using the orthonormality of the spin-weighted spherical harmonics with respect to the inner product (2.7),

$$({}_1Y_{lm}, F_{+1}^{(l+3)}) = (l+1)\sqrt{l(l+1)}\, a_{l+2,lm}$$

thus showing that

$$\int_{S^2} \overline{{}_1Y_{lm}}\, F_{+1}^{(l+3)} \mathrm{d}\Omega = \text{const.} \tag{3.77}$$

The infinite set of conserved quantities (3.77) was obtained by Newman and Penrose (1968), who showed that a similar result holds for massless fields of any spin and that, in the nonlinear Einstein–Maxwell theory, for an asymptotically flat space-time, there also exist some absolutely conserved quantities (see also Penrose and Rindler 1986). These conserved quantities, defined at future null infinity, characterize the time profile of the incoming radiation at the past null infinity. In the nonlinear Einstein–Maxwell theory only six of these electromagnetic and ten gravitational Newman–Penrose quantities remain conserved.

Polarization

The polarization of the radiation can be also readily determined from the spin-weighted components $F_{\pm 1}$. In fact, if F_s has a time dependence of the form

$$F_s(t) = F_s(0)\mathrm{e}^{\mathrm{i}\omega t},$$

where $F_s(0)$ is the value of F_s at $t = 0$, comparison with (2.14) shows that the time evolution of F_s amounts to rotating the vectors $\mathbf{e}_\theta$ and $\mathbf{e}_\phi$, about $\mathbf{e}_r$, with an angular velocity $-\omega/s$ or, equivalently, to rotating $\mathbf{F}$ about $\mathbf{e}_r$ with an angular velocity ω/s. Therefore, if F_s, with $s \neq 0$, is proportional to $\mathrm{e}^{\mathrm{i}\omega t}$ or to $\mathrm{e}^{-\mathrm{i}\omega t}$, the field has circular polarization; while the presence of both factors, $\mathrm{e}^{\mathrm{i}\omega t}$ y $\mathrm{e}^{-\mathrm{i}\omega t}$, means that the radiation has elliptic polarization.

If in the radiation zone F_{-1} is proportional to $\mathrm{e}^{\mathrm{i}\omega t}$, the outgoing radiation has right circular polarization (negative helicity) if $\omega > 0$ or left circular polarization (positive helicity) if $\omega < 0$. Since F_{+1} has spin-weight opposite to that of F_{-1} and corresponds to waves propagating in the direction $-\mathbf{e}_r$, the foregoing conclusions are equally valid for F_{+1}; that is, if F_{+1} is proportional to $\mathrm{e}^{\mathrm{i}\omega t}$, the ingoing radiation has right or left circular polarization according to whether ω is positive or negative, respectively.

Expansion of a plane wave

The electric field of a circularly polarized plane wave propagating in the direction $\mathbf{e}_z$ of unit amplitude is

$$\mathbf{E} = \mathrm{Re}\,[\mathrm{e}^{\mathrm{i}(kz-\omega t)}(\mathbf{e}_x \pm \mathrm{i}\mathbf{e}_y)] = \cos(kz-\omega t)\,\mathbf{e}_x \mp \sin(kz-\omega t)\,\mathbf{e}_y,$$

where the sign $\pm$ corresponds to the helicity of the wave. Using the equation $\nabla \times \mathbf{E} = -(1/c)\,\partial_t \mathbf{B}$ we find that $\mathbf{B} = \pm\sin(kz - \omega t)\,\mathbf{e}_x + \cos(kz - \omega t)\,\mathbf{e}_y$, therefore, $\mathbf{F} = \mathrm{e}^{\pm\mathrm{i}(kz-\omega t)}(\mathbf{e}_x + \mathrm{i}\mathbf{e}_y)$ and $\mathbf{r}\cdot\mathbf{F} = (x + \mathrm{i}y)\mathrm{e}^{\pm\mathrm{i}(kz-\omega t)} = (\mp\mathrm{i}/k)[x\partial_z - z\partial_x + \mathrm{i}(y\partial_z - z\partial_y)]\mathrm{e}^{\pm\mathrm{i}(kz-\omega t)} = (\pm\mathrm{i}/k)(L_x + \mathrm{i}L_y)\mathrm{e}^{\pm\mathrm{i}(kz-\omega t)}$. Hence, using the well-known expansion

$$\mathrm{e}^{\mathrm{i}kz} = \sum_{j=0}^{\infty} \mathrm{i}^j (2j+1) j_j(kr) P_j(\cos\theta) = \sum_{j=0}^{\infty} \mathrm{i}^j \sqrt{4\pi(2j+1)}\, j_j(kr) Y_{j0}(\theta,\phi), \tag{3.78}$$

and (3.31) we obtain

$$\begin{aligned} \mathbf{r}\cdot\mathbf{F} &= \pm\frac{\mathrm{i}}{k}\sum_{j=0}^{\infty}\sqrt{4\pi(2j+1)j(j+1)}\,(\pm\mathrm{i})^j j_j(kr) Y_{j,1}\mathrm{e}^{\mp\mathrm{i}\omega t} \\ &= L^2\left(\frac{1}{k}\sum_{j=1}^{\infty}\left[\frac{4\pi(2j+1)}{j(j+1)}\right]^{1/2}(\pm\mathrm{i})^{j+1} j_j(kr) Y_{j,1}\mathrm{e}^{\mp\mathrm{i}\omega t}\right). \end{aligned}$$

On the other hand, $\mathbf{r}\cdot\mathbf{F} = L^2\chi$ [see (3.57)], hence we can take

$$\chi = \frac{1}{k}\sum_{j=1}^{\infty}\left[\frac{4\pi(2j+1)}{j(j+1)}\right]^{1/2}(\pm\mathrm{i})^{j+1} j_j(kr) Y_{j,1}\mathrm{e}^{\mp\mathrm{i}\omega t}. \tag{3.79}$$

Substituting the expression (3.79) into (3.59) or (3.62) one obtains the multipole expansion of the electromagnetic field corresponding to a circularly polarized plane wave (*cf.* Jackson 1975, Sect. 16.8).

The potential, χ, corresponding to a circularly polarized plane wave propagating in an arbitrary direction, with polar and azimuth angles θ_1 and ϕ_1, can now be obtained by calculating the effect on (3.79) of the rotation with Euler angles $(\phi_1, \theta_1, 0)$, which takes the vector $\mathbf{e}_z$ into the new direction of propagation. In this manner, from (3.79), (2.45), and (2.51) we have

$$\begin{aligned} \chi &= \frac{1}{k}\sum_{j=1}^{\infty}\left[\frac{4\pi(2j+1)}{j(j+1)}\right]^{1/2}(\pm\mathrm{i})^{j+1} j_j(kr)\,\mathcal{R}(\phi_1,\theta_1,0)Y_{j,1}\mathrm{e}^{\mp\mathrm{i}\omega t} \\ &= \frac{1}{k}\sum_{j=1}^{\infty}\left[\frac{4\pi(2j+1)}{j(j+1)}\right]^{1/2}(\pm\mathrm{i})^{j+1} j_j(kr)\sum_{m=-j}^{j} D^j_{m,1}(\phi_1,\theta_1,0)\,Y_{jm}\mathrm{e}^{\mp\mathrm{i}\omega t} \\ &= -\frac{4\pi}{k}\sum_{j=1}^{\infty}\sum_{m=-j}^{j}\frac{(\pm\mathrm{i})^{j+1}}{\sqrt{j(j+1)}}\,\overline{{}_{-1}Y_{jm}(\theta_1,\phi_1)}\, j_j(kr)\,Y_{jm}\mathrm{e}^{\mp\mathrm{i}\omega t}. \end{aligned} \tag{3.80}$$

(Alternatively, (3.80) can be derived from (3.79) making use of the addition theorem (2.59).) Equation (3.80) shows that the coefficients in the expansion of the Debye potential χ in terms of the separable solutions of the wave equation $j_j(kr)Y_{jm}(\theta, \phi)e^{\mp i\omega t}$, are, essentially, the spherical harmonics with spin weight -1, evaluated in the direction of the propagation of the wave.

EXAMPLE. Scattering of a plane wave by a sphere.

Since the electromagnetic field is given by a single scalar potential, the boundary conditions satisfied by the electromagnetic field lead to boundary conditions for the scalar potential. For instance, we shall consider the scattering of a monochromatic plane wave of frequency ω by a perfectly conducting sphere; the tangential components of the (total) electric field must vanish at the surface of the sphere. Assuming that the origin of the system of coordinates is the center of the sphere, $E_\theta = 0 = E_\phi$ at $r = a$, where a is the radius of the sphere. From (3.60) it follows that these conditions amount to

$$(F_{+1} - \overline{F_{-1}})\big|_{r=a} = 0. \tag{3.81}$$

Since the conditions (3.81) involve complex conjugation, it is convenient to write the potential for the total electromagnetic field in the form

$$\chi = \psi_1 e^{-i\omega t} + \psi_2 e^{i\omega t}, \tag{3.82}$$

where ψ_1 and ψ_2 are solutions of the scalar Helmholtz equation. Then, making use of (3.62) we find that (3.81) is satisfied for all values of t if and only if

$$(\psi_1 - \overline{\psi_2})\big|_{r=a} = 0, \qquad \frac{1}{r}\partial_r r(\psi_1 + \overline{\psi_2})\bigg|_{r=a} = 0. \tag{3.83}$$

Assuming that the incident wave is given by (3.79) with the upper signs and making use of the relation $j_j(x) = \frac{1}{2}\big(h_j^{(1)}(x) + h_j^{(2)}(x)\big)$, and the fact that the scattered field must correspond to outgoing waves, in order to satisfy (3.83) we look for solutions of the Helmholtz equation of the form

$$\begin{aligned}\psi_1 &= \frac{1}{2k}\sum_{j=1}^{\infty}\left[\frac{4\pi(2j+1)}{j(j+1)}\right]^{1/2} i^{j+1}[(1+A_j)h_j^{(1)}(kr) + h_j^{(2)}(kr)]Y_{j,1},\\ \psi_2 &= \frac{1}{2k}\sum_{j=1}^{\infty}\left[\frac{4\pi(2j+1)}{j(j+1)}\right]^{1/2} (-i)^{j+1}B_j h_j^{(2)}(kr)Y_{j,-1},\end{aligned} \tag{3.84}$$

where A_j and B_j are complex constants, which determine the scattered field. Then, substituting (3.84) into (3.83), owing to the linear independence of the spherical harmonics, we obtain

$$1+A_j+\overline{B_j} = -\frac{h_j^{(2)}(ka)}{h_j^{(1)}(ka)}, \qquad 1+A_j-\overline{B_j} = -\frac{d[rh_j^{(2)}(kr)]/dr}{d[rh_j^{(1)}(kr)]/dr}\bigg|_{r=a}, \tag{3.85}$$

which have modulus equal to 1. Since the Wronskian of $h_j^{(1)}$ and $h_j^{(2)}$ is different from zero, (3.85) implies that the B_j are all different from zero, which means that the scattered field will be elliptically polarized.

According to (3.62), (3.65), (3.82), and (3.84), the radiative component of the scattered field at large distances from the sphere is given by

$$F_{-1}^{\text{sc}} \sim -\frac{\mathrm{i}}{kr}\sum_{j=1}^{\infty}\sqrt{2\pi(2j+1)}[A_j e^{\mathrm{i}(kr-\omega t)}\,{}_{-1}Y_{j,1} - B_j e^{-\mathrm{i}(kr-\omega t)}\,{}_{-1}Y_{j,-1}],$$

therefore, using (3.67), the time-averaged energy flux of the scattered field is

$$\left\langle \frac{\mathrm{d}^2 E^{\text{sc}}}{\mathrm{d}t\,\mathrm{d}\Omega} \right\rangle = \frac{c}{4k^2}\left\{ \left| \sum_{j=1}^{\infty}\sqrt{2j+1}\,A_j\,{}_{-1}Y_{j,1} \right|^2 + \left| \sum_{j=1}^{\infty}\sqrt{2j+1}\,B_j\,{}_{-1}Y_{j,-1} \right|^2 \right\}.$$

Since the energy flux of the incident wave is $c/4\pi$, the differential scattering cross section is

$$\frac{\mathrm{d}\sigma}{\mathrm{d}\Omega} = \frac{\pi}{k^2}\left\{ \left| \sum_{j=1}^{\infty}\sqrt{2j+1}\,A_j\,{}_{-1}Y_{j,1} \right|^2 + \left| \sum_{j=1}^{\infty}\sqrt{2j+1}\,B_j\,{}_{-1}Y_{j,-1} \right|^2 \right\}$$

and owing to the orthonormality of the spin-weighted spherical harmonics,

$$\sigma = \frac{\pi}{k^2}\sum_{j=1}^{\infty}(2j+1)\left[|A_j|^2 + |B_j|^2\right].$$

3.3 The equation for elastic waves in an isotropic medium

The equations for the elastic waves in an isotropic medium (see, *e.g.*, Landau and Lifshitz (1975), Chap. III) are given by

$$(1-2\sigma)\nabla^2\mathbf{u} + \nabla(\nabla\cdot\mathbf{u}) - \frac{2(1+\sigma)(1-2\sigma)\rho}{E}\partial_t^2\mathbf{u} = 0, \qquad (3.86)$$

where $\mathbf{u}$ is the displacement vector, σ is the Poisson ratio, E is the Young modulus and ρ is the mass density. Making use of (3.5), (3.6), and (3.8) we find that the vector equation (3.86) is equivalent to

$$(1-2\sigma)\left(\frac{1}{r}\partial_r^2(ru_{+1}) + \frac{1}{r^2}\eth\bar{\eth}u_{+1} + \frac{2}{r^2}\eth u_0\right) + \frac{1}{r^3}\partial_r r^2\eth u_0$$

$$
\begin{aligned}
& -\frac{1}{2r^2}\eth\eth u_{-1} + \frac{1}{2r^2}\eth\bar{\eth} u_{+1} - \kappa\, \partial_t^2 u_{+1} = 0, \\
(1-2\sigma)&\left(\frac{1}{r}\partial_r^2(r u_{-1}) + \frac{1}{r^2}\bar{\eth}\eth u_{-1} - \frac{2}{r^2}\bar{\eth} u_0\right) - \frac{1}{r^3}\partial_r r^2 \bar{\eth} u_0 \\
& + \frac{1}{2r^2}\bar{\eth}\eth u_{-1} - \frac{1}{2r^2}\bar{\eth}\bar{\eth} u_{+1} - \kappa\, \partial_t^2 u_{-1} = 0, \qquad (3.87) \\
(1-2\sigma)&\left(\partial_r \frac{1}{r^2}\partial_r(r^2 u_0) + \frac{1}{r^2}\bar{\eth}\eth u_0 + \frac{1}{r^2}\eth u_{-1} - \frac{1}{r^2}\bar{\eth} u_{+1}\right) \\
& + \partial_r \frac{1}{r^2}\partial_r(r^2 u_0) - \frac{1}{2}\partial_r \frac{1}{r}\eth u_{-1} + \frac{1}{2}\partial_r \frac{1}{r}\bar{\eth} u_{+1} - \kappa\, \partial_t^2 u_0 = 0,
\end{aligned}
$$

where the u_s are the spin-weighted components of $\mathbf{u}$ and

$$\kappa \equiv \frac{2(1+\sigma)(1-2\sigma)\rho}{E}.$$

The system of equations (3.87) admits separable solutions of the form

$$u_s = g_s(r)\, {}_sY_{jm}(\theta, \phi)\, \mathrm{e}^{-\mathrm{i}\omega t}, \qquad (s = 1, -1, 0), \tag{3.88}$$

where j is a nonnegative integer and m is an integer such that $|m| \leqslant j$. By substituting (3.88) into (3.87), using (2.27), we find that, if $j \neq 0$, the radial functions g_s are determined by

$$
\begin{aligned}
(1-2\sigma)&\left(\frac{1}{r}\frac{\mathrm{d}^2}{\mathrm{d}r^2}(r g_{\pm 1}) - \frac{\mu^2}{r^2} g_{\pm 1} + \frac{2\mu}{r^2} g_0\right) + \frac{\mu}{r^3}\frac{\mathrm{d}}{\mathrm{d}r}(r^2 g_0) \\
& - \frac{\mu^2}{2r^2}(g_1 + g_{-1}) + \kappa\,\omega^2 g_{\pm 1} = 0, \\
(1-2\sigma)&\left(\frac{\mathrm{d}}{\mathrm{d}r}\frac{1}{r^2}\frac{\mathrm{d}}{\mathrm{d}r}(r^2 g_0) - \frac{\mu^2}{r^2} g_0 + \frac{\mu}{r^2}(g_1 + g_{-1})\right) \\
& + \frac{\mathrm{d}}{\mathrm{d}r}\frac{1}{r^2}\frac{\mathrm{d}}{\mathrm{d}r}(r^2 g_0) - \frac{\mu}{2}\frac{\mathrm{d}}{\mathrm{d}r}\frac{1}{r}(g_1 + g_{-1}) + \kappa\,\omega^2 g_0 = 0,
\end{aligned}
\tag{3.89}
$$

where

$$\mu \equiv \sqrt{j(j+1)} \tag{3.90}$$

(the case with $j = 0$ will be considered below). Making use of the definitions

$$M \equiv \tfrac{1}{2}(g_1 - g_{-1}), \qquad H \equiv \tfrac{1}{2}(g_1 + g_{-1}), \tag{3.91}$$

the set of equations (3.89) can be rewritten as

$$\frac{\mathrm{d}^2 M}{\mathrm{d}r^2} + \frac{2}{r}\frac{\mathrm{d}M}{\mathrm{d}r} + \left(k_t^2 - \frac{\mu^2}{r^2}\right) M = 0, \tag{3.92}$$

where

$$k_t^2 \equiv \frac{\kappa\,\omega^2}{1-2\sigma} = \frac{2(1+\sigma)\omega^2\rho}{E},$$

and

$$(1-2\sigma)\left(\frac{1}{r}\frac{\mathrm{d}^2}{\mathrm{d}r^2}(rH) - \frac{\mu^2}{r^2}H + \frac{2\mu}{r^2}g_0\right) + \frac{\mu}{r^3}\frac{\mathrm{d}}{\mathrm{d}r}(r^2 g_0) - \frac{\mu^2}{r^2}H + \kappa\,\omega^2 H = 0, \tag{3.93}$$

$$2(1-\sigma)\frac{\mathrm{d}}{\mathrm{d}r}\frac{1}{r^2}\frac{\mathrm{d}}{\mathrm{d}r}(r^2 g_0) + \kappa\,\omega^2 g_0 - \frac{(1-2\sigma)\mu^2}{r^2}g_0 + \frac{2(1-2\sigma)\mu}{r^2}H - \mu\frac{\mathrm{d}}{\mathrm{d}r}\left(\frac{H}{r}\right) = 0. \tag{3.94}$$

Since $\mu^2 = j(j+1)$ [see (3.90)], the solution of (3.92) can be written as

$$M(r) = a_1\, h_j^{(1)}(k_t r) + a_2\, h_j^{(2)}(k_t r),$$

where a_1 and a_2 are constants.

In order to solve (3.93) and (3.94), we introduce the auxiliary functions (Torres del Castillo and Quintero-Téllez 1999)

$$v \equiv -\frac{\mu}{r}g_0 + \frac{1}{r}\frac{\mathrm{d}}{\mathrm{d}r}(rH), \qquad w \equiv \frac{\mu}{r}H - \frac{1}{r^2}\frac{\mathrm{d}}{\mathrm{d}r}(r^2 g_0). \tag{3.95}$$

(Note that, according to (3.6), (3.88), and (3.91), $\nabla\cdot\mathbf{u} = \sqrt{2}\,w(r)\,Y_{jm}(\theta,\phi) \times e^{-i\omega t}$; similarly, v and M are related to the radial part of $\mathbf{r}\cdot\nabla\times\nabla\times\mathbf{u}$ and $\mathbf{r}\cdot\nabla\times\mathbf{u}$, respectively.) Then, by means of a straightforward computation, from (3.93) and (3.94) one finds that

$$\frac{\mathrm{d}^2 v}{\mathrm{d}r^2} + \frac{2}{r}\frac{\mathrm{d}v}{\mathrm{d}r} + \left(k_t^2 - \frac{\mu^2}{r^2}\right)v = 0, \qquad \frac{\mathrm{d}^2 w}{\mathrm{d}r^2} + \frac{2}{r}\frac{\mathrm{d}w}{\mathrm{d}r} + \left(k_l^2 - \frac{\mu^2}{r^2}\right)w = 0, \tag{3.96}$$

with

$$k_l^2 \equiv \frac{\kappa\,\omega^2}{2(1-\sigma)} = \frac{(1+\sigma)(1-2\sigma)\omega^2\rho}{(1-\sigma)E}.$$

(Equations (3.92) and (3.96) also follow directly from the fact that $\mathbf{r}\cdot\nabla\times\mathbf{u}$, $\mathbf{r}\cdot\nabla\times\nabla\times\mathbf{u}$ and $\nabla\cdot\mathbf{u}$ obey wave equations as a consequence of (3.86).) Hence,

$$v(r) = b_1\, h_j^{(1)}(k_t r) + b_2\, h_j^{(2)}(k_t r), \qquad w(r) = c_1\, h_j^{(1)}(k_l r) + c_2\, h_j^{(2)}(k_l r),$$

where b_1, b_2, c_1, and c_2 are constants.

On the other hand, eliminating H from (3.95), one finds that

$$\left(\frac{\mathrm{d}^2}{\mathrm{d}r^2} + \frac{2}{r}\frac{\mathrm{d}}{\mathrm{d}r} - \frac{\mu^2}{r^2}\right)(r g_0) = \mu v - \frac{1}{r}\frac{\mathrm{d}}{\mathrm{d}r}(r^2 w). \tag{3.97}$$

Making use of (3.96), the right-hand side of (3.97) can be expressed as

$$-\frac{\mu}{k_t^2}\left(\frac{\mathrm{d}^2 v}{\mathrm{d}r^2}+\frac{2}{r}\frac{\mathrm{d}v}{\mathrm{d}r}-\frac{\mu^2}{r^2}v\right)+\frac{1}{k_l^2}\frac{1}{r}\frac{\mathrm{d}}{\mathrm{d}r}r^2\left(\frac{\mathrm{d}^2 w}{\mathrm{d}r^2}+\frac{2}{r}\frac{\mathrm{d}w}{\mathrm{d}r}-\frac{\mu^2}{r^2}w\right)$$
$$=\left(\frac{\mathrm{d}^2}{\mathrm{d}r^2}+\frac{2}{r}\frac{\mathrm{d}}{\mathrm{d}r}-\frac{\mu^2}{r^2}\right)\left(-\frac{\mu}{k_t^2}v+\frac{1}{k_l^2}r\frac{\mathrm{d}w}{\mathrm{d}r}\right)$$

thus,

$$\left(\frac{\mathrm{d}^2}{\mathrm{d}r^2}+\frac{2}{r}\frac{\mathrm{d}}{\mathrm{d}r}-\frac{\mu^2}{r^2}\right)\left(rg_0+\frac{\mu}{k_t^2}v-\frac{1}{k_l^2}r\frac{\mathrm{d}w}{\mathrm{d}r}\right)=0,$$

therefore,

$$g_0=-\frac{\mu}{k_t^2}\frac{v}{r}+\frac{1}{k_l^2}\frac{\mathrm{d}w}{\mathrm{d}r}+D_1 r^{j-1}+D_2 r^{-j-2}, \tag{3.98}$$

where D_1 and D_2 are constants. Substituting (3.98) into the second equation in (3.95), using (3.96), we find that

$$H=\frac{\mu}{k_l^2}\frac{w}{r}-\frac{1}{k_t^2 r}\frac{\mathrm{d}}{\mathrm{d}r}(rv)+(j+1)\frac{D_1}{\mu}r^{j-1}-j\frac{D_2}{\mu}r^{-j-2}. \tag{3.99}$$

Substituting (3.98) and (3.99) into (3.93) and (3.94), it follows that if $\omega \neq 0$, then D_1 and D_2 must vanish, hence

$$g_0=-\frac{\mu}{k_t^2}\frac{v}{r}+\frac{1}{k_l^2}\frac{\mathrm{d}w}{\mathrm{d}r}, \qquad H=\frac{\mu}{k_l^2}\frac{w}{r}-\frac{1}{k_t^2 r}\frac{\mathrm{d}}{\mathrm{d}r}(rv) \tag{3.100}$$

and, from (3.88), (3.91), (3.100), and (2.27) we obtain

$$\begin{aligned} u_0 &= \frac{1}{\sqrt{2}}\partial_r\psi_1-\frac{1}{\sqrt{2}r}\eth\bar{\eth}\psi_3,\\ u_{+1} &= \frac{1}{\sqrt{2}r}\eth\psi_1-\frac{\mathrm{i}}{\sqrt{2}}\eth\psi_2+\frac{1}{\sqrt{2}r}\partial_r r\eth\psi_3,\\ u_{-1} &= -\frac{1}{\sqrt{2}r}\bar{\eth}\psi_1-\frac{\mathrm{i}}{\sqrt{2}}\bar{\eth}\psi_2-\frac{1}{\sqrt{2}r}\partial_r r\bar{\eth}\psi_3 \end{aligned} \tag{3.101}$$

[*cf.* (3.24)], where

$$\begin{aligned} \psi_1 &\equiv \frac{\sqrt{2}}{k_l^2}w(r)\,Y_{jm}\,\mathrm{e}^{-\mathrm{i}\omega t},\\ \psi_2 &\equiv \frac{\mathrm{i}\sqrt{2}}{\mu}M(r)\,Y_{jm}\,\mathrm{e}^{-\mathrm{i}\omega t},\\ \psi_3 &\equiv -\frac{\sqrt{2}}{\mu k_t^2}v(r)\,Y_{jm}\,\mathrm{e}^{-\mathrm{i}\omega t}. \end{aligned} \tag{3.102}$$

According to (3.5) and (3.7), (3.101) amount to the simple expression

$$\mathbf{u} = -\nabla\psi_1 + \mathbf{r} \times \nabla\psi_2 + \nabla \times (\mathbf{r} \times \nabla\psi_3) \tag{3.103}$$

[*cf.* (3.26)] and, by virtue of (3.92) and (3.96), the scalar potentials (3.102) obey the wave equations

$$\nabla^2\psi_1 - \frac{1}{v_l^2}\partial_t^2\psi_1 = 0, \qquad \nabla^2\psi_{2,3} - \frac{1}{v_t^2}\partial_t^2\psi_{2,3} = 0, \tag{3.104}$$

where

$$v_l \equiv \frac{\omega}{k_l} = \sqrt{\frac{(1-\sigma)E}{(1+\sigma)(1-2\sigma)\rho}}, \qquad v_t \equiv \frac{\omega}{k_t} = \sqrt{\frac{E}{2(1+\sigma)\rho}}. \tag{3.105}$$

In the case of the separable solutions (3.88) with $j = 0$ (*i.e.*, $\mu = 0$), the only nonvanishing spin-weighted component of the displacement vector is u_0, which is a function of r and t only, and from (3.87) one obtains

$$\partial_r \frac{1}{r^2}\partial_r(r^2 u_0) + k_l^2 u_0 = 0.$$

Therefore, using the recurrence relations for the spherical Bessel functions, we have

$$u_0 = \big(a\, h_1^{(1)}(k_l r) + b\, h_1^{(2)}(k_l r)\big)\mathrm{e}^{-\mathrm{i}\omega t} = -\partial_r \frac{1}{k_l}\big(a\, h_0^{(1)}(k_l r) + b\, h_0^{(2)}(k_l r)\big)\mathrm{e}^{-\mathrm{i}\omega t},$$

which is of the form (3.101) with $\psi_1 = -\big(a\, h_0^{(1)}(k_l r) + b\, h_0^{(2)}(k_l r)\big)\mathrm{e}^{-\mathrm{i}\omega t}/k_l$ and $\psi_2 = \psi_3 = 0$, and these potentials also satisfy the wave equations (3.104). Thus, owing to the completeness of the spin-weighted spherical harmonics and the linearity of (3.104) and (3.103), it follows that the most general solution of (3.86) can be expressed in the form (3.103), where the scalar potentials ψ_i are solutions of the wave equations (3.104).

Equation (3.103) can be also written as

$$\mathbf{u} = -\nabla\psi_1 - \nabla \times (\psi_2\mathbf{r}) - \nabla \times \nabla \times (\psi_3\mathbf{r}), \tag{3.106}$$

which shows that the displacement vector $\mathbf{u}$ is the sum of an elastic wave with vanishing curl propagating with velocity v_l [(3.105)] and an elastic wave with vanishing divergence propagating with velocity v_t (*cf.* Landau and Lifshitz 1975). If the potentials ψ_i are real, then $\mathbf{u}$ is also real.

The potentials corresponding to plane waves, for instance, can be obtained in the following way. For the longitudinal plane wave propagating in an arbitrary direction $\mathbf{k}$, $\mathbf{u} = \mathbf{k}e^{i(\mathbf{k}\cdot\mathbf{r}-\omega t)}$, with $|\mathbf{k}| = k_l$, we have $\mathbf{u} = -i\nabla e^{i(\mathbf{k}\cdot\mathbf{r}-\omega t)}$, which is of the form (3.103) with $\psi_2 = 0 = \psi_3$ and

$$\psi_1 = ie^{i(\mathbf{k}\cdot\mathbf{r}-\omega t)} = 4\pi \sum_{j=0}^{\infty} \sum_{m=-j}^{j} i^{j+1}\overline{Y_{jm}(\theta_1, \phi_1)}\, j_j(k_l r) Y_{jm} e^{-i\omega t},$$

where θ_1 and ϕ_1 are the polar and azimuth angles of $\mathbf{k}$, respectively. For the circularly polarized transverse plane wave propagating along the z-axis, $\mathbf{u} = \cos(k_t z - \omega t)\mathbf{e}_x \mp \sin(k_t z - \omega t)\mathbf{e}_y$, we have $\nabla \cdot \mathbf{u} = 0$, $\nabla \times \mathbf{u} = \pm k_t \mathbf{u}$, and $\mathbf{r} \cdot \mathbf{u} = \operatorname{Re}[(x + iy)e^{\pm i(k_t z - \omega t)}] = \operatorname{Re}[(\pm i/k_t)(L_x + iL_y)e^{\pm i(k_t z-\omega t)}]$. Then from (3.103) and (3.104) it follows that $\nabla \cdot \mathbf{u} = -\nabla^2\psi_1$, $\mathbf{r} \cdot \mathbf{u} = -L^2\psi_3$, and $\mathbf{r} \cdot \nabla \times \mathbf{u} = -L^2\psi_2$; therefore, we can take $\psi_1 = 0$, $\psi_2 = \pm k_t \psi_3$, and

$$\psi_3 = -\operatorname{Re}\frac{1}{k_t}\sum_{j=1}^{\infty}\left[\frac{4\pi(2j+1)}{j(j+1)}\right]^{1/2} (\pm i)^{j+1} j_j(k_t r) Y_{j,1} e^{\mp i\omega t}$$

[*cf.* (3.79)]. Hence, if the wave propagates in an arbitrary direction, with polar and azimuth angles θ_1 and ϕ_1, the potential ψ_3 is [*cf.* (3.80)]

$$\psi_3 = \operatorname{Re}\frac{4\pi}{k_t}\sum_{j=1}^{\infty}\sum_{m=-j}^{j}\frac{(\pm i)^{j+1}}{\sqrt{j(j+1)}}\,\overline{{}_{-1}Y_{jm}(\theta_1, \phi_1)}\, j_j(k_t r)\, Y_{jm} e^{\mp i\omega t}.$$

3.4 The Weyl neutrino equation

The Weyl equation for the massless neutrino can be written as

$$i\sigma \cdot \nabla\psi = \frac{i}{c}\partial_t \psi, \tag{3.107}$$

where ψ is a two-component spinor field. The spinor ψ can be written as a linear combination of the spinors o and $\widehat{o}$ defined in (2.6),

$$\psi = \psi_- o + \psi_+ \widehat{o},$$

where $\psi_- = \psi^A \widehat{o}_A$, $\psi_+ = -\psi^A o_A$. The components ψ_+ and ψ_- have spin weight 1/2 and $-1/2$, respectively. A straightforward computation, making use of (1.16), (3.4), (2.6), and (2.17) shows that

$$\sigma \cdot \nabla\psi = \left[\frac{1}{r}\partial_r(r\psi_-) - \frac{1}{r}\bar{\eth}\psi_+\right] o + \left[-\frac{1}{r}\partial_r(r\psi_+) - \frac{1}{r}\eth\psi_-\right]\widehat{o}. \tag{3.108}$$

Hence, the Weyl equation (3.107) amounts to

$$\begin{aligned} \frac{1}{r}\partial_r(r\psi_-) - \frac{1}{c}\partial_t\psi_- &= \frac{1}{r}\bar{\eth}\psi_+, \\ -\frac{1}{r}\partial_r(r\psi_+) - \frac{1}{c}\partial_t\psi_+ &= \frac{1}{r}\eth\psi_-. \end{aligned} \tag{3.109}$$

This system of equations can be solved by separation of variables looking for solutions of the form

$$\psi_\pm = f_\pm(r)\, {}_{\pm\frac{1}{2}}Y_{jm}(\theta, \phi) e^{-i\omega t}, \tag{3.110}$$

where j and m are half-integers with $j \geqslant 1/2$ and $m = -j, -j+1, \ldots, j$. In effect, according to (2.27), $\eth\, {}_{-\frac{1}{2}}Y_{jm} = (j+\frac{1}{2})\, {}_{\frac{1}{2}}Y_{jm}$ and $\bar{\eth}\, {}_{\frac{1}{2}}Y_{jm} = -(j+\frac{1}{2})\, {}_{-\frac{1}{2}}Y_{jm}$, hence, substituting (3.110) into (3.109) we obtain

$$\begin{aligned} \frac{1}{r}\frac{d}{dr}(rf_-) + ikf_- &= -\frac{j+\frac{1}{2}}{r} f_+, \\ -\frac{1}{r}\frac{d}{dr}(rf_+) + ikf_+ &= \frac{j+\frac{1}{2}}{r} f_-. \end{aligned} \tag{3.111}$$

Making use of the recurrence relations (3.21) one can verify that the solution of (3.111) can be expressed as

$$f_\pm(r) = A\big(\pm h^{(1)}_{j-\frac{1}{2}}(kr) + i h^{(1)}_{j+\frac{1}{2}}(kr)\big) + B\big(\pm h^{(2)}_{j-\frac{1}{2}}(kr) + i h^{(2)}_{j+\frac{1}{2}}(kr)\big), \tag{3.112}$$

where A and B are arbitrary constants.

Substituting (3.112) into (3.110), with the aid of (3.64) we find that when $r \to \infty$,

$$\begin{aligned} \psi_+ &\sim \Bigg[A(-i)^{j+1/2} \frac{e^{i(kr-\omega t)}}{kr} \left(2 + \frac{(j+\frac{1}{2})^2}{ikr} \right) \\ &\qquad - B(j+\tfrac{1}{2}) i^{j-1/2} \frac{e^{-i(kr+\omega t)}}{(kr)^2} \Bigg] {}_{\frac{1}{2}}Y_{jm}, \\ \psi_- &\sim \Bigg[-B i^{j+1/2} \frac{e^{-i(kr+\omega t)}}{kr} \left(2 + \frac{(j+\frac{1}{2})^2}{ikr} \right) \\ &\qquad + A(j+\tfrac{1}{2})(-i)^{j-1/2} \frac{e^{i(kr-\omega t)}}{(kr)^2} \Bigg] {}_{-\frac{1}{2}}Y_{jm}. \end{aligned} \tag{3.113}$$

Hence, for outgoing waves, ψ_+, which is the amplitude of probability for the spin to be in the direction $-\mathbf{e}_r$, is dominant at large distances. Similarly, for ingoing waves, ψ_-, which is the amplitude of probability for the spin to be in the outward direction $\mathbf{e}_r$, is dominant at large distances [*cf.* (3.66) and (3.68)].

As in the case of the source-free electromagnetic field, any solution of the Weyl equation can be written in terms of a single potential. We begin by noting that, owing to (3.111), the separable solutions given by (3.110) and (3.112) can be also expressed in the form

$$\psi_+ = \left(\partial_r - \frac{1}{c}\partial_t \right) \chi, \qquad \psi_- = \frac{1}{r}\bar{\eth}\chi \tag{3.114}$$

[*cf.* (3.62)], with

$$\chi = -(j+\tfrac{1}{2})^{-1} r f_-(r)\, {}_{\frac{1}{2}}Y_{jm}(\theta,\phi) \mathrm{e}^{-\mathrm{i}\omega t}, \tag{3.115}$$

which satisfies the linear partial differential equation

$$\frac{1}{r}\left(\partial_r + \frac{1}{c}\partial_t\right) r \left(\partial_r - \frac{1}{c}\partial_t\right)\chi + \frac{1}{r^2}\eth\bar{\eth}\chi = 0 \tag{3.116}$$

[this last equation can be obtained by substituting (3.114) into (3.109)]. By virtue of the completeness of the spin-weighted spherical harmonics and the linearity of the differential operators in (3.107), (3.114), and (3.116), it follows that any solution of the Weyl neutrino equation can be expressed in the form (3.114), where χ is a solution of (3.116). (The solutions of (3.107) can also be written in the form $\psi_+ = (1/r)\eth\chi'$, $\psi_- = -(\partial_r + (1/c)\,\partial_t)\chi'$, where χ' is a potential with spin weight $-1/2$ that obeys a condition analogous to (3.116).)

For example, in the case of the plane wave

$$\psi^A = \widehat{\kappa}^A \mathrm{e}^{\mathrm{i}(\mathbf{k}\cdot\mathbf{r}-\omega t)},$$

where $k_i = -\sigma_{iAB}\widehat{\kappa}^A\kappa^B$ (see Section 1.3), using (3.78) and the addition theorem for the spherical harmonics (2.60) we have

$$\psi_+ = -\psi^A o_A = -\widehat{\kappa}^A o_A \sum_{j=0}^{\infty}\sum_{m=-j}^{j} 4\pi \mathrm{i}^j j_j(kr)\, \overline{Y_{jm}(\theta_1,\phi_1)}\, Y_{jm}\mathrm{e}^{-\mathrm{i}\omega t}, \tag{3.117}$$

where θ_1 and ϕ_1 are the polar and azimuth angles of $\mathbf{k}$, respectively.

It will be shown later that the spherical harmonics with spin weight 1/2 are related to the ordinary spherical harmonics by [see (3.152) and (3.153)]

$$\begin{aligned} {}_{\frac{1}{2}}Y_{jm} &= -\mathrm{i}\sqrt{\tfrac{j+m+1}{j+1}}\, o^1 Y_{j+\frac{1}{2},m+\frac{1}{2}} - \mathrm{i}\sqrt{\tfrac{j-m+1}{j+1}}\, o^2 Y_{j+\frac{1}{2},m-\frac{1}{2}}, \\ {}_{\frac{1}{2}}Y_{j+1,m} &= -\mathrm{i}\sqrt{\tfrac{j-m+1}{j+1}}\, o^1 Y_{j+\frac{1}{2},m+\frac{1}{2}} + \mathrm{i}\sqrt{\tfrac{j+m+1}{j+1}}\, o^2 Y_{j+\frac{1}{2},m-\frac{1}{2}}, \end{aligned} \tag{3.118}$$

with o^A given by (2.6), therefore,

$$\begin{aligned} o^1 Y_{jm} &= \frac{\mathrm{i}}{2}\sqrt{\tfrac{j+m}{j+\frac{1}{2}}}\, {}_{\frac{1}{2}}Y_{j-\frac{1}{2},m-\frac{1}{2}} + \frac{\mathrm{i}}{2}\sqrt{\tfrac{j-m+1}{j+\frac{1}{2}}}\, {}_{\frac{1}{2}}Y_{j+\frac{1}{2},m-\frac{1}{2}}, \\ o^2 Y_{jm} &= \frac{\mathrm{i}}{2}\sqrt{\tfrac{j-m}{j+\frac{1}{2}}}\, {}_{\frac{1}{2}}Y_{j-\frac{1}{2},m+\frac{1}{2}} - \frac{\mathrm{i}}{2}\sqrt{\tfrac{j+m+1}{j+\frac{1}{2}}}\, {}_{\frac{1}{2}}Y_{j+\frac{1}{2},m+\frac{1}{2}}. \end{aligned} \tag{3.119}$$

Thus, recalling that $\widehat{\kappa}^1 = -\overline{\kappa^2}$, $\widehat{\kappa}^2 = \overline{\kappa^1}$, and making use of (3.119) and (3.118), from (3.117) we have

$$\psi_+ = 4\pi \sum_{j=0}^{\infty}\sum_{m=-j}^{j} \mathrm{i}^j j_j(kr)\overline{Y_{jm}(\theta_1,\phi_1)}(\widehat{\kappa}^2 o^1 - \widehat{\kappa}^1 o^2) Y_{jm}\mathrm{e}^{-\mathrm{i}\omega t}$$

$$= 2\pi \mathrm{i} \sum_{j=0}^{\infty} \sum_{m=-j}^{j} \mathrm{i}^j j_j(kr) \overline{Y_{jm}(\theta_1, \phi_1)}$$

$$\times \left\{ \widehat{\kappa}^2 \left(\sqrt{\frac{j+m}{j+\frac{1}{2}}}\, {}_{\frac{1}{2}}Y_{j-\frac{1}{2}, m-\frac{1}{2}} + \sqrt{\frac{j-m+1}{j+\frac{1}{2}}}\, {}_{\frac{1}{2}}Y_{j+\frac{1}{2}, m-\frac{1}{2}} \right) \right.$$

$$\left. - \widehat{\kappa}^1 \left(\sqrt{\frac{j-m}{j+\frac{1}{2}}}\, {}_{\frac{1}{2}}Y_{j-\frac{1}{2}, m+\frac{1}{2}} - \sqrt{\frac{j+m+1}{j+\frac{1}{2}}}\, {}_{\frac{1}{2}}Y_{j+\frac{1}{2}, m+\frac{1}{2}} \right) \right\} \mathrm{e}^{-\mathrm{i}\omega t}$$

$$= 2\pi \mathrm{i} \sum_{j=\frac{1}{2}}^{\infty} \sum_{m=-j}^{j} {}_{\frac{1}{2}}Y_{jm} \left\{ \mathrm{i}^{j+\frac{1}{2}} j_{j+\frac{1}{2}}(kr) \left(\sqrt{\frac{j+m+1}{j+1}}\, \overline{\kappa^1}\, \overline{Y_{j+\frac{1}{2}, m+\frac{1}{2}}(\theta_1, \phi_1)} \right. \right.$$

$$\left. + \sqrt{\frac{j-m+1}{j+1}}\, \overline{\kappa^2}\, \overline{Y_{j+\frac{1}{2}, m-\frac{1}{2}}(\theta_1, \phi_1)} \right)$$

$$+ \mathrm{i}^{j-\frac{1}{2}} j_{j-\frac{1}{2}}(kr) \left(\sqrt{\frac{j-m}{j}}\, \overline{\kappa^1}\, \overline{Y_{j-\frac{1}{2}, m+\frac{1}{2}}(\theta_1, \phi_1)} \right.$$

$$\left. \left. - \sqrt{\frac{j+m}{j}}\, \overline{\kappa^2}\, \overline{Y_{j-\frac{1}{2}, m-\frac{1}{2}}(\theta_1, \phi_1)} \right) \right\} \mathrm{e}^{-\mathrm{i}\omega t}$$

$$= 2\pi \sqrt{k} \sum_{j=\frac{1}{2}}^{\infty} \sum_{m=-j}^{j} {}_{\frac{1}{2}}Y_{jm} \left(\mathrm{i}^{j+\frac{1}{2}} j_{j+\frac{1}{2}}(kr)\, \overline{{}_{\frac{1}{2}}Y_{jm}(\theta_1, \phi_1)} \right.$$

$$\left. + \mathrm{i}^{j-\frac{1}{2}} j_{j-\frac{1}{2}}(kr)\, \overline{{}_{\frac{1}{2}}Y_{jm}(\theta_1, \phi_1)} \right) \mathrm{e}^{-\mathrm{i}\omega t}$$

$$= 2\pi \sqrt{k} \sum_{j=\frac{1}{2}}^{\infty} \sum_{m=-j}^{j} \mathrm{i}^{j-\frac{1}{2}}\, \overline{{}_{\frac{1}{2}}Y_{jm}(\theta_1, \phi_1)} \left(j_{j-\frac{1}{2}}(kr) + \mathrm{i} j_{j+\frac{1}{2}}(kr) \right) {}_{\frac{1}{2}}Y_{jm} \mathrm{e}^{-\mathrm{i}\omega t}$$

$$= \left(\partial_r - \frac{1}{c} \partial_t \right) \left[-2\pi \sqrt{k} \sum_{j=\frac{1}{2}}^{\infty} \sum_{m=-j}^{j} \frac{\mathrm{i}^{j-\frac{1}{2}}}{j+\frac{1}{2}}\, \overline{{}_{\frac{1}{2}}Y_{jm}(\theta_1, \phi_1)} \right.$$

$$\left. \times\, r \left(-j_{j-\frac{1}{2}}(kr) + \mathrm{i} j_{j+\frac{1}{2}}(kr) \right) {}_{\frac{1}{2}}Y_{jm} \mathrm{e}^{-\mathrm{i}\omega t} \right],$$

hence, we can take

$$\chi = -2\pi \sqrt{k} \sum_{j=\frac{1}{2}}^{\infty} \sum_{m=-j}^{j} \frac{\mathrm{i}^{j-\frac{1}{2}}}{j+\frac{1}{2}}\, \overline{{}_{\frac{1}{2}}Y_{jm}(\theta_1, \phi_1)}$$

$$\times\, r \left(-j_{j-\frac{1}{2}}(kr) + \mathrm{i} j_{j+\frac{1}{2}}(kr) \right) {}_{\frac{1}{2}}Y_{jm} \mathrm{e}^{-\mathrm{i}\omega t}$$

[*cf.* (3.80)], which is a series in the functions (3.115), whose coefficients are, essentially, the spherical harmonics with spin weight 1/2 evaluated in the direction of the propagation of the wave. (Note that, in this derivation, the explicit form of the separable solutions [(3.110) and (3.112)] was not required.) An alternative procedure, that readily yields the multipole expansion of a plane wave of arbitrary spin, is given in Torres del Castillo and Hernández-Moreno (2002).

3.5 The Dirac equation

The Dirac equation is given by

$$\begin{aligned} \mathrm{i}\hbar\partial_t u &= -\mathrm{i}\hbar c\boldsymbol{\sigma}\cdot\nabla v + Mc^2 u, \\ \mathrm{i}\hbar\partial_t v &= -\mathrm{i}\hbar c\boldsymbol{\sigma}\cdot\nabla u - Mc^2 v. \end{aligned} \tag{3.120}$$

The two-component spinors u and v appearing in (3.120) can be written as

$$u = u_- o + u_+ \widehat{o}, \qquad v = v_- o + v_+ \widehat{o}$$

and with the aid of (3.108) we find that the set of equations (3.120) is equivalent to

$$\begin{aligned} \frac{1}{c}\partial_t u_- &= -\frac{1}{r}\partial_r(r v_-) + \frac{1}{r}\overline{\eth} v_+ - \frac{\mathrm{i}Mc}{\hbar}u_-, \\ \frac{1}{c}\partial_t u_+ &= \frac{1}{r}\eth v_- + \frac{1}{r}\partial_r(r v_+) - \frac{\mathrm{i}Mc}{\hbar}u_+, \\ \frac{1}{c}\partial_t v_- &= -\frac{1}{r}\partial_r(r u_-) + \frac{1}{r}\overline{\eth} u_+ + \frac{\mathrm{i}Mc}{\hbar}v_-, \\ \frac{1}{c}\partial_t v_+ &= \frac{1}{r}\eth u_- + \frac{1}{r}\partial_r(r u_+) + \frac{\mathrm{i}Mc}{\hbar}v_+. \end{aligned} \tag{3.121}$$

This system of equations admits separable solutions of the form

$$\begin{aligned} u_- &= g(r)\,{}_{-\frac{1}{2}}Y_{jm}(\theta,\phi)\mathrm{e}^{-\mathrm{i}Et/\hbar}, \\ u_+ &= G(r)\,{}_{\frac{1}{2}}Y_{jm}(\theta,\phi)\mathrm{e}^{-\mathrm{i}Et/\hbar}, \\ v_- &= f(r)\,{}_{-\frac{1}{2}}Y_{jm}(\theta,\phi)\mathrm{e}^{-\mathrm{i}Et/\hbar}, \\ v_+ &= F(r)\,{}_{\frac{1}{2}}Y_{jm}(\theta,\phi)\mathrm{e}^{-\mathrm{i}Et/\hbar}, \end{aligned} \tag{3.122}$$

where j is a half-integer greater than or equal to 1/2 and m is a half-integer such that $-j \leqslant m \leqslant j$. Substitution of (3.122) into (3.121) gives us the ordinary differential equations

$$\begin{aligned}
\frac{1}{r}\frac{\mathrm{d}}{\mathrm{d}r}(rf) + (j+\tfrac{1}{2})\frac{F}{r} + \frac{\mathrm{i}Mc}{\hbar}g &= \frac{\mathrm{i}E}{\hbar c}g,\\
-\frac{1}{r}\frac{\mathrm{d}}{\mathrm{d}r}(rF) - (j+\tfrac{1}{2})\frac{f}{r} + \frac{\mathrm{i}Mc}{\hbar}G &= \frac{\mathrm{i}E}{\hbar c}G,\\
\frac{1}{r}\frac{\mathrm{d}}{\mathrm{d}r}(rg) + (j+\tfrac{1}{2})\frac{G}{r} - \frac{\mathrm{i}Mc}{\hbar}f &= \frac{\mathrm{i}E}{\hbar c}f,\\
-\frac{1}{r}\frac{\mathrm{d}}{\mathrm{d}r}(rG) - (j+\tfrac{1}{2})\frac{g}{r} - \frac{\mathrm{i}Mc}{\hbar}F &= \frac{\mathrm{i}E}{\hbar c}F.
\end{aligned} \tag{3.123}$$

It is convenient to make use of the following combinations of the radial functions f, F, g, and G,

$$A_\pm \equiv \frac{1}{\sqrt{2}}(G \pm g), \qquad B_\pm \equiv -\frac{\mathrm{i}}{\sqrt{2}}(F \mp f), \tag{3.124}$$

since equations (3.123) are then equivalent to

$$\begin{aligned}
\frac{1}{r}\frac{\mathrm{d}}{\mathrm{d}r}(rA_\pm) \pm (j+\tfrac{1}{2})\frac{A_\pm}{r} &= \frac{E+Mc^2}{\hbar c}B_\pm,\\
-\frac{1}{r}\frac{\mathrm{d}}{\mathrm{d}r}(rB_\pm) \pm (j+\tfrac{1}{2})\frac{B_\pm}{r} &= \frac{E-Mc^2}{\hbar c}A_\pm.
\end{aligned} \tag{3.125}$$

Equations (3.125) and (3.132) below are the same radial equations that are obtained by means of the methods usually employed (*cf.*, for instance, Rose 1961, Messiah 1962, Davydov 1988).

By combining (3.125) one obtains the decoupled equations

$$\left[\frac{\mathrm{d}^2}{\mathrm{d}r^2} + \frac{2}{r}\frac{\mathrm{d}}{\mathrm{d}r} + k^2 - \frac{(j+\frac{1}{2})(j+\frac{1}{2}\pm 1)}{r^2}\right]A_\pm = 0,$$

where $k \equiv p/\hbar$ with $p = \sqrt{E^2 - M^2c^4}/c$, whose regular solution is a multiple of the spherical Bessel function $j_{j\pm\frac{1}{2}}(kr)$,

$$A_\pm(r) = a_\pm j_{j\pm\frac{1}{2}}(kr), \tag{3.126}$$

where $a_\pm$ are arbitrary constants. Substituting this expression into (3.125), making use of the recurrence relations for the spherical Bessel functions, we find that

$$B_\pm(r) = \pm a_\pm \frac{pc}{E+Mc^2} j_{j\mp\frac{1}{2}}(kr). \tag{3.127}$$

Hence, from (3.122) and (3.124) it follows that the system of equations (3.121) has separable solutions of the form

$$\begin{pmatrix} u_- \\ u_+ \\ v_- \\ v_+ \end{pmatrix} = \begin{pmatrix} A_+(r) X^m_{j+\frac{1}{2}} \\ \mathrm{i} B_+(r) X^m_{-j-\frac{1}{2}} \end{pmatrix} \mathrm{e}^{-\mathrm{i}Et/\hbar} + \begin{pmatrix} A_-(r) X^m_{-j-\frac{1}{2}} \\ \mathrm{i} B_-(r) X^m_{j+\frac{1}{2}} \end{pmatrix} \mathrm{e}^{-\mathrm{i}Et/\hbar}, \tag{3.128}$$

with

$$X^m_{j+\frac{1}{2}} \equiv \frac{1}{\sqrt{2}} \begin{pmatrix} -\frac{1}{2}Y_{jm} \\ \frac{1}{2}Y_{jm} \end{pmatrix}, \qquad X^m_{-j-\frac{1}{2}} \equiv \frac{1}{\sqrt{2}} \begin{pmatrix} -({}_{-\frac{1}{2}}Y_{jm}) \\ {}_{\frac{1}{2}}Y_{jm} \end{pmatrix} \tag{3.129}$$

and the functions $A_\pm$, $B_\pm$, given by (3.126) and (3.127). The factors $1/\sqrt{2}$ included in (3.129) are normalization factors; since the ${}_sY_{jm}$ are orthonormal we obtain

$$\int_0^{2\pi} \int_0^{\pi} (X^m_\kappa)^\dagger X^{m'}_{\kappa'} \sin\theta \, \mathrm{d}\theta \, \mathrm{d}\phi = \delta_{\kappa\kappa'} \delta_{mm'}. \tag{3.130}$$

Expansion of a plane wave

As shown in Section 1.3, any plane wave solution of the Dirac equation with nonvanishing wave vector can be expressed in the form

$$\begin{pmatrix} u \\ v \end{pmatrix} = \begin{pmatrix} a_-\kappa + a_+\widehat{\kappa} \\ \frac{pc}{E+Mc^2}(a_-\kappa - a_+\widehat{\kappa}) \end{pmatrix} \mathrm{e}^{\mathrm{i}(\mathbf{p}\cdot\mathbf{r}-Et)/\hbar}, \tag{3.131}$$

with $\mathbf{p} = \hbar\mathbf{k}$, $k_i = -\sigma_{iAB}\widehat{\kappa}^A\kappa^B$ and where a_+ and a_- are two arbitrary complex numbers. The field (3.131) is the superposition, with amplitudes a_- and a_+, of two plane waves with the spin of the particle aligned in the direction of $\mathbf{k}$ and $-\mathbf{k}$, respectively. Then, for instance, the spin-weighted component u_+ of (3.131) is given by

$$u_+ = -u^A o_A = -(a_-\kappa^A o_A + a_+\widehat{\kappa}^A o_A)\mathrm{e}^{\mathrm{i}(\mathbf{p}\cdot\mathbf{r}-Et)/\hbar}.$$

The expansion of the term $-a_+\widehat{\kappa}^A o_A \mathrm{e}^{\mathrm{i}(\mathbf{p}\cdot\mathbf{r}-Et)/\hbar}$ can be obtained from the results of the preceding section and by means of an analogous computation, making use of (3.152) and (3.153), one finds the expansion of $-\kappa^A o_A \mathrm{e}^{\mathrm{i}(\mathbf{p}\cdot\mathbf{r}-Et)\hbar}$; thus

$$\begin{aligned} u_+ &= 2\pi\sqrt{k} \sum_{j=\frac{1}{2}}^{\infty} \sum_{m=-j}^{j} \mathrm{i}^{j-\frac{1}{2}} \big[\mathrm{i}(a_+ \overline{{}_{\frac{1}{2}}Y_{jm}(\theta_1,\phi_1)} + a_- \overline{{}_{-\frac{1}{2}}Y_{jm}(\theta_1,\phi_1)}) j_{j+\frac{1}{2}}(kr) \\ &\quad + (a_+ \overline{{}_{\frac{1}{2}}Y_{jm}(\theta_1,\phi_1)} - a_- \overline{{}_{-\frac{1}{2}}Y_{jm}(\theta_1,\phi_1)}) j_{j-\frac{1}{2}}(kr)\big] \, {}_{\frac{1}{2}}Y_{jm} \mathrm{e}^{-\mathrm{i}Et/\hbar} \end{aligned}$$

and by comparing with (3.126)–(3.129) we conclude that, with respect to the basis $\{o, \widehat{o}\}$, the plane wave (3.131) has the expansion

$$\begin{pmatrix} u_- \\ u_+ \\ v_- \\ v_+ \end{pmatrix} = 2\pi\sqrt{2k}\sum_{j=\frac{1}{2}}^{\infty}\sum_{m=-j}^{j} \mathrm{i}^{j-\frac{1}{2}}\, \mathrm{e}^{-\mathrm{i}Et/\hbar}$$

$$\times \Big[\mathrm{i}(a_+ \overline{{}_{\frac{1}{2}}Y_{jm}(\theta_1, \phi_1)} + a_- \overline{{}_{-\frac{1}{2}}Y_{jm}(\theta_1, \phi_1)}) \begin{pmatrix} j_{j+\frac{1}{2}}(kr) X^m_{j+\frac{1}{2}} \\ \frac{\mathrm{i}pc}{E+Mc^2} j_{j-\frac{1}{2}}(kr) X^m_{-j-\frac{1}{2}} \end{pmatrix}$$

$$+ (a_+ \overline{{}_{\frac{1}{2}}Y_{jm}(\theta_1, \phi_1)} - a_- \overline{{}_{-\frac{1}{2}}Y_{jm}(\theta_1, \phi_1)}) \begin{pmatrix} j_{j-\frac{1}{2}}(kr) X^m_{-j-\frac{1}{2}} \\ -\frac{\mathrm{i}pc}{E+Mc^2} j_{j+\frac{1}{2}}(kr) X^m_{j+\frac{1}{2}} \end{pmatrix}\Big].$$

(The components of this expansion with respect to the canonical basis are obtained by simply replacing the spinors $X^m_{\pm(j+\frac{1}{2})}$ by $\chi^m_{\pm(j+\frac{1}{2})}$, defined in (3.149).)

Particle in a Coulomb field

In the case of an electron in the electromagnetic field produced by a point charge $-Ze$, placed at the origin, the Dirac equation (1.70) is modified following the minimal coupling rule: $-\mathrm{i}\hbar\nabla \mapsto -\mathrm{i}\hbar\nabla - \frac{e}{c}\mathbf{A}$, $\mathrm{i}\hbar\partial_t \mapsto \mathrm{i}\hbar\partial_t - e\varphi$, where φ and $\mathbf{A}$ are the potentials of the electromagnetic field and e is the electron charge. Choosing the potentials as $\varphi = -Ze/r$, $\mathbf{A} = \mathbf{0}$, the interaction is obtained by simply replacing $\mathrm{i}\hbar\partial_t$ by $\mathrm{i}\hbar\partial_t + Ze^2/r$; hence, for a solution of the form (3.122), we only need to substitute E by $E + Ze^2/r$ in (3.123) and making use of the definitions (3.124) one finds the radial equations

$$\begin{aligned} \frac{1}{r}\frac{\mathrm{d}}{\mathrm{d}r}(rA_\pm) \pm (j+\tfrac{1}{2})\frac{A_\pm}{r} &= \frac{1}{\hbar c}\left(E + \frac{Ze^2}{r} + Mc^2\right) B_\pm, \\ -\frac{1}{r}\frac{\mathrm{d}}{\mathrm{d}r}(rB_\pm) \pm (j+\tfrac{1}{2})\frac{B_\pm}{r} &= \frac{1}{\hbar c}\left(E + \frac{Ze^2}{r} - Mc^2\right) A_\pm. \end{aligned} \tag{3.132}$$

Equations (3.132) can be written as

$$\begin{aligned} \frac{\mathrm{d}R_1}{\mathrm{d}r} + \frac{\kappa}{r}R_1 &= \frac{1}{\hbar c}\left(E + \frac{Ze^2}{r} + Mc^2\right) R_2, \\ -\frac{\mathrm{d}R_2}{\mathrm{d}r} + \frac{\kappa}{r}R_2 &= \frac{1}{\hbar c}\left(E + \frac{Ze^2}{r} - Mc^2\right) R_1, \end{aligned} \tag{3.133}$$

where $R_1 = rA_+$ and $R_2 = rB_+$ when $\kappa = j + \frac{1}{2}$, and $R_1 = rA_-$ and $R_2 = rB_-$ when $\kappa = -(j+\frac{1}{2})$. Introducing the definitions

$$\mu \equiv \frac{Mc^2 + E}{\hbar c}, \qquad \nu \equiv \frac{Mc^2 - E}{\hbar c}, \qquad \rho \equiv \frac{\sqrt{M^2c^4 - E^2}}{\hbar c} r = \sqrt{\mu\nu}\, r, \tag{3.134}$$

where it is assumed that $E < Mc^2$, which corresponds to bound states, equations (3.133) amount to

$$\begin{aligned} \frac{\mathrm{d}R_1}{\mathrm{d}\rho} + \frac{\kappa}{\rho}R_1 &= \left(\frac{Z\alpha}{\rho} + \sqrt{\frac{\mu}{\nu}}\right) R_2, \\ -\frac{\mathrm{d}R_2}{\mathrm{d}\rho} + \frac{\kappa}{\rho}R_2 &= \left(\frac{Z\alpha}{\rho} - \sqrt{\frac{\nu}{\mu}}\right) R_1, \end{aligned} \tag{3.135}$$

where $\alpha \equiv e^2/\hbar c$ is the fine structure constant. These equations can be solved looking for series solutions of the form

$$R_1(\rho) = \mathrm{e}^{-\rho}\sum_{\lambda=0}^{\infty} a_\lambda \rho^{s+\lambda}, \qquad R_2(\rho) = \mathrm{e}^{-\rho}\sum_{\lambda=0}^{\infty} b_\lambda \rho^{s+\lambda} \tag{3.136}$$

(with $a_0, b_0 \neq 0$). Substituting (3.136) into (3.135) one obtains

$$\begin{aligned} Z\alpha b_0 - (s+\kappa)a_0 &= 0, \\ (s-\kappa)b_0 + Z\alpha a_0 &= 0 \end{aligned} \tag{3.137}$$

and

$$\begin{aligned} \sqrt{\frac{\mu}{\nu}}\, b_{\lambda-1} + a_{\lambda-1} &= -Z\alpha b_\lambda + (s+\lambda+\kappa)a_\lambda, \\ \sqrt{\frac{\nu}{\mu}}\, a_{\lambda-1} + b_{\lambda-1} &= Z\alpha a_\lambda + (s+\lambda-\kappa)b_\lambda. \end{aligned} \tag{3.138}$$

Since a_0 and b_0 are different from zero, from (3.137) it follows that

$$s = \sqrt{\kappa^2 - Z^2\alpha^2} = \sqrt{(j+\tfrac{1}{2})^2 - Z^2\alpha^2} \tag{3.139}$$

and from equations (3.138) one obtains the relation

$$\left(s+\lambda+\kappa - \sqrt{\frac{\mu}{\nu}}\, Z\alpha\right) a_\lambda = \left(\sqrt{\frac{\mu}{\nu}}(s+\lambda-\kappa) + Z\alpha\right) b_\lambda. \tag{3.140}$$

In order for the solutions (3.136) to be well behaved when $\rho \to \infty$, the expressions (3.136) must contain a finite number of terms, N. Making $a_{N+1} = 0$, from (3.140) we see that $b_{N+1} = 0$ and from (3.138) we obtain

$$b_N = -\sqrt{\frac{\nu}{\mu}}\, a_N. \tag{3.141}$$

Substituting (3.141) into (3.140) with $\lambda = N$ we can then conclude that $Z\alpha(\mu - \nu)/\sqrt{\mu\nu} = 2(s+N)$; thus, making use of the definitions (3.134) and

(3.139) it follows that

$$E = Mc^2 \left[1 + \left(\frac{Z\alpha}{N + \sqrt{(j+\frac{1}{2})^2 - Z^2\alpha^2}} \right)^2 \right]^{-1/2}, \tag{3.142}$$

where N can take the values $0, 1, 2, \ldots$, while j takes the values $\frac{1}{2}, \frac{3}{2}, \frac{5}{2}, \ldots$. For each couple of values of N and j, with $N = 1, 2, \ldots$ and $j = \frac{1}{2}, \frac{3}{2}, \ldots$, equations (3.135) have solutions for $\kappa = \pm(j + \frac{1}{2})$. However, when $N = 0$, equations (3.135) have solutions only if $\kappa = -(j + \frac{1}{2})$ (see below).

The solutions of (3.135) can be expressed in terms of associated Laguerre polynomials (Davis 1939). Taking into account (3.136), we write

$$R_1 = \sqrt{\mu}\, e^{-\rho} \rho^s (P + Q), \qquad R_2 = \sqrt{\nu}\, e^{-\rho} \rho^s (P - Q).$$

Substituting these expressions into (3.135) one obtains

$$\begin{aligned} \left(\rho \frac{\mathrm{d}}{\mathrm{d}\rho} - 2\rho + s + \frac{Z\alpha}{2} \frac{\mu - \nu}{\sqrt{\mu\nu}}\right) P &= -\left(\kappa + \frac{Z\alpha}{2} \frac{\mu + \nu}{\sqrt{\mu\nu}}\right) Q, \\ \left(\rho \frac{\mathrm{d}}{\mathrm{d}\rho} + s - \frac{Z\alpha}{2} \frac{\mu - \nu}{\sqrt{\mu\nu}}\right) Q &= -\left(\kappa - \frac{Z\alpha}{2} \frac{\mu + \nu}{\sqrt{\mu\nu}}\right) P, \end{aligned}$$

which can also be written as

$$\begin{aligned} \left(\rho \frac{\mathrm{d}}{\mathrm{d}\rho} - 2\rho + 2s + N\right) P &= -\left(\kappa + \sqrt{\kappa^2 + (2s + N)N}\right) Q, \\ \left(\rho \frac{\mathrm{d}}{\mathrm{d}\rho} - N\right) Q &= -\left(\kappa - \sqrt{\kappa^2 + (2s + N)N}\right) P \end{aligned} \tag{3.143}$$

and by combining these equations one finds that P and Q obey the decoupled equations

$$\begin{aligned} \rho^2 \frac{\mathrm{d}^2 P}{\mathrm{d}\rho^2} + [(2s + 1)\rho - 2\rho^2] \frac{\mathrm{d}P}{\mathrm{d}\rho} + 2(N - 1)\rho P &= 0, \\ \rho^2 \frac{\mathrm{d}^2 Q}{\mathrm{d}\rho^2} + [(2s + 1)\rho - 2\rho^2] \frac{\mathrm{d}Q}{\mathrm{d}\rho} + 2N\rho Q &= 0. \end{aligned}$$

Hence, $P(\rho)$ and $Q(\rho)$ are proportional to $L_{N-1}^{2s}(2\rho)$ and $L_N^{2s}(2\rho)$, respectively, where L_n^p denotes the associated Laguerre polynomials (the subscript n corresponds to the degree of the polynomial L_n^p).

In the case where $N = 0$, $Q(\rho)$ is a constant and $P(\rho)$ must be equal to zero; hence, from the first equation in (3.143) it follows that κ must be negative. If we take $Q(\rho) = L_N^{2s}(2\rho)$, then, using the recurrence relations for the associated Laguerre polynomials, from (3.143) we obtain $P(\rho) = -[(\kappa + \sqrt{\kappa^2 + (2s+N)N})/N]L^{2s}_{N-1}(2\rho)$.

Characterization of the separable solutions

In the case of a two-component spinor field, $\psi(x_i)$, the components of the image of ψ under the rotation $\mathcal{R}$ defined by a SU(2) matrix (Q^A_B) are given by

$$[\mathcal{R}\psi(x_i)]^A \equiv Q^A_B \psi^B(\tilde{a}_{im}x_m),$$

where (a_{ij}) is the SO(3) matrix corresponding to (Q^A_B). Then, making use of (1.15) one finds that for the rotations about $\mathbf{n}$, $\mathrm{d}Q/\mathrm{d}\alpha|_{\alpha=0} = -\frac{1}{2}\mathrm{i}n_k\sigma_k$, therefore

$$\begin{aligned} \frac{\mathrm{d}}{\mathrm{d}\alpha}[Q^A_B\psi^B(\tilde{a}_{im}x_m)]\bigg|_{\alpha=0} &= -\tfrac{1}{2}\mathrm{i}n_k\sigma_k{}^A{}_B\psi^B(x_i) + \varepsilon_{imp}n_p x_m \partial_i \psi^A(x_j) \\ &= -\mathrm{i}n_k(J_k\psi)^A, \end{aligned}$$

where now

$$(J_k\psi)^A = L_k\psi^A + \tfrac{1}{2}\sigma_k{}^A{}_B\psi^B. \tag{3.144}$$

By expressing the spinor field ψ in terms of o and $\widehat{o}$ in the form

$$\psi^A = \psi_- o^A + \psi_+ \widehat{o}^A,$$

where $\psi_- = \psi^A\widehat{o}_A$, $\psi_+ = -\psi^A o_A$, from (3.144) we obtain

$$\begin{aligned} (J_k\psi)^A &= (L_k\psi_-)o^A + (L_k\psi_+)\widehat{o}^A + \psi_-(L_k o^A + \tfrac{1}{2}\sigma_k{}^A{}_B o^B) \\ &\quad + \psi_+(L_k\widehat{o}^A + \tfrac{1}{2}\sigma_k{}^A{}_B\widehat{o}^B). \end{aligned} \tag{3.145}$$

Making use of (1.16), (2.6), and (3.39) we find

$$\begin{aligned} L_1 o^A + \tfrac{1}{2}\sigma_1{}^A{}_B o^B &= \frac{1}{2}\frac{\cos\phi}{\sin\theta}o^A, \\ L_2 o^A + \tfrac{1}{2}\sigma_2{}^A{}_B o^B &= \frac{1}{2}\frac{\sin\phi}{\sin\theta}o^A, \\ L_3 o^A + \tfrac{1}{2}\sigma_3{}^A{}_B o^B &= 0 \end{aligned} \tag{3.146}$$

and, noting that (1.61) and (1.68) give $\overline{\sigma_k{}^A{}_B o^B} = -\overline{\sigma_k{}^{AB}o_B} = -\sigma_{kAB}\widehat{o}^B$, we find that

$$\begin{aligned} L_1 \widehat{o}^A + \tfrac{1}{2}\sigma_1{}^A{}_B \widehat{o}^B &= -\frac{1}{2}\frac{\cos\phi}{\sin\theta}\widehat{o}^A, \\ L_2 \widehat{o}^A + \tfrac{1}{2}\sigma_2{}^A{}_B \widehat{o}^B &= -\frac{1}{2}\frac{\sin\phi}{\sin\theta}\widehat{o}^A, \\ L_3 \widehat{o}^A + \tfrac{1}{2}\sigma_3{}^A{}_B \widehat{o}^B &= 0, \end{aligned} \tag{3.147}$$

hence, substituting (3.146) and (3.147) into (3.145) we have

$$(J_k\psi)^A = (J_k^{(-1/2)}\psi_-)\,o^A + (J_k^{(1/2)}\psi_+)\,\widehat{o}^A,$$

where we have made use of the definition (3.41).

It can be seen that $\mathrm{i}J_k\psi$ is the Lie derivative of ψ with respect to $\mathrm{i}L_k$, which is a Killing vector field of the standard metric of three-dimensional Euclidean space (see Section 6.1).

The fact that the radial equations (3.123) can be partially decoupled, reducing to two independent sets of equations [(3.125) and (3.132)], is related to the existence of an operator, K, that commutes with the Dirac Hamiltonian, J^2 and J_3 (*cf.* Rose 1961, Messiah 1962, Davydov 1988). With respect to the basis induced by $\{o, \widehat{o}\}$, K is given by

$$K \equiv \begin{pmatrix} -Q & 0 \\ 0 & Q \end{pmatrix},$$

with

$$Q \equiv \begin{pmatrix} 0 & -\bar{\eth} \\ \eth & 0 \end{pmatrix}. \qquad (3.148)$$

The spinor fields $X^m_{\pm(j+\frac{1}{2})}$ defined by (3.129) are eigenfunctions of Q,

$$QX^m_{\pm(j+\frac{1}{2})} = \pm(j+\tfrac{1}{2})X^m_{\pm(j+\frac{1}{2})},$$

therefore, the first term on the right-hand side of (3.128) is an eigenfunction of K with eigenvalue $-j-\frac{1}{2}$, while the second term is an eigenfunction of K with eigenvalue $j+\frac{1}{2}$.

From the relation $u = u_-o + u_+\widehat{o}$, we obtain

$$\begin{pmatrix} u^1 \\ u^2 \end{pmatrix} = u_-\begin{pmatrix} o^1 \\ o^2 \end{pmatrix} + u_+\begin{pmatrix} \widehat{o}^1 \\ \widehat{o}^2 \end{pmatrix} = \begin{pmatrix} o^1 & \widehat{o}^1 \\ o^2 & \widehat{o}^2 \end{pmatrix}\begin{pmatrix} u_- \\ u_+ \end{pmatrix}.$$

Hence,

$$\begin{pmatrix} u_- \\ u_+ \end{pmatrix} = \begin{pmatrix} \widehat{o}^2 & -\widehat{o}^1 \\ -o^2 & o^1 \end{pmatrix}\begin{pmatrix} u^1 \\ u^2 \end{pmatrix} \equiv \Lambda\begin{pmatrix} u^1 \\ u^2 \end{pmatrix}$$

and, with respect to the canonical basis, Q is given by

$$\Lambda^{-1}Q\Lambda = -(I + \boldsymbol{\sigma}\cdot\mathbf{L}).$$

Since $J^2 = (\mathbf{L}+\mathbf{S})\cdot(\mathbf{L}+\mathbf{S}) = L^2 + 2\mathbf{L}\cdot\mathbf{S} + \mathbf{S}^2 = L^2 + \boldsymbol{\sigma}\cdot\mathbf{L} + \frac{3}{4}I$ for a spin-1/2 field, $L^2 = J^2 + \frac{1}{4}I - (I + \boldsymbol{\sigma}\cdot\mathbf{L})$, which implies that the spinor fields $X^m_{j+\frac{1}{2}}$ and

$X^m_{-(j+\frac{1}{2})}$, being eigenfunctions of J^2, J_3, and Q, are also eigenfunctions of L^2 with eigenvalue $l(l+1)$, where $l = j + \frac{1}{2}$ and $l = j - \frac{1}{2}$, respectively. Thus,

$$\chi^m_{\pm(j+\frac{1}{2})} \equiv \Lambda^{-1} X^m_{\pm(j+\frac{1}{2})} \tag{3.149}$$

are normalized eigenfunctions of J^2, J_3, L^2, and $-\boldsymbol{\sigma}\cdot\mathbf{L} - I$, with eigenvalues $j(j+1)$, m, $l(l+1)$, with $l = j \pm \frac{1}{2}$, and $\pm(j+\frac{1}{2})$, respectively, with respect to the canonical basis. Hence, for instance, $\chi^m_{j+\frac{1}{2}}$ can be expressed as

$$\chi^m_{j+\frac{1}{2}} = \begin{pmatrix} c_1 Y_{j+\frac{1}{2}, m-\frac{1}{2}} \\ c_2 Y_{j+\frac{1}{2}, m+\frac{1}{2}} \end{pmatrix},$$

since $J_3 = L_3 + S_3$ and the spinors constituting the canonical basis are eigenfunctions of S_3 with eigenvalues 1/2 and $-1/2$, while the ordinary spherical harmonics Y_{lm} are eigenfunctions of L_3 with eigenvalue m and, in the present case, $l = j + \frac{1}{2}$. The constants c_1 and c_2 are restricted by the condition that $\chi^m_{j+\frac{1}{2}}$ be an eigenfunction of

$$-\boldsymbol{\sigma}\cdot\mathbf{L} - I = \begin{pmatrix} -L_3 - 1 & -L_- \\ -L_+ & L_3 - 1 \end{pmatrix}$$

with eigenvalue $j + \frac{1}{2}$; in this way, making use of (3.31), one obtains the relation $c_1\sqrt{j+m+1} + c_2\sqrt{j-m+1} = 0$ which together with the normalization condition $|c_1|^2 + |c_2|^2 = 1$, determine c_1 and c_2 up to a common phase factor. By evaluating (3.149) at $\theta = 0$ one concludes that

$$\chi^m_{j+\frac{1}{2}} = \begin{pmatrix} \mathrm{i}\sqrt{\frac{j-m+1}{2(j+1)}}\, Y_{j+\frac{1}{2}, m-\frac{1}{2}} \\ -\mathrm{i}\sqrt{\frac{j+m+1}{2(j+1)}}\, Y_{j+\frac{1}{2}, m+\frac{1}{2}} \end{pmatrix}. \tag{3.150}$$

In a similar manner, it follows that

$$\chi^m_{-j-\frac{1}{2}} = \begin{pmatrix} -\mathrm{i}\sqrt{\frac{j+m}{2j}}\, Y_{j-\frac{1}{2}, m-\frac{1}{2}} \\ -\mathrm{i}\sqrt{\frac{j-m}{2j}}\, Y_{j-\frac{1}{2}, m+\frac{1}{2}} \end{pmatrix}. \tag{3.151}$$

From (3.149) we obtain the relation

$$X^m_{\pm(j+\frac{1}{2})} = \Lambda \chi^m_{\pm(j+\frac{1}{2})} = \begin{pmatrix} \widehat{o}^2 & -\widehat{o}^1 \\ -o^2 & o^1 \end{pmatrix} \chi^m_{\pm(j+\frac{1}{2})},$$

which is explicitly given by

$$\begin{aligned} {}_{-\frac{1}{2}}Y_{jm} &= \mathrm{i}\sqrt{\tfrac{j+m+1}{j+1}}\, \widehat{o}^1 Y_{j+\frac{1}{2}, m+\frac{1}{2}} + \mathrm{i}\sqrt{\tfrac{j-m+1}{j+1}}\, \widehat{o}^2 Y_{j+\frac{1}{2}, m-\frac{1}{2}}, \\ {}_{\frac{1}{2}}Y_{jm} &= -\mathrm{i}\sqrt{\tfrac{j+m+1}{j+1}}\, o^1 Y_{j+\frac{1}{2}, m+\frac{1}{2}} - \mathrm{i}\sqrt{\tfrac{j-m+1}{j+1}}\, o^2 Y_{j+\frac{1}{2}, m-\frac{1}{2}}, \end{aligned} \tag{3.152}$$

and

$$
\begin{aligned}
{}_{-\frac{1}{2}}Y_{jm} &= -\mathrm{i}\sqrt{\tfrac{j-m}{j}}\,\widehat{o}^1 Y_{j-\frac{1}{2},m+\frac{1}{2}} + \mathrm{i}\sqrt{\tfrac{j+m}{j}}\,\widehat{o}^2 Y_{j-\frac{1}{2},m-\frac{1}{2}}, \\
{}_{\frac{1}{2}}Y_{jm} &= -\mathrm{i}\sqrt{\tfrac{j-m}{j}}\,o^1 Y_{j-\frac{1}{2},m+\frac{1}{2}} + \mathrm{i}\sqrt{\tfrac{j+m}{j}}\,o^2 Y_{j-\frac{1}{2},m-\frac{1}{2}}.
\end{aligned}
\tag{3.153}
$$

3.6 The spin-2 Helmholtz equation

A spin-2 field corresponds to a symmetric, traceless two-index tensor field, t_{ij}. The five independent components of t_{ij} with respect to the orthonormal basis $\{\mathbf{e}_r, \mathbf{e}_\theta, \mathbf{e}_\phi\}$ can be combined into the five spin-weighted components

$$
\begin{aligned}
t_{\pm 2} &\equiv \tfrac{1}{2}(t_{\theta\theta} - t_{\phi\phi} \pm 2\mathrm{i}t_{\theta\phi}) = \tfrac{1}{2}(t_{rr} + 2t_{\theta\theta} \pm 2\mathrm{i}t_{\theta\phi}), \\
t_{\pm 1} &\equiv \mp\tfrac{1}{2}(t_{\theta r} \pm \mathrm{i}t_{\phi r}), \\
t_0 &\equiv \tfrac{1}{2}t_{rr}.
\end{aligned}
\tag{3.154}
$$

Thus, the field t_{ij} is real if and only if

$$
\overline{t_s} = (-1)^s t_{-s}.
$$

The Helmholtz equation for a symmetric, traceless tensor field t_{ij},

$$
\nabla^2 t_{ij} + k^2 t_{ij} = 0, \tag{3.155}
$$

written in terms of the spin-weighted components (3.154), is given by

$$
\begin{aligned}
\frac{1}{r^2}\partial_r(r^2\partial_r t_{+2}) + \frac{1}{r^2}\eth\bar{\eth} t_{+2} + \frac{4}{r^2}\eth t_{+1} + k^2 t_{+2} &= 0, \\
\frac{1}{r^2}\partial_r(r^2\partial_r t_{+1}) - \frac{4}{r^2}t_{+1} + \frac{1}{r^2}\eth\bar{\eth} t_{+1} - \frac{1}{r^2}\bar{\eth} t_{+2} + \frac{3}{r^2}\eth t_0 + k^2 t_{+1} &= 0, \\
\frac{1}{r^2}\partial_r(r^2\partial_r t_0) - \frac{6}{r^2}t_0 + \frac{1}{r^2}\eth\bar{\eth} t_0 + \frac{2}{r^2}(\eth t_{-1} - \bar{\eth} t_{+1}) + k^2 t_0 &= 0, \\
\frac{1}{r^2}\partial_r(r^2\partial_r t_{-1}) - \frac{4}{r^2}t_{-1} + \frac{1}{r^2}\bar{\eth}\eth t_{-1} + \frac{1}{r^2}\eth t_{-2} - \frac{3}{r^2}\bar{\eth} t_0 + k^2 t_{-1} &= 0, \\
\frac{1}{r^2}\partial_r(r^2\partial_r t_{-2}) + \frac{1}{r^2}\bar{\eth}\eth t_{-2} - \frac{4}{r^2}\bar{\eth} t_{-1} + k^2 t_{-2} &= 0.
\end{aligned}
\tag{3.156}
$$

Equations (3.156) admit separable solutions of the form

$$
t_s = \left[\frac{(j-|s|)!}{(j+|s|)!}\right]^{1/2} g_s(r)\,{}_sY_{jm}(\theta,\phi), \tag{3.157}
$$

where j is an integer greater than 1 and the constant factors have been introduced for convenience. Substituting (3.157) into (3.156) we obtain the system of ordinary differential equations

$$
\begin{aligned}
\left[\frac{\mathrm{d}^2}{\mathrm{d}r^2} + \frac{2}{r}\frac{\mathrm{d}}{\mathrm{d}r} - \frac{(j-1)(j+2)}{r^2} + k^2\right] g_{\pm 2} + \frac{4(j-1)(j+2)}{r^2} g_{\pm 1} &= 0, \\
\left[\frac{\mathrm{d}^2}{\mathrm{d}r^2} + \frac{2}{r}\frac{\mathrm{d}}{\mathrm{d}r} - \frac{j(j+1)+4}{r^2} + k^2\right] g_{\pm 1} + \frac{1}{r^2} g_{\pm 2} + \frac{3j(j+1)}{r^2} g_0 &= 0, \quad (3.158) \\
\left[\frac{\mathrm{d}^2}{\mathrm{d}r^2} + \frac{2}{r}\frac{\mathrm{d}}{\mathrm{d}r} - \frac{j(j+1)+6}{r^2} + k^2\right] g_0 + \frac{2}{r^2}(g_{-1} + g_1) &= 0.
\end{aligned}
$$

Combining (3.158) we find that the functions $g_2 - g_{-2} - 2(j+2)(g_1 - g_{-1})$, $g_2 - g_{-2} + 2(j-1)(g_1 - g_{-1})$, $g_2 + g_{-2} - 4(j+2)(g_1 + g_{-1}) + 6(j+1)(j+2)g_0$, $g_2 + g_{-2} - 2(g_1 + g_{-1}) - 2j(j+1)g_0$, and $g_2 + g_{-2} + 4(j-1)(g_1 + g_{-1}) + 6j(j-1)g_0$ obey decoupled equations (Torres del Castillo and Rojas-Marcial 1993) whose solutions are spherical Bessel functions provided $k \neq 0$. Thus, from (3.157) we obtain

$$
\begin{aligned}
t_{\pm 2} = {} & \tfrac{1}{2}[(j-1)j(j+1)(j+2)]^{1/2}\{a j_{j+2}(kr) + b n_{j+2}(kr) \\
& - 2[c j_j(kr) + d n_j(kr)] + e j_{j-2}(kr) + f n_{j-2}(kr) \\
& \pm 2[-A j_{j+1}(kr) - B n_{j+1}(kr) + C j_{j-1}(kr) + D n_{j-1}(kr)]\}\, {}_{\pm 2}Y_{jm}, \\
t_{\pm 1} = {} & \tfrac{1}{2}[j(j+1)]^{1/2}\{-(j+2)[a j_{j+2}(kr) + b n_{j+2}(kr)] \\
& + c j_j(kr) + d n_j(kr) + (j-1)[e j_{j-2}(kr) + f n_{j-2}(kr)] \\
& \pm (j+2)[A j_{j+1}(kr) + B n_{j+1}(kr)] \qquad (3.159) \\
& \pm (j-1)[C j_{j-1}(kr) + D n_{j-1}(kr)]\}\, {}_{\pm 1}Y_{jm}, \\
t_0 = {} & \{\tfrac{1}{2}(j+1)(j+2)[a j_{j+2}(kr) + b n_{j+2}(kr)] \\
& + \tfrac{1}{3}j(j+1)[c j_j(kr) + d n_j(kr)] \\
& + \tfrac{1}{2}j(j-1)[e j_{j-2}(kr) + f n_{j-2}(kr)]\}\, Y_{jm},
\end{aligned}
$$

where $a, b, c, d, e, f, A, B, C$, and D are arbitrary constants.

The cases where $j = 1$ or $j = 0$ must be treated separately since ${}_sY_{jm} = 0$ for $j < |s|$. It turns out that, also in these cases, the separable solutions of (3.155) are given by (3.159).

As in the case of the vector Helmholtz equation, the fact that the radial equations reduce to a set of uncoupled second-order differential equations is related to the existence of an operator that commutes with J^2, J_3 and ∇^2. Indeed, the separable solution (3.159) can be rewritten in the form

$$
\begin{aligned}
\begin{pmatrix} t_{-2} \\ t_{-1} \\ t_0 \\ t_{+1} \\ t_{+2} \end{pmatrix}
&= \frac{f_{j+2}(kr)}{2\sqrt{(2j+1)(2j+3)}}
\begin{pmatrix} \sqrt{j(j-1)}\,{}_{-2}Y_{jm} \\ -\sqrt{j(j+2)}\,{}_{-1}Y_{jm} \\ \sqrt{(j+1)(j+2)}\,Y_{jm} \\ -\sqrt{j(j+2)}\,{}_{1}Y_{jm} \\ \sqrt{j(j-1)}\,{}_{2}Y_{jm} \end{pmatrix} \\
&\quad + \frac{f_{j}(kr)}{\sqrt{\tfrac{2}{3}(2j-1)(2j+3)}}
\begin{pmatrix} -\sqrt{(j-1)(j+2)}\,{}_{-2}Y_{jm} \\ \tfrac{1}{2}\,{}_{-1}Y_{jm} \\ \tfrac{1}{3}\sqrt{j(j+1)}\,Y_{jm} \\ \tfrac{1}{2}\,{}_{1}Y_{jm} \\ -\sqrt{(j-1)(j+2)}\,{}_{2}Y_{jm} \end{pmatrix} \\
&\quad + \frac{f_{j-2}(kr)}{2\sqrt{(2j-1)(2j+1)}}
\begin{pmatrix} \sqrt{(j+1)(j+2)}\,{}_{-2}Y_{jm} \\ \sqrt{(j-1)(j+1)}\,{}_{-1}Y_{jm} \\ \sqrt{j(j-1)}\,Y_{jm} \\ \sqrt{(j-1)(j+1)}\,{}_{1}Y_{jm} \\ \sqrt{(j+1)(j+2)}\,{}_{2}Y_{jm} \end{pmatrix} \\
&\quad + \frac{f_{j+1}(kr)}{\sqrt{2(2j+1)}}
\begin{pmatrix} \sqrt{j-1}\,{}_{-2}Y_{jm} \\ -\tfrac{1}{2}\sqrt{j+2}\,{}_{-1}Y_{jm} \\ 0 \\ \tfrac{1}{2}\sqrt{j+2}\,{}_{1}Y_{jm} \\ -\sqrt{j-1}\,{}_{2}Y_{jm} \end{pmatrix} \\
&\quad + \frac{f_{j-1}(kr)}{\sqrt{2(2j+1)}}
\begin{pmatrix} -\sqrt{j+2}\,{}_{-2}Y_{jm} \\ -\tfrac{1}{2}\sqrt{j-1}\,{}_{-1}Y_{jm} \\ 0 \\ \tfrac{1}{2}\sqrt{j-1}\,{}_{1}Y_{jm} \\ \sqrt{j+2}\,{}_{2}Y_{jm} \end{pmatrix},
\end{aligned}
\tag{3.160}
$$

where f_l is a spherical Bessel function of order l. Each of the five terms in (3.160) is an eigenvector of the operator

$$
K = \begin{pmatrix}
0 & -2\bar{\eth} & 0 & 0 & 0 \\
\tfrac{1}{2}\eth & -3 & -\tfrac{3}{2}\bar{\eth} & 0 & 0 \\
0 & \eth & -4 & -\bar{\eth} & 0 \\
0 & 0 & \tfrac{3}{2}\eth & -3 & -\tfrac{1}{2}\bar{\eth} \\
0 & 0 & 0 & 2\eth & 0
\end{pmatrix},
$$

with eigenvalues $-2(j+2)$, -1, $2(j-1)$, $-(j+2)$, and $j-1$, respectively. With respect to the Cartesian basis $\{\mathbf{e}_x, \mathbf{e}_y, \mathbf{e}_z\}$, K corresponds to $2I + \mathbf{L}\cdot\mathbf{S}$, where $\mathbf{L}$ and $\mathbf{S}$ are the orbital and spin angular momentum operators. Hence,

$L^2 = J^2 - 2I - 2K$ and therefore each term in (3.160) is an eigenfunction of L^2 with eigenvalue $l(l+1)$, where $l = j+2, j, j-2, j+1$, and $j-1$, respectively; the index of each spherical Bessel function appearing in (3.160) coincides with the value of l of the eigenfunction of L^2 multiplying it. The parity of each term in (3.160) is $(-1)^l$ (assuming again that under the inversion $\mathbf{e}_r$ and $\mathbf{e}_\phi$ are left unchanged and $\mathbf{e}_\theta$ changes sign).

The divergence of a second-rank, symmetric, traceless tensor field, t, is the vector field, $\operatorname{div} t$, whose Cartesian components are given by $(\operatorname{div} t)_i = \partial_j t_{ij}$. Then the components of $\operatorname{div} t$ are given by

$$(\operatorname{div} t)_s = \frac{1}{\sqrt{2}}\left[\frac{1}{r}\eth t_{s-1} - \frac{2}{r^3}\partial_r(r^3 t_s) - \frac{1}{r}\bar{\eth} t_{s+1}\right] \tag{3.161}$$

[see (6.65)]. Substituting (3.159) into (3.161), using the recurrence relations for the spin-weighted spherical harmonics and for the spherical Bessel functions [(2.27) and (3.21)], we find that the divergence of the separable solution of the Helmholtz equation given by (3.159) vanishes if and only if

$$\begin{aligned} a &= \frac{j(2j-1)}{3(j+2)(2j+1)}c, & e &= \frac{(j+1)(2j+3)}{3(j-1)(2j+1)}c, & A &= \frac{j-1}{j+2}C,\\ b &= \frac{j(2j-1)}{3(j+2)(2j+1)}d, & f &= \frac{(j+1)(2j+3)}{3(j-1)(2j+1)}d, & B &= \frac{j-1}{j+2}D. \end{aligned} \tag{3.162}$$

Substituting (3.162) into (3.159) and making use of the recurrence relations for the Bessel functions we obtain

$$\begin{aligned} t_{+2} &= -\frac{\mathrm{i}}{r^2}\partial_r r^2 \eth\eth\psi_1 + \frac{1}{2k}\left(\frac{1}{r^2}\partial_r^2 r^2 - k^2\right)\eth\eth\psi_2,\\ t_{+1} &= \frac{\mathrm{i}}{2r}\bar{\eth}\eth\eth\psi_1 - \frac{1}{2kr^2}\partial_r r\bar{\eth}\eth\eth\psi_2,\\ t_0 &= \frac{1}{2kr^2}\bar{\eth}\bar{\eth}\eth\eth\psi_2,\\ t_{-1} &= \frac{\mathrm{i}}{2r}\eth\bar{\eth}\bar{\eth}\psi_1 + \frac{1}{2kr^2}\partial_r r\eth\bar{\eth}\bar{\eth}\psi_2,\\ t_{-2} &= \frac{\mathrm{i}}{r^2}\partial_r r^2\bar{\eth}\bar{\eth}\psi_1 + \frac{1}{2k}\left(\frac{1}{r^2}\partial_r^2 r^2 - k^2\right)\bar{\eth}\bar{\eth}\psi_2, \end{aligned} \tag{3.163}$$

where

$$\begin{aligned} \psi_1 &\equiv \frac{\mathrm{i}(2j+1)}{k(j+2)}[Cj_j(kr) + Dn_j(kr)]\,Y_{jm},\\ \psi_2 &\equiv \frac{(2j-1)(2j+3)}{3k(j-1)(j+2)}[cj_j(kr) + dn_j(kr)]\,Y_{jm}. \end{aligned}$$

The scalar potentials ψ_1 and ψ_2 are solutions of the scalar Helmholtz equation. On the other hand, from (3.159) and (3.161) we find that if $k \neq 0$ and $\operatorname{div} t = 0$ then, necessarily, $t = 0$.

The components (3.163) can be also written in terms of certain tensor operators, U_{ij} and V_{ij} (Campbell and Morgan 1971), whose Cartesian components are defined by

$$U_{jk}(\psi) \equiv \mathrm{i}L_j X_k \psi + \mathrm{i}L_k X_j \psi, \qquad V_{jk}(\psi) \equiv \varepsilon_{jlm} \partial_l U_{mk}(\psi),$$

where

$$\mathbf{X} \equiv \mathrm{i}\nabla \times \mathbf{L} - \nabla.$$

For a well-behaved function ψ, $U_{jk}(\psi)$ and $V_{jk}(\psi)$ are symmetric, traceless, divergenceless tensor fields. By computing the spin-weighted components of $U_{jk}(\psi)$ and $V_{jk}(\psi)$ we find that the expressions (3.163) are equivalent to

$$t_{ij} = U_{ij}(\psi_1) + \frac{1}{k} V_{ij}(\psi_2). \tag{3.164}$$

By virtue of the completeness of the spin-weighted spherical harmonics and the linearity of the differential operators appearing in (3.163) and of the scalar Helmholtz equation, any divergenceless solution of the spin-2 Helmholtz equation (3.155) can be expressed in the form (3.163) or (3.164), where ψ_1 and ψ_2 are solutions of the scalar Helmholtz equation. If ψ_1 and ψ_2 are real, then the tensor field t is real.

3.7 Linearized Einstein theory

The Einstein field equations linearized about the Minkowski metric are obtained assuming that in some coordinate system the metric of the space-time can be expressed in the form

$$g_{\alpha\beta} = \eta_{\alpha\beta} + h_{\alpha\beta}, \tag{3.165}$$

where $(\eta_{\alpha\beta}) = \operatorname{diag}(-1, 1, 1, 1)$, and the Greek indices run from 0 to 3. The curvature tensor of the metric (3.165) to first order in the metric perturbation $h_{\alpha\beta}$ is

$$K_{\alpha\beta\gamma\delta} = \tfrac{1}{2}\{\partial_\alpha \partial_\delta h_{\beta\gamma} - \partial_\alpha \partial_\gamma h_{\beta\delta} + \partial_\beta \partial_\gamma h_{\alpha\delta} - \partial_\beta \partial_\delta h_{\alpha\gamma}\}, \tag{3.166}$$

with the indices being lowered or raised by means of $\eta_{\alpha\beta}$ and its inverse $\eta^{\alpha\beta}$. The tensor field (3.166) possesses the symmetries of the curvature tensor

$$K_{\alpha\beta\gamma\delta} = -K_{\beta\alpha\gamma\delta} = -K_{\alpha\beta\delta\gamma} = K_{\gamma\delta\alpha\beta}, \tag{3.167}$$

$$K_{\alpha\beta\gamma\delta} + K_{\alpha\delta\beta\gamma} + K_{\alpha\gamma\delta\beta} = 0 \tag{3.168}$$

and it also satisfies the differential identities

$$\partial_\alpha K_{\beta\gamma\delta\epsilon} + \partial_\epsilon K_{\beta\gamma\alpha\delta} + \partial_\delta K_{\beta\gamma\epsilon\alpha} = 0. \tag{3.169}$$

In terms of the right dual of $K_{\alpha\beta\gamma\delta}$,

$$K^*_{\alpha\beta\gamma\delta} \equiv \tfrac{1}{2} K_{\alpha\beta}{}^{\rho\sigma} \varepsilon_{\rho\sigma\gamma\delta}, \tag{3.170}$$

where $\varepsilon_{\alpha\beta\gamma\delta}$ is completely anti-symmetric with $\varepsilon_{0123} = 1$, (3.168) and (3.169) can be written as

$$K^*_{\alpha\beta\gamma}{}^{\beta} = 0 \tag{3.171}$$

and

$$\partial^\gamma K^*_{\alpha\beta\gamma\delta} = 0, \tag{3.172}$$

respectively. From (3.167) and (3.170) it follows that

$$K^*_{\alpha\beta\gamma\delta} = -K^*_{\beta\alpha\gamma\delta} = -K^*_{\alpha\beta\delta\gamma},$$

which are analogous to the first two equations in (3.167); however, $K^*_{\alpha\beta\gamma\delta}$ may not possess all the symmetries of $K_{\alpha\beta\gamma\delta}$ [(3.167) and (3.168)]. In fact, from (3.170) one finds that

$$K^*_{\alpha\beta\gamma\delta} - K^*_{\gamma\delta\alpha\beta} = \tfrac{1}{2}(\varepsilon_{\alpha\beta\delta\rho} K_\gamma{}^\rho + \varepsilon_{\beta\alpha\gamma\rho} K_\delta{}^\rho + \varepsilon_{\gamma\beta\delta\rho} K_\alpha{}^\rho + \varepsilon_{\delta\alpha\gamma\rho} K_\beta{}^\rho), \tag{3.173}$$

where

$$K_{\alpha\beta} \equiv K^\gamma{}_{\alpha\gamma\beta}, \tag{3.174}$$

which is a symmetric tensor owing to (3.167). Similarly, one finds that

$$K^*_{\alpha\beta\gamma\delta} + K^*_{\alpha\delta\beta\gamma} + K^*_{\alpha\gamma\delta\beta} = -\varepsilon_{\beta\gamma\delta\rho} K_\alpha{}^\rho$$

and from the identities (3.169) it follows that

$$\partial^\gamma K_{\alpha\beta\gamma\delta} = \partial_\alpha K_{\beta\delta} - \partial_\beta K_{\alpha\delta}. \tag{3.175}$$

The linearized Einstein vacuum field equations are given by $K_{\alpha\beta} = 0$, and from (3.167), (3.168) and (3.171)–(3.175) one finds that the tensor field $K^*_{\alpha\beta\gamma\delta}$ satisfies the same relations as $K_{\alpha\beta\gamma\delta}$ if and only if the linearized Einstein vacuum field equations hold.

Thus, when $K_{\alpha\beta} = 0$, all the components of $K_{\alpha\beta\gamma\delta}$ can be expressed in terms of the tensor fields

$$E_{ij} \equiv K_{0i0j}, \qquad B_{ij} \equiv -K^*_{0i0j}. \tag{3.176}$$

Owing to (3.167), (3.171), (3.173), and (3.174), the fields E_{ij} and B_{ij} are symmetric and trace-free. Furthermore, (3.172) and (3.175) amount to the equations

$$\partial_i E_{ij} = 0, \qquad \partial_i B_{ij} = 0, \tag{3.177}$$

and

$$\frac{1}{c}\partial_t E_{ij} = \varepsilon_{ikm}\partial_k B_{mj}, \qquad \frac{1}{c}\partial_t B_{ij} = -\varepsilon_{ikm}\partial_k E_{mj}, \tag{3.178}$$

which are analogous to the source-free Maxwell equations (the minus sign in the definition of B_{ij} in (3.176) was included in order to obtain this analogy).

If the tensor fields E_{ij} and B_{ij} have a harmonic time dependence with frequency ω, from (3.177) and (3.178) it follows that they are divergenceless solutions of the spin-2 Helmholtz equation (3.155) with $k = \omega/c$; therefore, there exist solutions to the scalar Helmholtz equation, ψ_1 and ψ_2, such that [see (3.164)]

$$\begin{aligned} E_{ij} &= \mathrm{Re}\left[\left(U_{ij}(\psi_1) + \frac{1}{k}V_{ij}(\psi_2)\right)\mathrm{e}^{-\mathrm{i}\omega t}\right] \\ &= \mathrm{Re}\left[\frac{\mathrm{i}}{\omega}\partial_t U_{ij}(\psi_1\mathrm{e}^{-\mathrm{i}\omega t}) + \frac{1}{k}V_{ij}(\psi_2\mathrm{e}^{-\mathrm{i}\omega t})\right] \\ &= \frac{1}{c}\partial_t U_{ij}(\chi_{\mathrm{M}}) - V_{ij}(\chi_{\mathrm{E}}), \end{aligned} \tag{3.179}$$

where $\chi_{\mathrm{M}} = \mathrm{Re}\,(\mathrm{i}/k)\psi_1\mathrm{e}^{-\mathrm{i}\omega t}$ and $\chi_{\mathrm{E}} = -\mathrm{Re}\,(1/k)\psi_2\mathrm{e}^{-\mathrm{i}\omega t}$ are solutions of the scalar wave equation. Then, from (3.178) it follows that

$$B_{ij} = -\frac{1}{c}\partial_t U_{ij}(\chi_{\mathrm{E}}) - V_{ij}(\chi_{\mathrm{M}}). \tag{3.180}$$

According to (3.163), given the tensor fields E_{ij} and B_{ij}, the scalar potentials χ_{E} and χ_{M} can be obtained from

$$x_i x_j E_{ij} = -\bar{\eth}\bar{\eth}\eth\eth\chi_{\mathrm{E}}, \qquad x_i x_j B_{ij} = -\bar{\eth}\bar{\eth}\eth\eth\chi_{\mathrm{M}}.$$

The metric perturbations, $h_{\alpha\beta}$, corresponding to the curvatures (3.179) and (3.180) are given by

$$\begin{aligned} h_{00} &= -2\left(\partial_r^2 + \frac{1}{c^2}\partial_t^2\right)r^2\chi_{\mathrm{E}}, \\ h_{0j} &= -4\frac{x_j}{r}\partial_r r^2\frac{1}{c}\partial_t\chi_{\mathrm{E}} + 2\mathrm{i}L_j\frac{1}{r}\partial_r r^2\chi_{\mathrm{M}}, \\ h_{jk} &= -2\delta_{jk}\left(\partial_r^2 - \frac{1}{c^2}\partial_t^2\right)r^2\chi_{\mathrm{E}} - 4x_j x_k\frac{1}{c^2}\partial_t^2\chi_{\mathrm{E}} + 4\mathrm{i}x_{(j}L_{k)}\frac{1}{c}\partial_t\chi_{\mathrm{M}}, \end{aligned} \tag{3.181}$$

where L_i are the Cartesian components of the operator $\mathbf{L}$ [(3.28)], modulo the gauge transformations

$$h_{\alpha\beta} \mapsto h_{\alpha\beta} - \partial_\alpha\xi_\beta - \partial_\beta\xi_\alpha,$$

where ξ_α is an arbitrary vector field (Torres del Castillo 1990b).

The complex traceless symmetric tensor field

$$F_{ij} \equiv E_{ij} + \mathrm{i} B_{ij}$$

is then given by

$$F_{ij} = -\frac{\mathrm{i}}{c}\partial_t U_{ij}(\chi) - V_{ij}(\chi), \tag{3.182}$$

where $\chi \equiv \chi_{\mathrm{E}} + \mathrm{i}\chi_{\mathrm{M}}$ is a solution of the wave equation, thus showing that any solution of the linearized Einstein vacuum field equations can be expressed in terms of a single complex scalar potential. The spin-weighted components of (3.182) are given by

$$\begin{aligned} F_{+2} &= -\frac{1}{2r^2}\left(\frac{1}{c}\partial_t + \partial_r\right)^2 r^2 \eth\eth\chi, \\ F_{+1} &= \frac{1}{2r^2}\left(\frac{1}{c}\partial_t + \partial_r\right) r\bar{\eth}\eth\eth\chi, \\ F_0 &= -\frac{1}{2r^2}\bar{\eth}\bar{\eth}\eth\eth\chi, \\ F_{-1} &= \frac{1}{2r^2}\left(\frac{1}{c}\partial_t - \partial_r\right) r\eth\bar{\eth}\bar{\eth}\chi, \\ F_{-2} &= -\frac{1}{2r^2}\left(\frac{1}{c}\partial_t - \partial_r\right)^2 r^2 \bar{\eth}\bar{\eth}\chi. \end{aligned} \tag{3.183}$$

For a wave with frequency ω, the vector field

$$S_i = \frac{c^7}{8\pi G\omega^2}\varepsilon_{ijk} E_{jm} B_{km}, \tag{3.184}$$

where G is Newton's constant of gravitation, is analogous to the Poynting vector of the electromagnetic field. In fact, from (3.178) one can verify that the continuity equation

$$\partial_i S_i + \partial_t \frac{c^6}{16\pi G\omega^2}(E_{jk}E_{jk} + B_{jk}B_{jk}) = 0$$

holds. However, it should be remarked that, even in the linearized theory, there is no completely satisfactory definition for the energy or the momentum of the gravitational field. In any case, from (3.154) one finds that the radial component of the vector field (3.184) is

$$S_r = \frac{c^7}{16\pi G\omega^2}(|F_{-2}|^2 - |F_{+2}|^2 + 2|F_{-1}|^2 - 2|F_{+1}|^2). \tag{3.185}$$

Making use of (3.65) and the formulas

$$\left(\frac{2\mathrm{i}}{kr^2}\partial_r r^2 + \frac{1}{k^2r^2}\partial_r^2 r^2 - 1\right) h_j^{(1)}(kr) \sim -4(-\mathrm{i})^{j+1}\frac{\mathrm{e}^{\mathrm{i}kr}}{kr}\left(1 + O\left(\frac{1}{kr}\right)\right),$$

$$\left(-\frac{2\mathrm{i}}{kr^2}\partial_r r^2 + \frac{1}{k^2r^2}\partial_r^2 r^2 - 1\right) h_j^{(1)}(kr) \sim -\frac{(j+2)!}{4(j-2)!}(-\mathrm{i})^{j+1} \times \frac{\mathrm{e}^{\mathrm{i}kr}}{(kr)^5}\left(1 + O\left(\frac{1}{kr}\right)\right),$$

which follow from (3.64), one finds that for outgoing waves F_{-2} is the dominant component and

$$F_s = O\left(\frac{1}{r^{3+s}}\right) \tag{3.186}$$

thus, assuming that (3.184) represents the energy flux of a wave and making use of (3.185), the outgoing energy flux per unit time and unit solid angle is

$$\frac{\mathrm{d}^2 E_{\text{out}}}{\mathrm{d}t\,\mathrm{d}\Omega} = \lim_{r\to\infty} \frac{c^7}{16\pi G\omega^2} r^2 |F_{-2}|^2. \tag{3.187}$$

Similarly, one finds that for ingoing waves, F_{+2} is the dominant component,

$$F_s = O\left(\frac{1}{r^{3-s}}\right), \tag{3.188}$$

and the ingoing energy flux per unit time and unit solid angle is

$$\frac{\mathrm{d}^2 E_{\text{in}}}{\mathrm{d}t\,\mathrm{d}\Omega} = \lim_{r\to\infty} \frac{c^7}{64\pi G\omega^2} r^2 |F_{+2}|^2. \tag{3.189}$$

Thus, in the radiation zone F_{-2} represents the outgoing field and F_{+2} represents the ingoing field. As in the case of the electromagnetic waves, in the linear approximation, the gravitational waves have two independent polarizations and, since F_{-2} and F_{+2} have a well-defined spin-weight, if in the radiation zone F_{-2} or F_{+2} is proportional to $\mathrm{e}^{\mathrm{i}\omega t}$, the radiation has right circular polarization (negative helicity) if $\omega > 0$ or left circular polarization (positive helicity) if $\omega < 0$.

As in the case of the electromagnetic field, we can consider solutions of the wave equation satisfying (3.71), which lead to the conserved quantities (3.76). On the other hand, from (3.183) we find that, in terms of the coordinates $u = ct - r, r, \theta, \phi$,

$$F_{+2} = -\frac{1}{2r^2}\partial_r^2 r^2 \eth\eth\chi,$$

hence

$$F_{+2} = -\frac{1}{2r^2}\sum_{n=5}^{N+2}\frac{F_{+2}^{(n)}(u,\theta,\phi)}{r^n} + O\left(\frac{1}{r^{N+3}}\right)$$

with

$$F^{(n)}_{+2} = (n-3)(n-4)\eth\eth f_{n-2}$$
$$= (n-3)(n-4)\sum_{l,m}\sqrt{\frac{(l+2)!}{(l-2)!}}\,a_{n-2,lm}(u)\,{}_2Y_{lm}(\theta,\phi).$$

Hence

$$({}_2Y_{lm}, F^{(l+4)}_{+2}) = l(l+1)\sqrt{\frac{(l+2)!}{(l-2)!}}\,a_{l+2,lm}$$

that is,

$$\int_{S^2} \overline{{}_2Y_{lm}}\, F^{(l+4)}_{+2}\, d\Omega = \text{const.}$$

which gives the Newman–Penrose (1968) conserved quantities for the linearized gravitational field.

For a circularly polarized plane wave propagating in the $\mathbf{e}_z$ direction, the Cartesian components of the "electric part" of the curvature is proportional to

$$(E_{ij}) = \text{Re}\begin{pmatrix} 1 & \pm i & 0 \\ \pm i & -1 & 0 \\ 0 & 0 & 0 \end{pmatrix} e^{i(kz-\omega t)}$$
$$= \begin{pmatrix} \cos(kz-\omega t) & \mp\sin(kz-\omega t) & 0 \\ \mp\sin(kz-\omega t) & -\cos(kz-\omega t) & 0 \\ 0 & 0 & 0 \end{pmatrix},$$

hence

$$(F_{ij}) = \begin{pmatrix} 1 & i & 0 \\ i & -1 & 0 \\ 0 & 0 & 0 \end{pmatrix} e^{\pm i(kz-\omega t)}$$

and therefore $x_i x_j F_{ij} = (x+iy)^2 e^{\pm i(kz-\omega t)} = (-1/k^2)(L_x + iL_y)^2 e^{\pm i(kz-\omega t)}$; thus, making use of (3.78) and (3.31), it follows that

$$x_i x_j F_{ij}$$
$$= -\frac{1}{k^2}\sum_{j=2}^{\infty}\left[\frac{4\pi(2j+1)(j+2)!}{(j-2)!}\right]^{1/2}(\pm i)^j j_j(kr) Y_{j,2} e^{\mp i\omega t}$$
$$= -\bar{\eth}\bar{\eth}\eth\eth\left(\frac{1}{k^2}\sum_{j=2}^{\infty}\left[\frac{4\pi(2j+1)(j-2)!}{(j+2)!}\right]^{1/2}(\pm i)^j j_j(kr) Y_{j,2} e^{\mp i\omega t}\right).$$

Since $x_i x_j F_{ij} = -\bar{\eth}\bar{\eth}\eth\eth\chi$, we can take

$$\chi = \frac{1}{k^2}\sum_{j=2}^{\infty}\left[\frac{4\pi(2j+1)(j-2)!}{(j+2)!}\right]^{1/2}(\pm i)^j j_j(kr) Y_{j,2} e^{\mp i\omega t}. \qquad (3.190)$$

Then, by means of the addition theorem (2.59), one finds that the potential corresponding to a circularly polarized plane wave propagating in the direction with polar and azimuth angles θ_1 and ϕ_1, is

$$\chi = \frac{4\pi}{k^2} \sum_{j=2}^{\infty} \sum_{m=-j}^{j} \left[\frac{(j-2)!}{(j+2)!} \right]^{1/2} (\pm \mathrm{i})^j \, \overline{{}_2Y_{jm}(\theta_1, \phi_1)} \, j_j(kr) Y_{jm} \mathrm{e}^{\mp \mathrm{i}\omega t}. \quad (3.191)$$

3.8 Magnetic monopole

In all the examples of the application of the spin-weighted spherical harmonics given in the preceding sections, we have found that a spin-s field has $2s+1$ components with spin weights $-s, -s+1, \ldots, s$. By contrast, the equation considered in this section governs a single scalar field and its solution is given in terms of spherical harmonics with a variable spin weight, which depends on the parameters contained in the equation. Following Cortés-Cuautli (1997), we shall solve the time-independent Schrödinger equation for a (spin-0) particle of mass M and electric charge e in the presence of the electromagnetic field produced by a magnetic monopole g and an electric charge $-Ze$ placed at the origin (see also Tamm 1931, Wu and Yang 1976). This equation is given by

$$-\frac{\hbar^2}{2M} \left(\nabla - \frac{\mathrm{i}e}{\hbar c} \mathbf{A} \right)^2 \psi + e\varphi \, \psi = E\psi, \quad (3.192)$$

where the electromagnetic potentials $\mathbf{A}$ and φ can be chosen as

$$\mathbf{A} = g \frac{(\mp 1 - \cos\theta)}{r \sin\theta} \mathbf{e}_\phi, \qquad \varphi = -\frac{Ze}{r}. \quad (3.193)$$

With the negative sign, the vector potential $\mathbf{A}$ is singular on the positive z axis, while with the positive sign, $\mathbf{A}$ diverges on the negative z axis. Thus, we shall consider both signs in (3.193) in order to find a well-behaved solution of the Schrödinger equation everywhere. As shown in Wu and Yang (1976), the solutions corresponding to these two choices of $\mathbf{A}$ can be joined to form a section on a line bundle provided that

$$\frac{eg}{\hbar c} = \frac{n}{2}, \quad (3.194)$$

where n is an integer. Condition (3.194) is the well-known Dirac quantization condition (Dirac 1931, 1948). It what follows, we will consider the wave function as an ordinary function, without stressing its relationship with a line bundle.

Making use of the expression for the Laplace operator in spherical coordinates

$$\nabla^2 = \frac{1}{r^2} \partial_r r^2 \partial_r + \frac{1}{r^2 \sin\theta} \partial_\theta \sin\theta \, \partial_\theta + \frac{1}{r^2 \sin^2\theta} \partial_\phi^2,$$

and the fact that the divergence of the vector potential (3.193) is equal to zero, the Schrödinger equation (3.192) takes the form

$$-\frac{\hbar^2}{2M}\left[\frac{1}{r^2}\partial_r r^2 \partial_r + \frac{1}{r^2}\left(\frac{1}{\sin\theta}\partial_\theta \sin\theta\, \partial_\theta + \frac{1}{\sin^2\theta} e^{\mp iq\phi}\, \partial_\phi^2 e^{\pm iq\phi} + 2iq\frac{\cos\theta}{\sin^2\theta} e^{\mp iq\phi}\, \partial_\phi e^{\pm iq\phi} - \frac{q^2}{\sin^2\theta}\right) + \frac{q^2}{r^2}\right]\psi - \frac{Ze^2}{r}\psi = E\psi, \tag{3.195}$$

where we have introduced the dimensionless quantity $q \equiv eg/\hbar c$ which, according to the Dirac quantization condition (3.194), can only take the values $q = n/2$, with $n = 0, \pm 1, \pm 2, \ldots$.

According to (2.23), (3.195) can be rewritten as

$$-\frac{\hbar^2}{2M}\left[\frac{1}{r^2}\partial_r r^2 \partial_r + \frac{1}{r^2}(\bar{\eth}\eth - q)\right] e^{\pm iq\phi}\psi - \left(\frac{Ze^2}{r} + E\right) e^{\pm iq\phi}\psi = 0, \tag{3.196}$$

provided we *assign* a spin weight q to the wave function ψ. In order to solve (3.196), we look for a separable solution of the form

$$\psi = R(r) e^{\mp iq\phi}\, {}_qY_{jm}(\theta, \phi), \tag{3.197}$$

with $j = |q|, |q| + 1, |q| + 2, \ldots$, and $-j \leqslant m \leqslant j$ [see (2.16)]. Substituting (3.197) into (3.196), with the aid of (2.22) we obtain the radial equation

$$-\frac{\hbar^2}{2M}\left[\frac{1}{r^2}\frac{d}{dr}r^2\frac{d}{dr}R(r) - \frac{1}{r^2}(j(j+1) - q^2)R(r)\right] - \left(\frac{Ze^2}{r} + E\right) R(r) = 0. \tag{3.198}$$

Thus, the only effect on the radial equation of the presence of the magnetic monopole is to replace the factor $l(l+1)$, where l is the orbital quantum number, by $j(j+1) - q^2$, and, by contrast with the quantum number l, j can take half-integral values. It should be clear that a similar result applies if one considers any central potential in place of the Coulomb potential (*cf.* Tamm 1931, Wu and Yang 1976). Hence, the solution of the radial equation (3.198) can be obtained from that corresponding to the hydrogen atom by simply replacing l by $-\frac{1}{2} + \sqrt{(j+\frac{1}{2})^2 - q^2}$ (which comes from the identification $l(l+1) = j(j+1) - q^2$). In this manner (assuming $E < 0$) we conclude that

$$R(\rho) = \rho^{-\frac{1}{2}+\sqrt{(j+\frac{1}{2})^2 - q^2}}\, e^{-\rho/2} L_{n_r}^{2\sqrt{(j+\frac{1}{2})^2 - q^2}}(\rho),$$

where L_n^p denotes the associated Laguerre polynomials and

$$\rho \equiv \left(\frac{8M|E|}{\hbar^2}\right)^{1/2} r.$$

The energy eigenvalues are given by

$$E = -\frac{MZ^2e^4}{2\hbar^2}\left[n_r + \tfrac{1}{2} + \sqrt{(j+\tfrac{1}{2})^2 - q^2}\right]^{-2}, \tag{3.199}$$

with $n_r = 0, 1, 2, \ldots$. Thus, by contrast with the hydrogen atom, the degeneracy of each energy level is $2j + 1$, since m does not enter into (3.199). In the case where q vanishes, (3.199) reduces to the well-known expression for the energy eigenvalues of the hydrogen atom, identifying $n_r + j + 1$ with the principal quantum number n and j with l. When $Z = 0$, the regular solution of (3.198) is proportional to the spherical Bessel function $j_l(kr)$, where $l = -\frac{1}{2} + \sqrt{(j+\frac{1}{2})^2 - q^2}$ and $k = \sqrt{2ME}/\hbar$.

In the present case, the spin weight, q, assigned to the wave function does not correspond to the behavior of ψ under rotations about $\mathbf{e}_r$, since ψ is a scalar function. However, as is known, the electromagnetic field of an electric charge e and a magnetic monopole g possesses the angular momentum $\mathbf{L}_{\text{emf}} = -(eg/c)(\mathbf{r}/r)$, where $\mathbf{r}$ is the vector going from the monopole to the electric charge (see, *e.g.*, Jackson 1975, Feynman 1987); therefore, the spin weight of ψ is related to the magnitude of the angular momentum of the electromagnetic field through $q = L_{\text{emf}}/\hbar$. A similar treatment can be applied in the case of the Dirac equation for a charged particle in the field of a magnetic monopole and an electric charge (Torres del Castillo and Cortés-Cuautli 1997).

The spin-weighted spherical harmonics are also useful in general relativity; in fact, these functions were introduced by Newman and Penrose (1966) in the study of the asymptotic behavior of the gravitational field (see also Walker 1983, Stewart 1990). Furthermore, the spin-weighted spherical harmonics appear in the solution by separation of variables of various nonscalar differential equations in spherically symmetric space-times (see, *e.g.*, Torres del Castillo 1996).

4
Spin-Weighted Cylindrical Harmonics

As shown in Chapter 3, the spin-weighted spherical harmonics are very useful in the solution of linear nonscalar equations and in the derivation of general expressions for the solutions of such equations. The usefulness of the spin-weighted spherical harmonics is related to the appearance of the operators $\eth$ and $\bar{\eth}$, when the equations are written in terms of spin-weighted components.

In this chapter, it will be shown that there are classes of functions similar to the spin-weighted spherical harmonics, adapted to the cylindrical coordinates (circular, parabolic, and elliptic) (Torres del Castillo 1992b, Torres del Castillo and Cartas-Fuentevilla 1994). The definition of these functions will be based on the appropriate definition of spin weight and of the corresponding raising and lowering spin weight operators. The spin-weighted cylindrical harmonics defined in this chapter might be called spin-weighted plane harmonics, since its definition is directly related to the Euclidean plane.

4.1 Definitions and basic properties

Let $\{\mathbf{e}_\rho, \mathbf{e}_\phi, \mathbf{e}_z\}$ be the orthonormal basis induced by the circular cylindrical coordinates (ρ, ϕ, z). A quantity η has spin weight s if under the rotation about $\mathbf{e}_z$ given by

$$\mathbf{e}_\rho + \mathrm{i}\mathbf{e}_\phi \mapsto \mathrm{e}^{\mathrm{i}\theta}(\mathbf{e}_\rho + \mathrm{i}\mathbf{e}_\phi) \tag{4.1}$$

it transforms according to

$$\eta \mapsto \mathrm{e}^{\mathrm{i}s\theta}\eta.$$

From this definition it follows that if η has spin weight s, then its complex conjugate, $\bar{\eta}$, has spin weight $-s$ and if κ has spin weight s', then $\eta\kappa$ has spin weight

$s + s'$. The vector fields $\mathbf{e}_z$ and $\mathbf{e}_\rho \pm \mathrm{i}\mathbf{e}_\phi$ have spin weight 0 and ± 1, respectively. Therefore, if $\mathbf{F}$ is an arbitrary vector field, the scalar fields

$$F_0 \equiv -\frac{1}{\sqrt{2}} \mathbf{F} \cdot \mathbf{e}_z, \qquad F_{\pm 1} \equiv \pm \frac{1}{\sqrt{2}} \mathbf{F} \cdot (\mathbf{e}_\rho \pm \mathrm{i}\mathbf{e}_\phi), \tag{4.2}$$

have spin weight 0 and ± 1, respectively. In terms of the spin-weighted components (4.2), the vector field $\mathbf{F}$ is expressed as

$$\mathbf{F} = -\sqrt{2}\, F_0\, \mathbf{e}_z - \frac{1}{\sqrt{2}} F_{-1}(\mathbf{e}_\rho + \mathrm{i}\mathbf{e}_\phi) + \frac{1}{\sqrt{2}} F_{+1}(\mathbf{e}_\rho - \mathrm{i}\mathbf{e}_\phi). \tag{4.3}$$

Similarly, the components of a traceless totally symmetric n-index tensor field can be combined into $2n + 1$ components with spin weight $-n, -n+1, \ldots, n$ (see Section 6.3).

We shall employ again the symbols $\eth$ and $\bar{\eth}$ to denote the spin weight raising and lowering operators. If η has spin weight s, $\eth\eta$ and $\bar{\eth}\eta$ will be defined by

$$\begin{aligned} \eth\eta &\equiv -\left(\partial_\rho + \frac{\mathrm{i}}{\rho}\partial_\phi - \frac{s}{\rho}\right)\eta = -\rho^s\left(\partial_\rho + \frac{\mathrm{i}}{\rho}\partial_\phi\right)(\rho^{-s}\eta), \\ \bar{\eth}\eta &\equiv -\left(\partial_\rho - \frac{\mathrm{i}}{\rho}\partial_\phi + \frac{s}{\rho}\right)\eta = -\rho^{-s}\left(\partial_\rho - \frac{\mathrm{i}}{\rho}\partial_\phi\right)(\rho^s\eta) \end{aligned} \tag{4.4}$$

(Torres del Castillo 1992b). Then $\eth\eta$ and $\bar{\eth}\eta$ have spin weight $s+1$ and $s-1$, respectively (see Section 6.3), and by means of a straightforward computation one finds that

$$\bar{\eth}\eth\eta = \eth\bar{\eth}\eta = \partial_\rho^2\eta + \frac{1}{\rho}\partial_\rho\eta + \frac{1}{\rho^2}\partial_\phi^2\eta + \frac{2\mathrm{i}s}{\rho^2}\partial_\phi\eta - \frac{s^2}{\rho^2}\eta. \tag{4.5}$$

Furthermore, $\overline{\eth\eta} = \bar{\eth}\bar{\eta}$, $\eth(\eta\kappa) = \eta\eth\kappa + \kappa\eth\eta$ and $\bar{\eth}(\eta\kappa) = \eta\bar{\eth}\kappa + \kappa\bar{\eth}\eta$.

In terms of the operators $\eth$ and $\bar{\eth}$, the gradient of a function f with spin weight 0 is given by

$$\nabla f = \partial_z f\, \mathbf{e}_z - \tfrac{1}{2}\bar{\eth} f\, (\mathbf{e}_\rho + \mathrm{i}\mathbf{e}_\phi) - \tfrac{1}{2}\eth f\, (\mathbf{e}_\rho - \mathrm{i}\mathbf{e}_\phi). \tag{4.6}$$

Similarly, the divergence and the curl of a vector field $\mathbf{F}$ are given by

$$\begin{aligned} \nabla \cdot \mathbf{F} &= -\sqrt{2}\,\partial_z F_0 + \frac{1}{\sqrt{2}}\eth F_{-1} - \frac{1}{\sqrt{2}}\bar{\eth} F_{+1}, \\ \nabla \times \mathbf{F} &= \frac{\mathrm{i}}{\sqrt{2}}(\eth F_{-1} + \bar{\eth} F_{+1})\, \mathbf{e}_z + \frac{\mathrm{i}}{\sqrt{2}}\left[\partial_z F_{-1} + \bar{\eth} F_0\right](\mathbf{e}_\rho + \mathrm{i}\mathbf{e}_\phi) \\ &\quad + \frac{\mathrm{i}}{\sqrt{2}}\left[\partial_z F_{+1} - \eth F_0\right](\mathbf{e}_\rho - \mathrm{i}\mathbf{e}_\phi). \end{aligned} \tag{4.7}$$

Then, using the identity $\nabla \times (\nabla \times \mathbf{F}) = \nabla(\nabla \cdot \mathbf{F}) - \nabla^2 \mathbf{F}$, and the expressions (4.6) and (4.7) it follows that

$$\nabla^2 \mathbf{F} = -\sqrt{2}\left[\partial_z^2 F_0 + \bar{\eth}\eth F_0\right]\mathbf{e}_z - \frac{1}{\sqrt{2}}\left[\partial_z^2 F_{-1} + \bar{\eth}\eth F_{-1}\right](\mathbf{e}_\rho + \mathrm{i}\mathbf{e}_\phi)$$
$$+ \frac{1}{\sqrt{2}}\left[\partial_z^2 F_{+1} + \eth\bar{\eth} F_{+1}\right](\mathbf{e}_\rho - \mathrm{i}\mathbf{e}_\phi). \qquad (4.8)$$

Using (4.6) and (4.7) and the commutativity of $\eth$ and $\bar{\eth}$ one finds that the Laplacian of a function of spin weight 0 is

$$\nabla^2 f = \partial_z^2 f + \bar{\eth}\eth f. \qquad (4.9)$$

We shall denote by ${}_sF_{\alpha m}$ a function of ρ and ϕ with spin weight s such that

$$\bar{\eth}\eth\, {}_sF_{\alpha m} = -\alpha^2\, {}_sF_{\alpha m}, \qquad (4.10)$$
$$-\mathrm{i}\partial_\phi\, {}_sF_{\alpha m} = m\, {}_sF_{\alpha m}, \qquad (4.11)$$

where α is a (real or complex) constant and m is an integer or a half-integer according to whether s is an integer or a half-integer. The solutions of (4.10) and (4.11) will be called *spin-weighted cylindrical harmonics*. Condition (4.11) implies that ${}_sF_{\alpha m}(\rho, \phi) = f(\rho)\mathrm{e}^{\mathrm{i}m\phi}$ and from (4.5) and (4.10) it follows that $f(\rho)$ must satisfy the equation

$$\rho^2 \frac{\mathrm{d}^2 f}{\mathrm{d}\rho^2} + \rho\frac{\mathrm{d}f}{\mathrm{d}\rho} + [\alpha^2\rho^2 - (m+s)^2]f = 0.$$

Therefore, if $\alpha \neq 0$, $f(\rho)$ is a linear combination of the Bessel functions $J_{m+s}(\alpha\rho)$ and $N_{m+s}(\alpha\rho)$, or of $H^{(1)}_{m+s}(\alpha\rho)$ and $H^{(2)}_{m+s}(\alpha\rho)$. We shall employ the notation

$${}_sZ_{\alpha m}(\rho, \phi) \equiv Z_{m+s}(\alpha\rho)\mathrm{e}^{\mathrm{i}m\phi} \qquad (\alpha \neq 0), \qquad (4.12)$$

where Z_ν is a Bessel function. Thus, when $\alpha \neq 0$, the solution of (4.10) and (4.11) is given by

$${}_sF_{\alpha m} = A\, {}_sJ_{\alpha m} + B\, {}_sN_{\alpha m} = C\, {}_sH^{(1)}_{\alpha m} + D\, {}_sH^{(2)}_{\alpha m}, \qquad (4.13)$$

where A, B, C, and D are constants.

In the case where $\alpha = 0$ and $m + s \neq 0$, $f(\rho)$ is a linear combination of ρ^{m+s} and ρ^{-m-s}. Hence,

$${}_sF_{0,m} = A\,\rho^{m+s}\mathrm{e}^{\mathrm{i}m\phi} + B\,\rho^{-m-s}\mathrm{e}^{\mathrm{i}m\phi} \qquad (m+s \neq 0). \qquad (4.14)$$

Finally, in the case where $\alpha = 0$ and $m + s = 0$,

$${}_sF_{0,-s} = A\mathrm{e}^{-\mathrm{i}s\phi} + B(\ln\rho)\mathrm{e}^{-\mathrm{i}s\phi}. \qquad (4.15)$$

Using the recurrence relations for the Bessel functions, (4.4) and (4.12) one finds that, for $\alpha \neq 0$,

$$\begin{aligned} ð\, {}_sZ_{\alpha m} &= \alpha\, {}_{s+1}Z_{\alpha m}, \\ \bar{ð}\, {}_sZ_{\alpha m} &= -\alpha\, {}_{s-1}Z_{\alpha m} \end{aligned} \tag{4.16}$$

[*cf.* (2.27)]. In the case where $\alpha = 0$ we obtain

$$ð(\rho^{m+s} e^{im\phi}) = 0, \qquad \bar{ð}(\rho^{m+s} e^{im\phi}) = -2(m+s)\rho^{m+s-1} e^{im\phi},$$
$$\bar{ð}(\rho^{-m-s} e^{im\phi}) = 0, \qquad ð(\rho^{-m-s} e^{im\phi}) = 2(m+s)\rho^{-m-s-1} e^{im\phi},$$

and

$$\begin{aligned} &ð(e^{-is\phi}) = 0, && \bar{ð}(e^{-is\phi}) = 0, \\ &ð(\ln\rho\, e^{-is\phi}) = -\rho^{-1} e^{-is\phi}, && \bar{ð}(\ln\rho\, e^{-is\phi}) = -\rho^{-1} e^{-is\phi}. \end{aligned} \tag{4.17}$$

The functions ${}_0Z_{\alpha m}$ are also related to the operators

$$P_1 \equiv -i\partial_x, \qquad P_2 \equiv -i\partial_y, \qquad L_3 \equiv -i(x\partial_y - y\partial_x), \tag{4.18}$$

where x and y are Cartesian coordinates on the plane; these operators correspond to the x- and y-components of the linear momentum and to the angular momentum about the origin. Alternatively, P_1, P_2, and L_3 are the generators of translations parallel to the x- and y-axis and of rotations about the origin, respectively; hence, they form a basis of the Lie algebra of the group of rigid motions of the plane. Instead of P_1 and P_2, it is convenient to make use of the nonhermitian operators

$$P_\pm \equiv P_1 \pm iP_2. \tag{4.19}$$

Then the basic commutation relations are given by

$$[P_+, P_-] = 0, \qquad [L_3, P_\pm] = \pm P_\pm, \tag{4.20}$$

which imply that

$$[L_3, P^2] = 0, \qquad [P_\pm, P^2] = 0, \tag{4.21}$$

where

$$P^2 \equiv {P_1}^2 + {P_2}^2 = P_+P_- = P_-P_+. \tag{4.22}$$

In terms of the polar coordinates ρ, ϕ, the operators defined above are given by

$$\begin{aligned} P_\pm &= -ie^{\pm i\phi}\left(\partial_\rho \pm \frac{i}{\rho}\partial_\phi\right), \\ L_3 &= -i\partial_\phi, \\ P^2 &= -\left(\partial_\rho^{\,2} + \frac{1}{\rho}\partial_\rho + \frac{1}{\rho^2}\partial_\phi^{\,2}\right). \end{aligned} \tag{4.23}$$

Hence, on functions with spin weight zero, $\bar{\eth}\eth = -P^2$, and from (4.10) and (4.11) we conclude that the ${}_0Z_{\alpha m}$ are eigenfunctions of P^2 and L_3,

$$\begin{aligned} P^2\, {}_0Z_{\alpha m} &= \alpha^2\, {}_0Z_{\alpha m}, \\ L_3\, {}_0Z_{\alpha m} &= m\, {}_0Z_{\alpha m}. \end{aligned} \tag{4.24}$$

The recurrence relations for the Bessel functions amount to

$$P_\pm\, {}_0Z_{\alpha m} = \pm \mathrm{i}\alpha\, {}_0Z_{\alpha, m\pm 1}. \tag{4.25}$$

4.2 Representation of the Euclidean group of the plane

The rigid motions of the Euclidean plane form the Euclidean group SE(2). Given a system of Cartesian coordinates on the plane, any rigid motion can be obtained by composing a rotation about the origin O through an angle β, followed by a translation over a distance R parallel to the resulting x'-axis and by a rotation about the new origin O'' through an angle γ. The resulting transformation will be denoted by $T(\beta, R, \gamma)$. Then, it can be shown that

$$T(\beta, R, \gamma) = T(\beta, 0, 0)\, T(0, R, 0)\, T(\gamma, 0, 0), \tag{4.26}$$

where, according to the above definition, $T(\beta, 0, 0)$ and $T(\gamma, 0, 0)$ are rotations about the origin O and $T(0, R, 0)$ is a translation parallel to the original x-axis. (Expression (4.26) is analogous to that for a rotation parametrized by the Euler angles, see (1.54).)

As we shall see, the functions ${}_0J_{\alpha m}$ form bases for linear (infinite-dimensional) representations of SE(2) in the same way as the spherical harmonics Y_{lm} form bases for representations of SO(3). Under a rigid motion of the plane, each function ${}_0J_{\alpha m}$ is transformed onto a series in ${}_0J_{\alpha m'}$. Since ${}_0J_{\alpha m}$ is an eigenfunction of the infinitesimal generator of rotations about the origin, it is also an eigenfunction of any rotation about the origin. The effect of the rotation $T(\gamma, 0, 0)$ on an arbitrary complex-valued function defined on the plane, $f(\rho, \phi)$, is given by $[T(\gamma, 0, 0) f](\rho, \phi) \equiv f(\rho, \phi - \gamma)$, therefore, according to (4.12),

$$T(\gamma, 0, 0)\, {}_0J_{\alpha m} = \mathrm{e}^{-\mathrm{i}m\gamma}\, {}_0J_{\alpha m}. \tag{4.27}$$

Similarly, if $\mathbf{r}$ is an arbitrary point of the plane, then $[T(0, R, 0) f](\mathbf{r}) \equiv f(\mathbf{r} - R\mathbf{e}_x)$. In order to find $T(0, R, 0)\, {}_0J_{\alpha m}$ we notice that if f is an analytic function, using Taylor's formula and (4.18)–(4.20), we have

$$\begin{aligned} f(\mathbf{r} + R\mathbf{e}_x) &= \sum_{s=0}^{\infty} \frac{R^s}{s!} (\partial_x)^s f(\mathbf{r}) = \sum_{s=0}^{\infty} \frac{1}{s!} \left(\frac{\mathrm{i}R}{2}\right)^s (P_+ + P_-)^s f(\mathbf{r}) \\ &= \sum_{s=0}^{\infty} \frac{1}{s!} \left(\frac{\mathrm{i}R}{2}\right)^s \sum_{r=0}^{s} \frac{s!}{r!(s-r)!} P_+^r P_-^{s-r} f(\mathbf{r}). \end{aligned} \tag{4.28}$$

Applying (4.28) to ${}_0Z_{\alpha m}$ and making use of (4.25) one finds

$$
\begin{aligned}
{}_0Z_{\alpha m}(\mathbf{r}+R\mathbf{e}_x) &= \sum_{s=0}^{\infty}\sum_{r=0}^{s}\frac{1}{r!(s-r)!}\left(\frac{\mathrm{i}R}{2}\right)^s (\mathrm{i}\alpha)^r(-\mathrm{i}\alpha)^{s-r}\,{}_0Z_{\alpha,m+2r-s}(\mathbf{r})\\
&= \sum_{m'=-\infty}^{\infty}\sum_{k=0}^{\infty}\frac{(-1)^k}{k!(m-m'+k)!}\left(\frac{\alpha R}{2}\right)^{m-m'+2k}\,{}_0Z_{\alpha m'}(\mathbf{r}).
\end{aligned}
\tag{4.29}
$$

In particular, for ${}_0Z_{\alpha m} = {}_0J_{\alpha m}$ and $\mathbf{r} = \mathbf{0}$, taking into account that $J_m(0) = \delta_{m0}$ and that the polar coordinates of $R\mathbf{e}_x$ are $\rho = R$ and $\phi = 0$, from (4.29) and (4.12), we obtain

$$
J_m(\alpha R) = \sum_{k=0}^{\infty}\frac{(-1)^k}{k!(m+k)!}\left(\frac{\alpha R}{2}\right)^{m+2k}, \tag{4.30}
$$

which is the series expansion for the Bessel functions of integral order. Then, (4.29) can be rewritten as

$$
{}_0Z_{\alpha m}(\mathbf{r}+R\mathbf{e}_x) = \sum_{m'=-\infty}^{\infty} J_{m-m'}(\alpha R)\,{}_0Z_{\alpha m'}(\mathbf{r}). \tag{4.31}
$$

If (ρ, ϕ) are the polar coordinates of $\mathbf{r}$, then the polar coordinates of the point $\mathbf{r}+R\mathbf{e}_x$ are (ρ', ϕ'), with $\rho' = \sqrt{\rho^2+R^2+2\rho R\cos\phi}$, $\tan\phi' = (\rho\sin\phi)/(\rho\cos\phi + R)$. Hence, (4.31) is equivalent to

$$
Z_m(\alpha\rho')\mathrm{e}^{\mathrm{i}m\phi'} = \sum_{m'=-\infty}^{\infty} J_{m-m'}(\alpha R) Z_{m'}(\alpha\rho)\mathrm{e}^{\mathrm{i}m'\phi}. \tag{4.32}
$$

Thus, in particular,

$$
J_m(\alpha\rho')\mathrm{e}^{\mathrm{i}m\phi'} = \sum_{m'=-\infty}^{\infty} J_{m-m'}(\alpha R) J_{m'}(\alpha\rho)\mathrm{e}^{\mathrm{i}m'\phi}, \tag{4.33}
$$

which is known as Neumann's addition theorem (see, *e.g.*, Hochstadt 1971). For $m = 0$, (4.33) reduces to

$$
J_0\left(\alpha\sqrt{\rho^2+R^2-2\rho R\cos\phi}\right) = \sum_{m'=-\infty}^{\infty} J_{m'}(\alpha R) J_{m'}(\alpha\rho)\mathrm{e}^{\mathrm{i}m'\phi}, \tag{4.34}
$$

where we have used the relation

$$
J_{-n}(x) = J_n(-x), \tag{4.35}
$$

which follows from (4.30). Equation (4.34) is known as Gegenbauer's addition theorem (alternative derivations of this theorem can be found, *e.g.*, in Lebedev 1965, Vilenkin 1968, Hochstadt 1971, Torres del Castillo 1992b).

From (4.31) and (4.35) it follows that

$$T(0, R, 0)\,{}_0J_{\alpha m} = \sum_{m'=-\infty}^{\infty} J_{m'-m}(\alpha R)\,{}_0J_{\alpha m'}, \tag{4.36}$$

therefore, (4.26), (4.27), and (4.36) yield

$$T(\beta, R, \gamma)\,{}_0J_{\alpha m} = \sum_{m'=-\infty}^{\infty} \mathrm{e}^{-\mathrm{i}m'\beta} J_{m'-m}(\alpha R)\mathrm{e}^{-\mathrm{i}m\gamma}\,{}_0J_{\alpha m'}. \tag{4.37}$$

For a fixed α, the matrix elements

$$D^{\alpha}_{km}(\beta, R, \gamma) \equiv \mathrm{e}^{-\mathrm{i}k\beta} J_{k-m}(\alpha R)\mathrm{e}^{-\mathrm{i}m\gamma}, \tag{4.38}$$

appearing in (4.37), give an infinite-dimensional representation of SE(2) (alternative derivations can be found, *e.g.*, in Vilenkin 1968, Miller 1977, Tung 1985, Torres del Castillo 1993). From (4.38) and the relations $J_{-m}(x) = (-1)^m J_m(x) = J_m(-x)$, one finds that

$$\overline{D^{\alpha}_{mm'}(\beta, R, \gamma)} = \mathrm{e}^{\mathrm{i}m\beta} J_{m'-m}(\alpha R)\mathrm{e}^{\mathrm{i}m'\gamma} = D^{\alpha}_{m'm}(-\gamma, -R, -\beta). \tag{4.39}$$

Taking into account that $[T(\beta, R, \gamma)]^{-1} = T(-\gamma, -R, -\beta)$ (as can be seen from (4.26)), (4.39) means that the representation given by the functions $D^{\alpha}_{m'm}$ is unitary.

The matrix elements $D^{\alpha}_{m'm}$ are related to the spin-weighted cylindrical harmonics in various ways. For instance, (4.12) and (4.38) yield

$$\begin{aligned} D^{\alpha}_{m'm}(\beta, R, \gamma) &= (-1)^{m'-m}\mathrm{e}^{-\mathrm{i}m\gamma}\,{}_mJ_{\alpha,-m'}(R, \beta) \\ &= \mathrm{e}^{-\mathrm{i}m'\beta}\,{}_{m'}J_{\alpha,-m}(R, \gamma). \end{aligned} \tag{4.40}$$

The analog of (2.61) is given by

$$\sum_{m=-\infty}^{\infty} \overline{{}_{s'}J_{\alpha m}(\rho, \phi)}\,{}_sJ_{\alpha m}(\rho, \phi) = \delta_{ss'}, \tag{4.41}$$

which can be derived from (4.33) or from (4.40), using the fact that, for each value of α, the functions $D^{\alpha}_{m'm}$ form a linear representation of SE(2).

Angular and linear momentum

For a vector field $\mathbf{F}$, the operator corresponding to the z-component of the total angular momentum is given by

$$\begin{aligned} J_3\mathbf{F} &= (-\mathrm{i}\mathbf{e}_z \cdot \mathbf{r} \times \nabla)\mathbf{F} + \mathrm{i}\mathbf{e}_z \times \mathbf{F} \\ &= -\mathrm{i}\partial_\phi \mathbf{F} + \mathrm{i}\mathbf{e}_z \times \mathbf{F} \end{aligned} \tag{4.42}$$

[*cf.* (3.37)]. From the relation $\mathbf{e}_\rho + \mathrm{i}\mathbf{e}_\phi = \mathrm{e}^{-\mathrm{i}\phi}(\mathbf{e}_x + \mathrm{i}\mathbf{e}_y)$ it follows that $\partial_\phi(\mathbf{e}_\rho + \mathrm{i}\mathbf{e}_\phi) = \mathbf{e}_\phi - \mathrm{i}\mathbf{e}_\rho = \mathbf{e}_z \times (\mathbf{e}_\rho + \mathrm{i}\mathbf{e}_\phi)$; therefore, expressing the vector field $\mathbf{F}$ in the form (4.3) we obtain

$$\begin{aligned} &J_3\left(-\sqrt{2}\,F_0\,\mathbf{e}_z - \frac{1}{\sqrt{2}}F_{-1}(\mathbf{e}_\rho + \mathrm{i}\mathbf{e}_\phi) + \frac{1}{\sqrt{2}}F_{+1}(\mathbf{e}_\rho - \mathrm{i}\mathbf{e}_\phi)\right) \\ &= -\sqrt{2}\,(-\mathrm{i}\partial_\phi F_0)\,\mathbf{e}_z - \frac{1}{\sqrt{2}}(-\mathrm{i}\partial_\phi F_{-1})(\mathbf{e}_\rho + \mathrm{i}\mathbf{e}_\phi) + \frac{1}{\sqrt{2}}(-\mathrm{i}\partial_\phi F_{+1})(\mathbf{e}_\rho - \mathrm{i}\mathbf{e}_\phi). \end{aligned}$$

Hence, defining the operators

$$J_3^{(s)}\eta \equiv -\mathrm{i}\partial_\phi \eta, \tag{4.43}$$

where s is the spin weight of η, we have

$$(J_3\mathbf{F})_s = J_3^{(s)}F_s. \tag{4.44}$$

In a similar manner, using the relations

$$\partial_x(\mathbf{e}_\rho + \mathrm{i}\mathbf{e}_\phi) = \frac{\mathrm{i}\sin\phi}{\rho}(\mathbf{e}_\rho + \mathrm{i}\mathbf{e}_\phi), \qquad \partial_y(\mathbf{e}_\rho + \mathrm{i}\mathbf{e}_\phi) = -\frac{\mathrm{i}\cos\phi}{\rho}(\mathbf{e}_\rho + \mathrm{i}\mathbf{e}_\phi)$$

and their complex conjugates, one finds that the operators $P_1 = -\mathrm{i}\partial_x$ and $P_2 = -\mathrm{i}\partial_y$, corresponding to the x- and y-components of the linear momentum, acting on a vector field $\mathbf{F}$ are given by

$$(P_k\mathbf{F})_s = P_k^{(s)}F_s, \qquad (k = 1, 2),$$

with

$$\begin{aligned} P_1^{(s)} &\equiv -\mathrm{i}\left(\partial_x - \mathrm{i}s\frac{\sin\phi}{\rho}\right) = -\mathrm{i}\left(\cos\phi\,\partial_\rho - \frac{\sin\phi}{\rho}\partial_\phi - \mathrm{i}s\frac{\sin\phi}{\rho}\right), \\ P_2^{(s)} &\equiv -\mathrm{i}\left(\partial_y + \mathrm{i}s\frac{\cos\phi}{\rho}\right) = -\mathrm{i}\left(\sin\phi\,\partial_\rho + \frac{\cos\phi}{\rho}\partial_\phi + \mathrm{i}s\frac{\cos\phi}{\rho}\right). \end{aligned} \tag{4.45}$$

The operators $J_3^{(s)}$, $P_1^{(s)}$, and $P_2^{(s)}$ do not change the spin weight and satisfy the commutation relations

$$[P_1^{(s)}, P_2^{(s)}] = 0, \qquad [J_3^{(s)}, P_k^{(s)}] = \mathrm{i}\varepsilon_{kl3}P_l^{(s)}$$

[*cf.* (4.20)]. These relations imply that $P_1^{(s)2} + P_2^{(s)2}$ commutes with $J_3^{(s)}$, $P_1^{(s)}$, and $P_2^{(s)}$; furthermore,

$$[J_3^{(s)}, P_1^{(s)} \pm \mathrm{i} P_2^{(s)}] = \pm(P_1^{(s)} \pm \mathrm{i} P_2^{(s)}).$$

By means of a straightforward computation one finds that

$$P_1^{(s)2} + P_2^{(s)2} = -\left(\partial_\rho^2 + \frac{1}{\rho}\partial_\rho + \frac{1}{\rho^2}\partial_\phi^2 + \frac{2\mathrm{i}s}{\rho^2}\partial_\phi - \frac{s^2}{\rho^2}\right) = -\bar{\eth}\eth \tag{4.46}$$

and

$$(P_1^{(s)} \pm \mathrm{i} P_2^{(s)})\,{}_sZ_{\alpha m} = \pm \mathrm{i}\alpha\,{}_sZ_{\alpha,m\pm1}.$$

Thus, equations (4.10) and (4.11) can be rewritten as

$$(P_1^{(s)2} + P_2^{(s)2})\,{}_sF_{\alpha m} = \alpha^2\,{}_sF_{\alpha m}, \qquad J_3^{(s)}\,{}_sF_{\alpha m} = m\,{}_sF_{\alpha m}.$$

For a two-component spinor field u, the operator corresponding to the z-component of the total angular momentum is given by

$$J_3 u = -\mathrm{i}\partial_\phi u + \tfrac{1}{2}\sigma_3 u$$

[see (3.144)]. Writing $u = u_- o + u_+ \widehat{o}$, one finds that

$$J_3 u = (-\mathrm{i}\partial_\phi u_-)o + (-\mathrm{i}\partial_\phi u_+)\widehat{o},$$

i.e., $(J_3 u)_\pm = J_3^{\pm 1/2} u_\pm$.

On the other hand, using the relations

$$\partial_x o = \frac{\mathrm{i}\sin\phi}{2\rho} o, \qquad \partial_y o = -\frac{\mathrm{i}\cos\phi}{2\rho} o, \qquad \partial_x \widehat{o} = -\frac{\mathrm{i}\sin\phi}{2\rho}\widehat{o}, \qquad \partial_y \widehat{o} = \frac{\mathrm{i}\cos\phi}{2\rho}\widehat{o},$$

it follows that, for an arbitrary two-component spinor field u,

$$P_k u \equiv -\mathrm{i}\partial_k u = -\mathrm{i}\partial_k(u_- o + u_+ \widehat{o}) = (P_k^{-1/2} u_-)o + (P_k^{1/2} u_+)\widehat{o},$$

$(k = 1, 2)$, with $P_k^{(s)}$ defined by (4.45).

The spinor fields $\mathrm{i}J_3 u$, $\mathrm{i}P_1 u$, and $\mathrm{i}P_2 u$ are the Lie derivatives of u with respect to ∂_ϕ, ∂_x, and ∂_y, respectively, which are Killing vector fields of the Euclidean space (see Section 6.1).

4.3 Applications

In this section we shall solve many of the equations considered in the preceding chapter, using the spin-weighted cylindrical harmonics. It will be shown that, also in this case, the use of spin-weighted quantities simplifies the solution of the linear nonscalar equations and allows us to find expressions for their solutions in terms of scalar potentials.

4.3.1 Solution of the vector Helmholtz equation

According to (4.8), the vector Helmholtz equation, $\nabla^2\mathbf{F} + k^2\mathbf{F} = 0$, amounts to the set of equations

$$\partial_z^2 F_s + \bar{\eth}\eth F_s + k^2 F_s = 0 \qquad (s = 0, \pm 1), \tag{4.47}$$

where the F_s are the (cylindrical) spin-weighted components of $\mathbf{F}$. Taking into account the fact that F_s has spin weight s, we look for solutions of (4.47) of the form

$$F_s = g_s(z)\, {}_sJ_{\alpha m}(\rho, \phi) + G_s(z)\, {}_sN_{\alpha m}(\rho, \phi), \tag{4.48}$$

where m is an integer. Substituting (4.48) into (4.47), with the aid of (4.10), we obtain

$$\frac{\mathrm{d}^2 g_s}{\mathrm{d}z^2} + (k^2 - \alpha^2) g_s = 0 \qquad (s = 0, \pm 1),$$

with identical equations for the functions G_s; hence, if $\alpha^2 \neq k^2$, $g_s(z) = A_s \mathrm{e}^{\gamma z} + B_s \mathrm{e}^{-\gamma z}$, with

$$\gamma^2 = \alpha^2 - k^2$$

and if $\alpha^2 = k^2$, $g_s(z) = A_s + B_s z$, where the A_s and B_s are arbitrary constants. Thus, assuming that α is different from zero, from (4.13) and (4.48) it follows that the vector Helmholtz equation admits separable solutions of the form

$$F_s = (A_s \mathrm{e}^{\gamma z} + B_s \mathrm{e}^{-\gamma z})\, {}_sJ_{\alpha m} + (C_s \mathrm{e}^{\gamma z} + D_s \mathrm{e}^{-\gamma z})\, {}_sN_{\alpha m} \tag{4.49}$$

and, if $\alpha = \pm k$,

$$F_s = (A_s + B_s z)\, {}_sJ_{\alpha m} + (C_s + D_s z)\, {}_sN_{\alpha m}. \tag{4.50}$$

From (4.7) and (4.16) we find that the divergence of the vector field (4.49) vanishes if and only if

$$\begin{aligned} \frac{\alpha}{2}(A_1 + A_{-1}) &= \gamma A_0, & \frac{\alpha}{2}(B_1 + B_{-1}) &= -\gamma B_0, \\ \frac{\alpha}{2}(C_1 + C_{-1}) &= \gamma C_0, & \frac{\alpha}{2}(D_1 + D_{-1}) &= -\gamma D_0. \end{aligned} \tag{4.51}$$

Introducing the constants

$$\begin{aligned} a_1 &\equiv \frac{\mathrm{i}}{\sqrt{2}\alpha}(A_1 - A_{-1}), & a_2 &\equiv \frac{\mathrm{i}}{\sqrt{2}\alpha}(C_1 - C_{-1}), \\ b_1 &\equiv \frac{\mathrm{i}}{\sqrt{2}\alpha}(B_1 - B_{-1}), & b_2 &\equiv \frac{\mathrm{i}}{\sqrt{2}\alpha}(D_1 - D_{-1}), \\ c_1 &\equiv \frac{k}{\sqrt{2}\alpha\gamma}(A_1 + A_{-1}), & c_2 &\equiv \frac{k}{\sqrt{2}\alpha\gamma}(C_1 + C_{-1}), \\ d_1 &\equiv \frac{-k}{\sqrt{2}\alpha\gamma}(B_1 + B_{-1}), & d_2 &\equiv \frac{-k}{\sqrt{2}\alpha\gamma}(D_1 + D_{-1}), \end{aligned}$$

and assuming that the conditions (4.51) hold, using (4.16), the components (4.49) can be written as

$$\begin{aligned} F_0 &= -\frac{1}{\sqrt{2}k}\bar{\eth}\eth\psi_2, \\ F_{+1} &= -\frac{\mathrm{i}}{\sqrt{2}}\eth\psi_1 + \frac{1}{\sqrt{2}k}\partial_z\eth\psi_2, \\ F_{-1} &= -\frac{\mathrm{i}}{\sqrt{2}}\bar{\eth}\psi_1 - \frac{1}{\sqrt{2}k}\partial_z\bar{\eth}\psi_2, \end{aligned} \tag{4.52}$$

where

$$\psi_1 = (a_1 e^{\gamma z} + b_1 e^{-\gamma z})\,{}_0J_{\alpha m} + (a_2 e^{\gamma z} + b_2 e^{-\gamma z})\,{}_0N_{\alpha m},$$

$$\psi_2 = (c_1 e^{\gamma z} + d_1 e^{-\gamma z})\,{}_0J_{\alpha m} + (c_2 e^{\gamma z} + d_2 e^{-\gamma z})\,{}_0N_{\alpha m},$$

which are solutions of the scalar Helmholtz equation.

Using (4.6) and (4.7) it can be verified that (4.52) amount to the simple expression

$$\mathbf{F} = \mathbf{e}_z \times \nabla\psi_1 + \frac{1}{k}\nabla \times (\mathbf{e}_z \times \nabla\psi_2) \tag{4.53}$$

[*cf.* (3.26)] or, equivalently,

$$\mathbf{F} = -\nabla \times (\psi_1 \mathbf{e}_z) - \frac{1}{k}\nabla \times \nabla \times (\psi_2 \mathbf{e}_z).$$

In a similar manner, one finds that if the divergence of the vector field given by (4.50) vanishes, then (4.53) also applies, with ψ_1 and ψ_2 being solutions of the scalar Helmholtz equation of the form $(a_1 + b_1 z)\,{}_0J_{\alpha m} + (a_2 + b_2 z)\,{}_0N_{\alpha m}$.

As a simple example of the application of the solutions (4.49) and (4.50) we shall solve the Maxwell–London equations for the case of an infinite superconducting cylinder of radius a placed in an originally uniform magnetic field perpendicular to the axis of the cylinder. We shall employ a system of cylindrical coordinates such that the z axis coincides with the axis of the cylinder and the angle ϕ is measured from the direction of the original magnetic field (hence, the original magnetic induction is of the form $B_b\mathbf{e}_x$). Outside the cylinder the magnetic induction and the magnetic field satisfy the equations $\nabla \cdot \mathbf{B} = 0$ and $\nabla \times \mathbf{H} = 0$, with $\mathbf{B} = \mathbf{H}$, therefore, there exists a magnetic scalar potential φ_M such that $\mathbf{B} = -\nabla\varphi_M$ and $\nabla^2\varphi_M = 0$. Solving the Laplace equation, taking into account that φ_M does not depend on z, that $\mathbf{B} \to B_b\mathbf{e}_x$ as $\rho \to \infty$ and that, owing to the symmetry under the reflection on the xz plane, φ_M must be an even function of ϕ, we obtain

$$\varphi_M = -B_b\,\rho\cos\phi + \sum_{m=1}^{\infty} b_m \rho^{-m}\cos m\phi,$$

where the b_m are some constants. From $\mathbf{B} = -\nabla\varphi_M$, (4.4) and (4.6) we find that, for $\rho \geqslant a$,

$$\begin{aligned} B_1 &= \frac{1}{\sqrt{2}}\eth\varphi_M = \frac{1}{\sqrt{2}}B_b e^{-i\phi} + \frac{1}{\sqrt{2}}\sum_{m=1}^{\infty} m b_m \rho^{-m-1} e^{im\phi}, \\ B_{-1} &= -\frac{1}{\sqrt{2}}\bar{\eth}\varphi_M = -\frac{1}{\sqrt{2}}B_b e^{i\phi} - \frac{1}{\sqrt{2}}\sum_{m=1}^{\infty} m b_m \rho^{-m-1} e^{-im\phi}. \end{aligned} \tag{4.54}$$

On the other hand, inside the superconductor, the magnetic induction is assumed to satisfy the equation $\nabla^2\mathbf{B} = \lambda^{-2}\mathbf{B}$, where λ is a constant, which is the vector Helmholtz equation with $k = 1/(i\lambda)$. The symmetry of the problem implies that the component B_0 must be equal to zero and that the remaining components must depend on ρ and ϕ only. Then, since $\nabla \cdot \mathbf{B} = 0$ and $B_{\pm 1}$ must be bounded at $\rho = 0$, from (4.50), (4.7), and (4.16) we obtain, for $\rho \leqslant a$,

$$\begin{aligned} B_1 &= \sum_{m=-\infty}^{\infty} a_m ({}_1J_{\alpha m}) = \sum_{m=-\infty}^{\infty} a_m J_{m+1}(\alpha\rho) e^{im\phi}, \\ B_{-1} &= -\sum_{m=-\infty}^{\infty} a_m ({}_{-1}J_{\alpha m}) = -\sum_{m=-\infty}^{\infty} a_m J_{m-1}(\alpha\rho) e^{im\phi}, \end{aligned} \tag{4.55}$$

where $\alpha = k = 1/(i\lambda)$ and the a_m are some constants. (Note that since ${}_sJ_{-\alpha,m} = (-1)^{m+s}{}_sJ_{\alpha m}$, it is not necessary to include terms with $\alpha = -1/(i\lambda)$ in (4.55).) By equating the corresponding components (4.54) and (4.55) at $\rho = a$ we find that the only nonvanishing coefficients are b_1, a_1, and a_{-1}, which are given by

$$\begin{aligned} a_1 &= a_{-1} = \frac{B_b}{\sqrt{2}\,J_0(a/i\lambda)} = \frac{B_b}{\sqrt{2}\,I_0(a/\lambda)}, \\ b_1 &= B_b a^2 \frac{J_2(a/i\lambda)}{J_0(a/i\lambda)} = -B_b a^2 \frac{I_2(a/\lambda)}{I_0(a/\lambda)}, \end{aligned}$$

where the I_ν are modified Bessel functions.

Vector plane harmonics

By analogy with the vector spherical harmonics, (3.29), we can define the *vector plane harmonics*

$$\mathbf{Q}_{\alpha m} \equiv \frac{1}{\alpha}\mathbf{M}\,{}_0J_{\alpha m} \tag{4.56}$$

where

$$\mathbf{M} \equiv -i\mathbf{e}_z \times \nabla.$$

The Cartesian components of the vector operator $\mathbf{M}$ are given by $\mathbf{M} = (-P_2, P_1, 0)$; therefore,

$$\mathbf{M} \cdot \mathbf{M} = {P_1}^2 + {P_2}^2 = P^2$$

and from (4.24) and (4.56) we have

$$\mathbf{M} \cdot \mathbf{Q}_{\alpha m} = \alpha \, {}_0J_{\alpha m}. \tag{4.57}$$

The vector plane harmonics (4.56) satisfy the orthogonality relations

$$\begin{aligned}\int \overline{\mathbf{Q}_{\alpha' m'}} \cdot \mathbf{Q}_{\alpha m} \, da &= \int \overline{{}_0J_{\alpha' m'}} \, {}_0J_{\alpha m} \, da \\ &= \int_0^{2\pi} e^{i(m-m')\phi} \, d\phi \int_0^\infty J_{m'}(\alpha' \rho) J_m(\alpha \rho) \rho \, d\rho \\ &= \frac{2\pi}{\alpha} \delta_{mm'} \delta(\alpha - \alpha'),\end{aligned} \tag{4.58}$$

which follow from the hermiticity of $\mathbf{M}$ and (4.57), and

$$\int \overline{\mathbf{Q}_{\alpha' m'}} \cdot \mathbf{e}_z \times \mathbf{Q}_{\alpha m} \, da = 0.$$

The spin-weighted components of $\mathbf{Q}_{\alpha m}$ are $(\mathbf{Q}_{\alpha m})_{\pm 1} = \mp ({}_{\pm 1}J_{\alpha m})/\sqrt{2}$ and $(\mathbf{Q}_{\alpha m})_0 = 0$. Thus, as a consequence of (4.41), the vector plane harmonics satisfy

$$\sum_{m=-\infty}^{\infty} \overline{{}_0J_{\alpha m}} \, \mathbf{Q}_{\alpha m} = 0, \qquad \sum_{m=-\infty}^{\infty} \overline{\mathbf{Q}_{\alpha m}} \cdot \mathbf{Q}_{\alpha m} = 1.$$

The vector plane harmonics are divergenceless

$$\nabla \cdot \mathbf{Q}_{\alpha m} = 0,$$

as can be seen by writing $\mathbf{Q}_{\alpha m} = i\nabla \times ({}_0J_{\alpha m}\mathbf{e}_z)$, and any divergenceless vector field $\mathbf{F}$ that is bounded for all finite values of ρ can be expressed in the form

$$\mathbf{F} = \int_0^\infty d\alpha \sum_{m=-\infty}^{\infty} [f_{\alpha m}(z)\mathbf{Q}_{\alpha m}(\rho, \phi) + \nabla \times (g_{\alpha m}(z)\mathbf{Q}_{\alpha m}(\rho, \phi))], \tag{4.59}$$

where the $f_{\alpha m}(z)$ and $g_{\alpha m}(z)$ are functions of z only (see, *e.g.*, Yoshida 1992). It can be readily seen that

$$\int \overline{f(z)\mathbf{Q}_{\alpha' m'}} \cdot \nabla \times (g(z)\mathbf{Q}_{\alpha m}) \, da = 0,$$

if $f(z)$ and $g(z)$ are arbitrary functions of z only. For a given divergenceless vector field $\mathbf{F}$, the coefficients $f_{\alpha m}(z)$ and $g_{\alpha m}(z)$ appearing in (4.59) can be determined making use of the relations

$$\begin{aligned} \mathbf{e}_z \cdot \mathbf{Q}_{\alpha m} &= 0, \\ \mathbf{e}_z \cdot \nabla \times \mathbf{Q}_{\alpha m} &= \mathrm{i}\alpha\, {}_0J_{\alpha m}, \\ \nabla \times \nabla \times \mathbf{Q}_{\alpha m} &= \alpha^2 \mathbf{Q}_{\alpha m}, \end{aligned}$$

and the orthogonality of the functions ${}_0J_{\alpha m}$ [see (4.58)].

The vector field $\mathbf{Q}_{\alpha m}$ is an eigenfunction of the z-component of the *total* angular momentum, J_3, with eigenvalue m. Indeed, for an arbitrary vector field $\mathbf{u}$, $J_3\mathbf{u} = -\mathrm{i}\mathbf{e}_z \cdot \mathbf{r} \times \nabla\mathbf{u} + \mathrm{i}\mathbf{e}_z \times \mathbf{u} = L_3\mathbf{u} + \mathrm{i}\mathbf{e}_z \times \mathbf{u}$ (see, *e.g.*, Section 3.1), therefore, using (4.20) we find the operator identity

$$\begin{aligned} J_3\mathbf{M} &= L_3(-P_2, P_1, 0) + \mathrm{i}\mathbf{e}_z \times (-P_2, P_1, 0) \\ &= (-L_3P_2 - \mathrm{i}P_1, L_3P_1 - \mathrm{i}P_2, 0) \\ &= (-P_2L_3, P_1L_3, 0) = \mathbf{M}L_3, \end{aligned}$$

which, together with (4.24) and (4.56), yield

$$J_3\mathbf{Q}_{\alpha m} = m\, \mathbf{Q}_{\alpha m}.$$

The operators P^2 and $\mathbf{M}$ commute, hence $\mathbf{Q}_{\alpha m}$ is also an eigenfunction of P^2 with eigenvalue α^2,

$$P^2\mathbf{Q}_{\alpha m} = \alpha^2\mathbf{Q}_{\alpha m}.$$

Eigenfunctions of the curl operator

As shown in Section 3.1, if the vector field $\mathbf{u}$ is an eigenfunction of the curl operator with a nonvanishing eigenvalue λ, $\nabla \times \mathbf{u} = \lambda\mathbf{u}$, then $\mathbf{u}$ is a divergenceless solution of the Helmholtz equation $\nabla^2\mathbf{u} + \lambda^2\mathbf{u} = 0$; hence, there exist two solutions of the scalar Helmholtz equation, ψ_1 and ψ_2, such that $\mathbf{u} = \mathbf{e}_z \times \nabla\psi_1 + \lambda^{-1}\nabla \times (\mathbf{e}_z \times \nabla\psi_2)$ and from the condition $\nabla \times \mathbf{u} = \lambda\mathbf{u}$ it follows that $\psi_1 = \psi_2$. Thus, if $\psi \equiv \lambda^{-1}\psi_1$ we conclude that the eigenfunctions of the curl operator with nonvanishing eigenvalue can be expressed in the form

$$\mathbf{u} = \lambda\, \mathbf{e}_z \times \nabla\psi + \nabla \times (\mathbf{e}_z \times \nabla\psi), \tag{4.60}$$

where ψ is a solution of the scalar Helmholtz equation $\nabla^2\psi + \lambda^2\psi = 0$.

The vector fields (4.60) corresponding to separable scalar potentials of the form $\psi = J_m(\alpha\rho)\mathrm{e}^{\mathrm{i}(m\phi - kz)} = {}_0J_{\alpha m}(\rho, \phi)\mathrm{e}^{-\mathrm{i}kz}$ with $\lambda = \pm(\alpha^2 + k^2)^{1/2}$, are known as Chandrasekhar–Kendall eigenfunctions (Chandrasekhar and Kendall 1957, Morse and Feshbach 1953).

The source-free electromagnetic field

The electric and magnetic fields in vacuum, in a source-free region, are divergenceless and, if it is assumed that they have a harmonic time dependence with frequency ω, satisfy the vector Helmholtz equation. Hence,

$$\begin{aligned}
\mathbf{E} &= \mathrm{Re}\left[\left(\mathbf{e}_z \times \nabla\psi_1 + \frac{1}{k}\nabla \times (\mathbf{e}_z \times \nabla\psi_2)\right)\mathrm{e}^{-\mathrm{i}\omega t}\right] \\
&= \mathrm{Re}\left[\frac{\mathrm{i}}{\omega}\partial_t(\mathbf{e}_z \times \nabla\psi_1 \mathrm{e}^{-\mathrm{i}\omega t}) + \frac{1}{k}\nabla \times (\mathbf{e}_z \times \nabla\psi_2 \mathrm{e}^{-\mathrm{i}\omega t})\right] \\
&= \frac{1}{c}\partial_t(\mathbf{e}_z \times \nabla\chi_\mathrm{M}) - \nabla \times (\mathbf{e}_z \times \nabla\chi_\mathrm{E}),
\end{aligned} \tag{4.61}$$

where $\chi_\mathrm{M} = \mathrm{Re}\,(\mathrm{i}/k)\psi_1\mathrm{e}^{-\mathrm{i}\omega t}$ and $\chi_\mathrm{E} = -\mathrm{Re}\,(1/k)\psi_2\mathrm{e}^{-\mathrm{i}\omega t}$ obey the wave equation, $\nabla^2\chi - (1/c^2)\,\partial_t^2\chi = 0$; then

$$\mathbf{B} = -\frac{1}{c}\partial_t(\mathbf{e}_z \times \nabla\chi_\mathrm{E}) - \nabla \times (\mathbf{e}_z \times \nabla\chi_\mathrm{M}). \tag{4.62}$$

If the scalar potentials χ_E and χ_M are real, the fields **E** and **B** are also real. These fields can be expressed in the usual way in terms of the electromagnetic potentials, φ and **A**, given by

$$\varphi = -\mathbf{e}_z \cdot \nabla\chi_\mathrm{E}, \qquad \mathbf{A} = \frac{1}{c}(\partial_t\chi_\mathrm{E})\,\mathbf{e}_z - \mathbf{e}_z \times \nabla\chi_\mathrm{M}.$$

The linearity of the wave equation and the fact that any electromagnetic field can be expressed as a superposition of monochromatic waves imply that any solution of the source-free Maxwell equations can be written in the form (4.61) and (4.62).

The spin-weighted components of the complex vector field $\mathbf{F} = \mathbf{E} + \mathrm{i}\mathbf{B}$ are

$$\begin{aligned}
F_{+1} &= -\frac{1}{\sqrt{2}}\left(\frac{1}{c}\partial_t + \partial_z\right)\eth\chi, \\
F_0 &= \frac{1}{\sqrt{2}}\bar{\eth}\eth\chi, \\
F_{-1} &= -\frac{1}{\sqrt{2}}\left(\frac{1}{c}\partial_t - \partial_z\right)\bar{\eth}\chi,
\end{aligned} \tag{4.63}$$

where $\chi \equiv \chi_\mathrm{E} + \mathrm{i}\chi_\mathrm{M}$ is a solution of the wave equation, showing again that any solution of the source-free Maxwell equations in vacuum can be written in terms of a single complex scalar potential.

4.3.2 Elastic waves in an isotropic elastic medium

Making use of (4.6)–(4.8) one finds that the spin-weighted components of the equations for the elastic waves in an isotropic medium [(3.86)],

$$(1-2\sigma)\nabla^2\mathbf{u} + \nabla(\nabla\cdot\mathbf{u}) - \frac{2(1+\sigma)(1-2\sigma)\rho}{E}\partial_t^2\mathbf{u} = 0, \tag{4.64}$$

are

$$\begin{aligned}
(1-2\sigma)\left(\eth\bar{\eth}u_{+1} + \partial_z^2 u_{+1}\right) + \eth\left(-\tfrac{1}{2}\eth u_{-1} + \tfrac{1}{2}\bar{\eth}u_{+1} + \partial_z u_0\right) - \kappa\,\partial_t^2 u_{+1} &= 0,\\
(1-2\sigma)\left(\bar{\eth}\eth u_{-1} + \partial_z^2 u_{-1}\right) + \bar{\eth}\left(\tfrac{1}{2}\eth u_{-1} - \tfrac{1}{2}\bar{\eth}u_{+1} - \partial_z u_0\right) - \kappa\,\partial_t^2 u_{-1} &= 0,\\
(1-2\sigma)\left(\bar{\eth}\eth u_0 + \partial_z^2 u_0\right) + \partial_z\left(-\tfrac{1}{2}\eth u_{-1} + \tfrac{1}{2}\bar{\eth}u_{+1} + \partial_z u_0\right) - \kappa\,\partial_t^2 u_0 &= 0,
\end{aligned} \tag{4.65}$$

where

$$\kappa = \frac{2(1+\sigma)(1-2\sigma)\rho}{E}.$$

This system of equations admits solutions of the form

$$u_k = g_k(z)\,{}_kJ_{\alpha m}(\rho,\phi)\,\mathrm{e}^{-\mathrm{i}\omega t} + G_k(z)\,{}_kN_{\alpha m}(\rho,\phi)\,\mathrm{e}^{-\mathrm{i}\omega t}, \qquad (k = 0, \pm 1), \tag{4.66}$$

where the g_k and G_k are functions to be determined, α is a constant different from zero, m is an integer and ω is also a constant. Substituting (4.66) into (4.65), making use of (4.10), (4.16) and the linear independence of ${}_sJ_{\alpha m}$ and ${}_sN_{\alpha m}$, one finds that the functions g_k must obey the system of ordinary differential equations

$$\begin{aligned}
(1-2\sigma)\left(\frac{\mathrm{d}^2 g_1}{\mathrm{d}z^2} - \alpha^2 g_1\right) - \frac{1}{2}\alpha^2 g_{-1} - \frac{1}{2}\alpha^2 g_1 + \alpha\frac{\mathrm{d}g_0}{\mathrm{d}z} + \kappa\,\omega^2 g_1 &= 0,\\
(1-2\sigma)\left(\frac{\mathrm{d}^2 g_{-1}}{\mathrm{d}z^2} - \alpha^2 g_{-1}\right) - \frac{1}{2}\alpha^2 g_{-1} - \frac{1}{2}\alpha^2 g_1 + \alpha\frac{\mathrm{d}g_0}{\mathrm{d}z} + \kappa\,\omega^2 g_{-1} &= 0,\\
(1-2\sigma)\left(\frac{\mathrm{d}^2 g_0}{\mathrm{d}z^2} - \alpha^2 g_0\right) - \frac{1}{2}\alpha\frac{\mathrm{d}g_{-1}}{\mathrm{d}z} - \frac{1}{2}\alpha\frac{\mathrm{d}g_1}{\mathrm{d}z} + \frac{\mathrm{d}^2 g_0}{\mathrm{d}z^2} + \kappa\,\omega^2 g_0 &= 0
\end{aligned} \tag{4.67}$$

and the functions G_k obey a system of the same form, with G_k in place of g_k.

Equations (4.67) can be rewritten as

$$\frac{\mathrm{d}^2 n}{\mathrm{d}z^2} + (k_t^2 - \alpha^2)n = 0, \tag{4.68}$$

$$(1-2\sigma)\frac{\mathrm{d}^2 H}{\mathrm{d}z^2} - 2\alpha^2(1-\sigma)H + \alpha\frac{\mathrm{d}g_0}{\mathrm{d}z} + \kappa\,\omega^2 H = 0, \tag{4.69}$$

$$2(1-\sigma)\frac{\mathrm{d}^2 g_0}{\mathrm{d}z^2} - \alpha^2(1-2\sigma)g_0 - \alpha\frac{\mathrm{d}H}{\mathrm{d}z} + \kappa\,\omega^2 g_0 = 0, \tag{4.70}$$

with

$$n \equiv \tfrac{1}{2}(g_1 - g_{-1}), \qquad H \equiv \tfrac{1}{2}(g_1 + g_{-1}) \tag{4.71}$$

and

$$k_t^2 = \frac{\kappa\,\omega^2}{1-2\sigma} = \frac{2(1+\sigma)\omega^2\rho}{E}. \tag{4.72}$$

By combining (4.69) and (4.70) one can obtain a decoupled fourth-order equation (with constant coefficients) for H that can be easily solved and then, using (4.69) and (4.70) again, one finds g_0. However, it is convenient to follow a different procedure, introducing the two auxiliary one-variable functions

$$v \equiv \frac{\mathrm{d}H}{\mathrm{d}z} - \alpha g_0, \qquad w \equiv \alpha H - \frac{\mathrm{d}g_0}{\mathrm{d}z}. \tag{4.73}$$

These combinations arise by considering the scalar functions $\mathbf{e}_z \cdot \nabla \times \nabla \times \mathbf{u}$ and $\nabla \cdot \mathbf{u}$, respectively; for instance, making use of (4.7) and (4.16) one finds that for a vector field with components (4.66), $\nabla \cdot \mathbf{u} = \sqrt{2}\big(\tfrac{1}{2}\alpha(g_1 + g_{-1}) - \mathrm{d}g_0/\mathrm{d}z\big)\,{}_0J_{\alpha m}\mathrm{e}^{-\mathrm{i}\omega t} + \sqrt{2}\big(\tfrac{1}{2}\alpha(G_1 + G_{-1}) - \mathrm{d}G_0/\mathrm{d}z\big)\,{}_0N_{\alpha m}\mathrm{e}^{-\mathrm{i}\omega t}$, and from (4.64) it follows that $\mathbf{e}_z \cdot \nabla \times \nabla \times \mathbf{u}$, $\nabla \cdot \mathbf{u}$ and $\mathbf{e}_z \cdot \nabla \times \mathbf{u}$ obey the scalar wave equation (the function $n(z)$, defined by (4.71), is related to $\mathbf{e}_z \cdot \nabla \times \mathbf{u}$). From (4.69), (4.70), and (4.73) it follows that

$$\frac{\mathrm{d}^2 v}{\mathrm{d}z^2} + (k_t^2 - \alpha^2)v = 0, \qquad \frac{\mathrm{d}^2 w}{\mathrm{d}z^2} + (k_l^2 - \alpha^2)w = 0, \tag{4.74}$$

with

$$k_l^2 = \frac{\kappa\,\omega^2}{2(1-\sigma)} = \frac{(1+\sigma)(1-2\sigma)\omega^2\rho}{(1-\sigma)E}. \tag{4.75}$$

The solutions of (4.74) are of the same form as those of (4.68). (Actually, (4.68) and (4.74) follow directly from the fact that $\mathbf{e}_z \cdot \nabla \times \mathbf{u}$, $\mathbf{e}_z \cdot \nabla \times \nabla \times \mathbf{u}$ and $\nabla \cdot \mathbf{u}$ obey scalar wave equations.)

On the other hand, from (4.73) and (4.74) we obtain

$$\left(\frac{\mathrm{d}^2}{\mathrm{d}z^2} - \alpha^2\right) g_0 = -\frac{\mathrm{d}w}{\mathrm{d}z} + \alpha v = \left(\frac{\mathrm{d}^2}{\mathrm{d}z^2} - \alpha^2\right)\left(\frac{1}{k_l^2}\frac{\mathrm{d}w}{\mathrm{d}z} - \frac{\alpha}{k_t^2}v\right)$$

hence,

$$g_0 = -\frac{\alpha}{k_t^2}v + \frac{1}{k_l^2}\frac{\mathrm{d}w}{\mathrm{d}z} + A\mathrm{e}^{\alpha z} + B\mathrm{e}^{-\alpha z}, \tag{4.76}$$

where A and B are some constants. Then, from (4.73) and (4.74) one has

$$H = \frac{\alpha}{k_l^2}w - \frac{1}{k_t^2}\frac{\mathrm{d}v}{\mathrm{d}z} + A\mathrm{e}^{\alpha z} - B\mathrm{e}^{-\alpha z}. \tag{4.77}$$

Substituting (4.76) and (4.77) into (4.69) and (4.70) one finds that if $\omega \neq 0$, then $A = B = 0$. Thus,

$$g_0 = -\frac{\alpha}{k_t^2} v + \frac{1}{k_l^2}\frac{dw}{dz}, \qquad H = \frac{\alpha}{k_l^2} w - \frac{1}{k_t^2}\frac{dv}{dz}, \tag{4.78}$$

with similar expressions for G_0 and $\frac{1}{2}(G_1 + G_{-1})$. Then, from (4.66), (4.71), (4.78), and (4.16) we find that the spin-weighted components of the displacement vector $\mathbf{u}$ can be expressed as

$$\begin{aligned} u_{+1} &= \frac{1}{\sqrt{2}}\eth\psi_1 - \frac{i}{\sqrt{2}}\eth\psi_2 + \frac{1}{\sqrt{2}}\partial_z\eth\psi_3, \\ u_{-1} &= -\frac{1}{\sqrt{2}}\bar{\eth}\psi_1 - \frac{i}{\sqrt{2}}\bar{\eth}\psi_2 - \frac{1}{\sqrt{2}}\partial_z\bar{\eth}\psi_3, \\ u_0 &= \frac{1}{\sqrt{2}}\partial_z\psi_1 - \frac{1}{\sqrt{2}}\eth\bar{\eth}\psi_3, \end{aligned} \tag{4.79}$$

with

$$\begin{aligned} \psi_1 &\equiv \frac{\sqrt{2}}{k_l^2}(w(z)\,{}_0J_{\alpha m} + W(z)\,{}_0N_{\alpha m})e^{-i\omega t}, \\ \psi_2 &\equiv \frac{i\sqrt{2}}{\alpha}(n(z)\,{}_0J_{\alpha m} + N(z)\,{}_0N_{\alpha m})e^{-i\omega t}, \\ \psi_3 &\equiv -\frac{\sqrt{2}}{\alpha k_t^2}(v(z)\,{}_0J_{\alpha m} + V(z)\,{}_0N_{\alpha m})e^{-i\omega t}, \end{aligned} \tag{4.80}$$

where the functions $W(z)$, $N(z)$ and $V(z)$ obey the same equations as $w(z)$, $n(z)$ and $v(z)$, respectively [(4.68) and (4.74)]. With the aid of (4.9), (4.10), (4.68), and (4.74) one finds that the three scalar potentials (4.80) satisfy the wave equations

$$\nabla^2\psi_1 - \frac{1}{v_l^2}\partial_t^2\psi_1 = 0, \qquad \nabla^2\psi_{2,3} - \frac{1}{v_t^2}\partial_t^2\psi_{2,3} = 0, \tag{4.81}$$

where

$$v_l = \frac{\omega}{k_l} = \sqrt{\frac{(1-\sigma)E}{(1+\sigma)(1-2\sigma)\rho}}, \qquad v_t = \frac{\omega}{k_t} = \sqrt{\frac{E}{2(1+\sigma)\rho}} \tag{4.82}$$

and from (4.6) it follows that (4.79) amount to

$$\mathbf{u} = -\nabla\psi_1 + \mathbf{e}_z \times \nabla\psi_2 + \nabla \times (\mathbf{e}_z \times \nabla\psi_3), \tag{4.83}$$

or, equivalently,

$$\mathbf{u} = -\nabla\psi_1 - \nabla \times (\psi_2\mathbf{e}_z) - \nabla \times \nabla \times (\psi_3\mathbf{e}_z). \tag{4.84}$$

In a similar manner, one can show that (4.65) admit separable solutions analogous to (4.66) in terms of the spin-weighted cylindrical harmonics with $\alpha = 0$ [(4.14) and (4.15)], which can also be written in the form (4.83) with the potentials ψ_i satisfying (4.81). By virtue of the completeness of the spin-weighted cylindrical harmonics and the linearity of (4.83) and (4.81), it follows that the general solution of (4.64) is given by (4.83) or (4.84), where the scalar potentials ψ_i are solutions of the wave equations (4.81).

Equation (4.84) shows that the displacement vector field, in effect, can be written as the sum of a vector field $(-\nabla\psi_1)$ with vanishing curl and a vector field $(-\nabla \times (\psi_2 \mathbf{e}_z) - \nabla \times \nabla \times (\psi_3 \mathbf{e}_z))$ with vanishing divergence (as assumed, *e.g.*, in Landau and Lifshitz 1975). It is easy to verify directly that (4.84) satisfies (4.64) provided that the scalar potentials ψ_i obey the corresponding wave equations [(4.81)]. If the potentials ψ_i are real, then the displacement vector field is also real. It should be remarked that the expressions (4.83) and (4.84) are not linked to a particular coordinate system, despite the fact that the circular cylindrical coordinates were employed to obtain these formulas; however, owing to the presence of the (constant) vector field $\mathbf{e}_z$, (4.83) and (4.84) are adapted to the Cartesian or the cylindrical coordinates (circular, parabolic or elliptic).

The solutions of (4.64) generated by the potential ψ_1 propagate with the velocity v_l, while those generated by ψ_2 or ψ_3 propagate with the velocity v_t. If the potentials ψ_i are plane waves, then the elastic waves generated by ψ_1 are longitudinal waves, whereas those generated by ψ_2 or ψ_3 are transverse. (This is the reason why the subscripts l and t have been employed in the definitions (4.72), (4.75), and (4.82).) In fact, if we substitute $\psi_1 = A\cos(\mathbf{k}\cdot\mathbf{r} - \omega t)$, with $|\mathbf{k}| = k_l$, into (4.83) we will obtain $\mathbf{u} = A\sin(\mathbf{k}\cdot\mathbf{r} - \omega t)\,\mathbf{k}$, which represents a longitudinal elastic wave (with $\mathbf{u}$ parallel to $\mathbf{k}$); on the other hand, $\psi_2 = A\cos(\mathbf{k}\cdot\mathbf{r} - \omega t)$, with $|\mathbf{k}| = k_t$, yields $\mathbf{u} = -A\sin(\mathbf{k}\cdot\mathbf{r} - \omega t)\,\mathbf{e}_z \times \mathbf{k}$, which satisfies $\mathbf{u}\cdot\mathbf{k} = 0$ and, hence, is a transverse wave. Similarly, if $\psi_3 = A\cos(\mathbf{k}\cdot\mathbf{r} - \omega t)$, with $|\mathbf{k}| = k_t$, then $\mathbf{u} = A\cos(\mathbf{k}\cdot\mathbf{r} - \omega t)\,(\mathbf{e}_z \times \mathbf{k}) \times \mathbf{k}$, which also satisfies $\mathbf{u}\cdot\mathbf{k} = 0$.

It may be noticed that, according to (4.12), a separable solution of the form $u_k = g_k(z)\,{}_kJ_{\alpha m}(\rho, \phi)\mathrm{e}^{-i\omega t}$ [see (4.66)] corresponds to

$$u_\rho = \frac{1}{\sqrt{2}}(u_{+1} - u_{-1}) = \frac{1}{\sqrt{2}}\big(g_1(z)J_{m+1}(\alpha\rho) - g_{-1}(z)J_{m-1}(\alpha\rho)\big)\mathrm{e}^{im\phi}\mathrm{e}^{-i\omega t}, \tag{4.85}$$

which is not separable since $g_1(z)$ and $g_{-1}(z)$ are not independent. It may be also noticed that the presence of Bessel functions of order $m+1$ and $m-1$ accompanying the factor $\mathrm{e}^{im\phi}$ in (4.85) arises in a natural way by expressing each spin-weighted component of $\mathbf{u}$ in terms of the spin-weighted harmonics of the corresponding weight [(4.66)].

4.3.3 Solution of the equations of equilibrium for an isotropic elastic medium

The equations of equilibrium for an isotropic elastic medium in the absence of body forces are (Landau and Lifshitz 1975)

$$(1-2\sigma)\nabla^2\mathbf{u} + \nabla(\nabla\cdot\mathbf{u}) = 0, \tag{4.86}$$

[*cf.* (4.64)]. Looking for solutions of (4.86) of the form

$$u_k = g_k(z)\,{}_kJ_{\alpha m}(\rho,\phi) + G_k(z)\,{}_kN_{\alpha m}(\rho,\phi), \qquad (k=0,\pm1), \tag{4.87}$$

with $\alpha \neq 0$, we begin by noticing that if $\mathbf{u}$ satisfies (4.86), then $\nabla\cdot\mathbf{u}$, $\mathbf{e}_z\cdot\nabla\times\mathbf{u}$, and $z\nabla\cdot\mathbf{u} + 2(1-2\sigma)\mathbf{e}_z\cdot\mathbf{u}$ obey the Laplace equation. For a vector field given by (4.87),

$$\begin{aligned}
\nabla\cdot\mathbf{u} &= \sqrt{2}\left(\frac{1}{2}\alpha(g_1+g_{-1}) - \frac{dg_0}{dz}\right){}_0J_{\alpha m} \\
&\quad + \sqrt{2}\left(\frac{1}{2}\alpha(G_1+G_{-1}) - \frac{dG_0}{dz}\right){}_0N_{\alpha m}, \\
\mathbf{e}_z\cdot\nabla\times\mathbf{u} &= -\frac{i\alpha}{\sqrt{2}}(g_1-g_{-1})\,{}_0J_{\alpha m} - \frac{i\alpha}{\sqrt{2}}(G_1-G_{-1})\,{}_0N_{\alpha m}, \\
z\nabla\cdot\mathbf{u} &+ 2(1-2\sigma)\mathbf{e}_z\cdot\mathbf{u} \\
&= \sqrt{2}\left(\frac{1}{2}\alpha z(g_1+g_{-1}) - z\frac{dg_0}{dz} - 2(1-2\sigma)g_0\right){}_0J_{\alpha m} \\
&\quad + \sqrt{2}\left(\frac{1}{2}\alpha z(G_1+G_{-1}) - z\frac{dG_0}{dz} - 2(1-2\sigma)G_0\right){}_0N_{\alpha m}.
\end{aligned}$$

Since these functions satisfy the Laplace equation, from (4.9) and (4.10) it follows that $n(z) \equiv \frac{1}{2}(g_1 - g_{-1})$, and

$$w(z) \equiv \frac{1}{2}\alpha(g_1+g_{-1}) - \frac{dg_0}{dz}, \qquad v(z) \equiv zw - 2(1-2\sigma)g_0 \tag{4.88}$$

obey the equations

$$\frac{d^2n}{dz^2} - \alpha^2 n = 0, \qquad \frac{d^2w}{dz^2} - \alpha^2 w = 0, \qquad \frac{d^2v}{dz^2} - \alpha^2 v = 0. \tag{4.89}$$

Then, from (4.88) and (4.89) we obtain

$$\begin{aligned}
\tfrac{1}{2}\alpha(g_1+g_{-1}) &= w + \frac{dg_0}{dz} = w + \frac{d}{dz}\frac{zw-v}{2(1-2\sigma)} \\
&= \frac{1}{2(1-2\sigma)}\left[(3-4\sigma)w - \frac{dv}{dz} + z\frac{dw}{dz}\right]
\end{aligned}$$

and

$$\begin{aligned} 2(1-2\sigma)g_0 &= zw - v = \frac{1}{\alpha^2}\left(z\frac{\mathrm{d}^2 w}{\mathrm{d}z^2} - \frac{\mathrm{d}^2 v}{\mathrm{d}z^2}\right) \\ &= \frac{1}{\alpha^2}\frac{\mathrm{d}}{\mathrm{d}z}\left(z\frac{\mathrm{d}w}{\mathrm{d}z} - \frac{\mathrm{d}v}{\mathrm{d}z} + (3-4\sigma)w\right) - \frac{4(1-\sigma)}{\alpha^2}\frac{\mathrm{d}w}{\mathrm{d}z}, \end{aligned}$$

with analogous expressions for the G_k.

Thus,

$$\begin{aligned} u_{+1} &= \frac{\mathrm{i}}{\sqrt{2}}\eth\psi_1 + \frac{1}{\sqrt{2}}\eth(\psi_2 + z\psi_3), \\ u_{-1} &= \frac{\mathrm{i}}{\sqrt{2}}\bar{\eth}\psi_1 - \frac{1}{\sqrt{2}}\bar{\eth}(\psi_2 + z\psi_3), \\ u_0 &= \frac{1}{\sqrt{2}}\partial_z(\psi_2 + z\psi_3) - 2\sqrt{2}\,(1-\sigma)\psi_3, \end{aligned} \tag{4.90}$$

where

$$\begin{aligned} \psi_1 &= \frac{\sqrt{2}}{\mathrm{i}\alpha}\big(n(z)\,{}_0J_{\alpha m} + N(z)\,{}_0N_{\alpha m}\big), \\ \psi_2 &= \frac{1}{\sqrt{2}\,(1-2\sigma)\alpha^2}\left\{\left[(3-4\sigma)w - \frac{\mathrm{d}v}{\mathrm{d}z}\right]{}_0J_{\alpha m}\right. \\ &\qquad \left. + \left[(3-4\sigma)W - \frac{\mathrm{d}V}{\mathrm{d}z}\right]{}_0N_{\alpha m}\right\}, \\ \psi_3 &= \frac{1}{\sqrt{2}\,(1-2\sigma)\alpha^2}\left(\frac{\mathrm{d}w}{\mathrm{d}z}\,{}_0J_{\alpha m} + \frac{\mathrm{d}W}{\mathrm{d}z}\,{}_0N_{\alpha m}\right) \end{aligned} \tag{4.91}$$

and N, W, and V are defined in terms of the G_k by the same formulas that define n, w, and v in terms of the g_k. As a consequence of (4.89) and of the corresponding equations for N, W, and V, the three scalar potentials (4.91) obey the Laplace equation,

$$\nabla^2\psi_1 = 0, \qquad \nabla^2\psi_2 = 0, \qquad \nabla^2\psi_3 = 0. \tag{4.92}$$

Making use of (4.6) and (4.7), one finds that (4.90) is equivalent to

$$\mathbf{u} = \nabla \times (\psi_1\mathbf{e}_z) - \nabla(\psi_2 + z\psi_3) + 4(1-\sigma)\psi_3\mathbf{e}_z \tag{4.93}$$

[*cf.* (4.84)]. In the case where $\alpha = 0$, one also obtains (4.92) and (4.93) but now the potentials ψ_1, ψ_2, ψ_3 are not independent (Torres del Castillo 1992c).

An expression analogous to (4.93) for the solutions of (4.86) in terms of *four* harmonic scalar potentials was obtained by Papkovich and Neuber (see, *e.g.*, Sokolnikoff 1956, Fung 1965, Timoshenko and Goodier 1970). The Papkovich–Neuber solution can be written in the form

$$\mathbf{u} = -\nabla(\phi_0 + x\phi_1 + y\phi_2 + z\phi_3) + 4(1-\sigma)(\phi_1\mathbf{e}_x + \phi_2\mathbf{e}_y + \phi_3\mathbf{e}_z). \tag{4.94}$$

The potential ϕ_0 can be omitted provided that the potentials are allowed to have singularities (the displacement vector $\mathbf{u}$ given by (4.94) may be well behaved even if the potentials have singularities). The right-hand side of (4.94) is left unchanged if ϕ_0 is set equal to zero and ϕ_i is replaced by $\phi_i + \partial f/\partial x_i$, $(i = 1, 2, 3)$, where

$$f = r^{4(1-\sigma)} \int \phi_0 r^{4\sigma-5} \mathrm{d}r,$$

the x_i are Cartesian coordinates and r is the usual radial coordinate. (The condition $\nabla^2\phi_0 = 0$ implies $\nabla^2 f = 0$ and, hence, $\nabla^2(\partial f/\partial x_i) = 0$.)

Another expression similar to (4.93) is given in Landau and Lifshitz (1975), where the solutions of (4.86) are written in terms of four arbitrary harmonic functions.

4.3.4 Solution of the Dirac equation

The orthonormal basis $\{\mathbf{e}_\rho, \mathbf{e}_\phi, \mathbf{e}_z\}$ can be considered as induced by the two-component spinor field

$$o \equiv \begin{pmatrix} \mathrm{e}^{-\mathrm{i}\phi/2} \\ 0 \end{pmatrix}, \tag{4.95}$$

by means of the relations

$$\mathbf{e}_\rho + \mathrm{i}\mathbf{e}_\phi = o^{\mathrm{t}}\varepsilon\sigma o, \qquad \mathbf{e}_z = o^\dagger \sigma o,$$

so that the rotation (4.1) is induced by the transformation

$$o \mapsto \mathrm{e}^{\mathrm{i}\theta/2} o$$

and therefore we shall assign to o the spin weight 1/2.

An arbitrary two-component spinor field u can be expressed in the form

$$u = u_- o + u_+ \widehat{o}, \tag{4.96}$$

where $u_\pm$ is a complex-valued function with spin weight $\pm 1/2$. Making use of (4.95) and (4.4) we find that

$$\sigma \cdot \nabla(u_- o + u_+ \widehat{o}) = (\partial_z u_- - \bar{\eth} u_+) o + (-\partial_z u_+ - \eth u_-)\widehat{o},$$

therefore, the Dirac equation (1.71) takes the form

$$\begin{aligned} \frac{1}{c}\partial_t u_- &= -\partial_z v_- + \bar{\eth} v_+ - \frac{\mathrm{i}Mc}{\hbar} u_-, \\ \frac{1}{c}\partial_t u_+ &= \eth v_- + \partial_z v_+ - \frac{\mathrm{i}Mc}{\hbar} u_+, \\ \frac{1}{c}\partial_t v_- &= -\partial_z u_- + \bar{\eth} u_+ + \frac{\mathrm{i}Mc}{\hbar} v_-, \\ \frac{1}{c}\partial_t v_+ &= \eth u_- + \partial_z u_+ + \frac{\mathrm{i}Mc}{\hbar} v_+. \end{aligned} \tag{4.97}$$

These equations admit separable solutions of the form

$$\begin{aligned} u_- &= {}_{-\frac{1}{2}}J_{\alpha m}(\rho,\phi)\, g(z)\mathrm{e}^{-\mathrm{i}Et/\hbar}, \\ u_+ &= {}_{\frac{1}{2}}J_{\alpha m}(\rho,\phi)\, G(z)\mathrm{e}^{-\mathrm{i}Et/\hbar}, \\ v_- &= {}_{-\frac{1}{2}}J_{\alpha m}(\rho,\phi)\, f(z)\mathrm{e}^{-\mathrm{i}Et/\hbar}, \\ v_+ &= {}_{\frac{1}{2}}J_{\alpha m}(\rho,\phi)\, F(z)\mathrm{e}^{-\mathrm{i}Et/\hbar}, \end{aligned} \tag{4.98}$$

where m is a half-integer and α, E are constants. The components $u_\pm$, $v_\pm$ are bounded everywhere only if α is real and different from zero (the spin-weighted functions ${}_{\pm\frac{1}{2}}N_{\alpha m}$ are not included in the solutions (4.98) because they diverge at $\rho = 0$). Substitution of (4.98) into (4.97) gives

$$\begin{aligned} \frac{\mathrm{d}A}{\mathrm{d}z} + \alpha A &= \frac{E+Mc^2}{\hbar c}C, & -\frac{\mathrm{d}C}{\mathrm{d}z} + \alpha C &= \frac{E-Mc^2}{\hbar c}A, \\ \frac{\mathrm{d}B}{\mathrm{d}z} - \alpha B &= \frac{E+Mc^2}{\hbar c}D, & -\frac{\mathrm{d}D}{\mathrm{d}z} - \alpha D &= \frac{E-Mc^2}{\hbar c}B, \end{aligned} \tag{4.99}$$

where

$$A \equiv \frac{1}{2}(G+g), \qquad B \equiv \frac{1}{2}(G-g), \qquad C \equiv \frac{1}{2\mathrm{i}}(F-f), \qquad D \equiv \frac{1}{2\mathrm{i}}(F+f).$$

Thus, $\mathrm{d}^2A/\mathrm{d}z^2 = -(k^2-\alpha^2)A$, where $k \equiv \sqrt{E^2 - M^2c^4}/\hbar c$; hence,

$$A = a_1\mathrm{e}^{\mathrm{i}\sqrt{k^2-\alpha^2}\,z} + a_2\mathrm{e}^{-\mathrm{i}\sqrt{k^2-\alpha^2}\,z}, \tag{4.100}$$

where a_1 and a_2 are two arbitrary constants. Substitution of (4.100) into the first equation (4.99) yields

$$C = a_1\frac{(\alpha + \mathrm{i}\sqrt{k^2-\alpha^2})\hbar c}{E+Mc^2}\mathrm{e}^{\mathrm{i}\sqrt{k^2-\alpha^2}\,z} + a_2\frac{(\alpha - \mathrm{i}\sqrt{k^2-\alpha^2})\hbar c}{E+Mc^2}\mathrm{e}^{-\mathrm{i}\sqrt{k^2-\alpha^2}\,z}. \tag{4.101}$$

The solutions (4.100) and (4.101) are bounded only if $|\alpha| \leqslant k$. Since the equations for B and D in (4.99) differ from those for A and C only by the sign of α, one obtains

$$\begin{aligned} B &= b_1\mathrm{e}^{\mathrm{i}\sqrt{k^2-\alpha^2}\,z} + b_2\mathrm{e}^{-\mathrm{i}\sqrt{k^2-\alpha^2}\,z}, \\ D &= b_1\frac{(-\alpha + \mathrm{i}\sqrt{k^2-\alpha^2})\hbar c}{E+Mc^2}\mathrm{e}^{\mathrm{i}\sqrt{k^2-\alpha^2}\,z} \\ &\quad + b_2\frac{(-\alpha - \mathrm{i}\sqrt{k^2-\alpha^2})\hbar c}{E+Mc^2}\mathrm{e}^{-\mathrm{i}\sqrt{k^2-\alpha^2}\,z}, \end{aligned}$$

where b_1 and b_2 are arbitrary constants. Therefore, the system of equations (4.97) admits separable solutions of the form

$$\begin{pmatrix} u_- \\ u_+ \\ v_- \\ v_+ \end{pmatrix} = \begin{pmatrix} A(z)X_{\alpha m} \\ \mathrm{i}C(z)X_{-\alpha m} \end{pmatrix} \mathrm{e}^{-\mathrm{i}Et/\hbar} + \begin{pmatrix} B(z)X_{-\alpha m} \\ \mathrm{i}D(z)X_{\alpha m} \end{pmatrix} \mathrm{e}^{-\mathrm{i}Et/\hbar}, \quad (4.102)$$

where

$$X_{\alpha m} \equiv \begin{pmatrix} -\frac{1}{2}J_{\alpha m} \\ \frac{1}{2}J_{\alpha m} \end{pmatrix}, \qquad X_{-\alpha m} \equiv \begin{pmatrix} -(_{-\frac{1}{2}}J_{\alpha m}) \\ \frac{1}{2}J_{\alpha m} \end{pmatrix}, \qquad (\alpha \neq 0). \quad (4.103)$$

The spin-weighted cylindrical harmonics with $\alpha = 0$, ${}_sF_{0m}$, are bounded only if $m = -s$ [see (4.14) and (4.15)]; hence, in addition to the solutions (4.98), there are two families of bounded separable solutions of (4.97) given by

$$m = \tfrac{1}{2}: \begin{cases} u_- = g(z)\mathrm{e}^{\mathrm{i}\phi/2}\mathrm{e}^{-\mathrm{i}Et/\hbar}, \\ u_+ = 0, \\ u_- = f(z)\mathrm{e}^{\mathrm{i}\phi/2}\mathrm{e}^{-\mathrm{i}Et/\hbar}, \\ u_- = 0, \end{cases} \quad (4.104)$$

and

$$m = -\tfrac{1}{2}: \begin{cases} u_- = 0, \\ u_+ = G(z)\mathrm{e}^{-\mathrm{i}\phi/2}\mathrm{e}^{-\mathrm{i}Et/\hbar}, \\ v_- = 0, \\ v_+ = F(z)\mathrm{e}^{-\mathrm{i}\phi/2}\mathrm{e}^{-\mathrm{i}Et/\hbar}. \end{cases} \quad (4.105)$$

Substituting (4.104) and (4.105) into (4.97), making use of (4.17), we obtain

$$g(z) = a_1\mathrm{e}^{\mathrm{i}kz} + a_2\mathrm{e}^{-\mathrm{i}kz}, \qquad f(z) = \frac{\hbar ck}{E + Mc^2}(a_1\mathrm{e}^{\mathrm{i}kz} - a_2\mathrm{e}^{-\mathrm{i}kz}),$$

$$G(z) = b_1\mathrm{e}^{\mathrm{i}kz} + b_2\mathrm{e}^{-\mathrm{i}kz}, \qquad F(z) = \frac{\hbar ck}{E + Mc^2}(-b_1\mathrm{e}^{\mathrm{i}kz} + b_2\mathrm{e}^{-\mathrm{i}kz}).$$

Thus, equations (4.97) admit bounded separable solutions of the form

$$\begin{pmatrix} u_- \\ u_+ \\ v_- \\ v_+ \end{pmatrix} = \begin{pmatrix} \zeta(z)X_{0m} \\ \xi(z)X_{0m} \end{pmatrix} \mathrm{e}^{-\mathrm{i}Et/\hbar} \qquad (m = \pm\tfrac{1}{2}),$$

where

$$X_{0,\frac{1}{2}} \equiv \begin{pmatrix} \mathrm{e}^{\mathrm{i}\phi/2} \\ 0 \end{pmatrix}, \qquad X_{0,-\frac{1}{2}} \equiv \begin{pmatrix} 0 \\ \mathrm{e}^{-\mathrm{i}\phi/2} \end{pmatrix}. \quad (4.106)$$

The solutions (4.104) and (4.105), corresponding to $\alpha = 0$, are superpositions of plane waves traveling along the z axis in the positive and negative directions.

The fact that the Dirac equation can be reduced to the two independent pairs of differential equations (4.99) is related to the existence of an operator, K, that commutes with the Dirac Hamiltonian, J_3 and $P_1{}^2 + P_2{}^2$. The two-component spinors $X_{\alpha m}$ defined by (4.103) and (4.106) satisfy

$$QX_{\kappa m} = \kappa X_{\kappa m},$$

where

$$Q \equiv \begin{pmatrix} 0 & -\bar{\eth} \\ \eth & 0 \end{pmatrix} \tag{4.107}$$

[*cf.* (3.148)], *i.e.*, $Q(u_- o + u_+ \widehat{o}) = -(\bar{\eth} u_+) o + (\eth u_-)\widehat{o}$. Then, letting

$$K \equiv \begin{pmatrix} -Q & 0 \\ 0 & Q \end{pmatrix}, \tag{4.108}$$

one finds that each term on the right-hand side of (4.102) is an eigenfunction of K with eigenvalue $-\alpha$ and α, respectively.

The 2×2 matrix (4.107) defines the operator Q with respect to the basis $\{o, \widehat{o}\}$. In order to find the expression for Q with respect to the canonical basis, we note that (4.96) gives $u^A = u_- o^A + u_+ \widehat{o}^A$; hence, making use of (4.95), we have

$$\begin{pmatrix} u^1 \\ u^2 \end{pmatrix} = \begin{pmatrix} u_- \mathrm{e}^{-\mathrm{i}\phi/2} \\ u_+ \mathrm{e}^{\mathrm{i}\phi/2} \end{pmatrix} = \begin{pmatrix} \mathrm{e}^{-\mathrm{i}\phi/2} & 0 \\ 0 & \mathrm{e}^{\mathrm{i}\phi/2} \end{pmatrix} \begin{pmatrix} u_- \\ u_+ \end{pmatrix}$$

and

$$\begin{pmatrix} u_- \\ u_+ \end{pmatrix} = \begin{pmatrix} \mathrm{e}^{\mathrm{i}\phi/2} & 0 \\ 0 & \mathrm{e}^{-\mathrm{i}\phi/2} \end{pmatrix} \begin{pmatrix} u^1 \\ u^2 \end{pmatrix} \equiv \Lambda \begin{pmatrix} u^1 \\ u^2 \end{pmatrix}.$$

Then, with respect to the canonical basis, Q corresponds to the operator

$$\Lambda^{-1} Q \Lambda = \sigma_1 P_2 - \sigma_2 P_1 \tag{4.109}$$

and K corresponds to

$$\begin{pmatrix} \sigma_2 & 0 \\ 0 & -\sigma_2 \end{pmatrix} P_1 - \begin{pmatrix} \sigma_1 & 0 \\ 0 & -\sigma_1 \end{pmatrix} P_2 = \gamma_5(\gamma_2 P_1 - \gamma_1 P_2),$$

where $\gamma_5 \equiv \mathrm{i}\gamma^0\gamma^1\gamma^2\gamma^3$ and the γ^μ are the Dirac matrices (see, *e.g.*, Bjorken and Drell 1964). Furthermore, (4.107) or (4.109) implies that $Q^2 = P_1{}^2 + P_2{}^2$ and therefore $K^2 = P_1{}^2 + P_2{}^2$.

4.3.5 Solution of the spin-2 Helmholtz equation

The components of a symmetric, traceless two-index tensor field, t_{ij}, with respect to the orthonormal basis $\{\mathbf{e}_\rho, \mathbf{e}_\phi, \mathbf{e}_z\}$ form the five spin-weighted combinations

$$\begin{aligned} t_{\pm 2} &\equiv \tfrac{1}{2}(t_{\rho\rho} - t_{\phi\phi} \pm 2\mathrm{i}t_{\rho\phi}) = \tfrac{1}{2}(t_{zz} + 2t_{\rho\rho} \pm 2\mathrm{i}t_{\rho\phi}), \\ t_{\pm 1} &\equiv \mp\tfrac{1}{2}(t_{\rho z} \pm \mathrm{i}t_{\phi z}), \\ t_0 &\equiv \tfrac{1}{2}t_{zz}. \end{aligned} \tag{4.110}$$

Then, the Helmholtz equation, $\nabla^2 t_{ij} + k^2 t_{ij} = 0$, is equivalent to the equations

$$\partial_z^2 t_s + \bar{\eth}\eth t_s + k^2 t_s = 0 \qquad (s = 0, \pm 1, \pm 2). \tag{4.111}$$

These equations admit separable solutions of the form

$$t_s = g_s(z)\, {}_sJ_{\alpha m}(\rho, \phi) + G_s(z)\, {}_sN_{\alpha m}(\rho, \phi) \tag{4.112}$$

[*cf.* (4.48)], where m is an integer and α is a real number different from zero. Any tensor field of the form (4.112) is an eigenfunction of J_3 and of ${P_1}^2 + {P_2}^2$, with eigenvalues m and α^2, respectively. Substituting (4.112) into (4.111) one finds that

$$\frac{\mathrm{d}^2 g_s}{\mathrm{d}z^2} + (k^2 - \alpha^2) g_s = 0 \qquad (s = 0, \pm 1, \pm 2),$$

with identical equations for the G_s; hence, making $\gamma^2 \equiv \alpha^2 - k^2$,

$$g_s(z) = \begin{cases} A_s \mathrm{e}^{\gamma z} + B_s \mathrm{e}^{-\gamma z}, & \text{if } \alpha^2 \neq k^2, \\ A_s + B_s z, & \text{if } \alpha^2 = k^2, \end{cases}$$

where the A_s and B_s are arbitrary constants. Thus, the spin-2 Helmholtz equation admits separable solutions of the form

$$t_s = (A_s \mathrm{e}^{\gamma z} + B_s \mathrm{e}^{-\gamma z})\, {}_sJ_{\alpha m} + (C_s \mathrm{e}^{\gamma z} + D_s \mathrm{e}^{-\gamma z})\, {}_sN_{\alpha m}, \tag{4.113}$$

if $\alpha^2 \neq k^2$, and

$$t_s = (A_s + B_s z)\, {}_sJ_{\alpha m} + (C_s + D_s z)\, {}_sN_{\alpha m}, \tag{4.114}$$

if $\alpha = \pm k$.

The spin-weighted components of the divergence of a symmetric, traceless, 2-index tensor field t_{ij} are given by

$$(\operatorname{div} t)_s = \frac{1}{\sqrt{2}}(\eth t_{s-1} - 2\partial_z t_s - \bar{\eth} t_{s+1}), \tag{4.115}$$

therefore, the tensor field (4.113) has vanishing divergence if and only if

$$\frac{\alpha}{2}(A_{s+1}+A_{s-1}) = \gamma A_s, \qquad \frac{\alpha}{2}(B_{s+1}+B_{s-1}) = -\gamma B_s,$$
$$\frac{\alpha}{2}(C_{s+1}+C_{s-1}) = \gamma C_s, \qquad \frac{\alpha}{2}(D_{s+1}+D_{s-1}) = -\gamma D_s, \tag{4.116}$$

$s = 0, \pm 1$. Introducing the combinations

$$a_1 \equiv \frac{\mathrm{i}}{\alpha^3}(A_1 - A_{-1}), \qquad a_2 \equiv \frac{\mathrm{i}}{\alpha^3}(C_1 - C_{-1}),$$
$$b_1 \equiv \frac{\mathrm{i}}{\alpha^3}(B_1 - B_{-1}), \qquad b_2 \equiv \frac{\mathrm{i}}{\alpha^3}(D_1 - D_{-1}),$$
$$a_3 \equiv \frac{k}{\gamma\alpha^3}(A_1 + A_{-1}), \qquad a_4 \equiv \frac{k}{\gamma\alpha^3}(C_1 + C_{-1}),$$
$$b_3 \equiv -\frac{k}{\gamma\alpha^3}(B_1 + B_{-1}), \qquad b_4 \equiv -\frac{k}{\gamma\alpha^3}(D_1 + D_{-1}),$$

and assuming that the conditions (4.116) hold, one finds that the components (4.113) can be written as

$$\begin{aligned}
t_{+2} &= -\mathrm{i}\partial_z \eth\eth\psi_1 + \frac{1}{2k}(\partial_z^2 - k^2)\eth\eth\psi_2,\\
t_{+1} &= \frac{\mathrm{i}}{2}\bar{\eth}\eth\eth\psi_1 - \frac{1}{2k}\partial_z\bar{\eth}\eth\eth\psi_2,\\
t_0 &= \frac{1}{2k}\bar{\eth}\bar{\eth}\eth\eth\psi_2,\\
t_{-1} &= \frac{\mathrm{i}}{2}\eth\bar{\eth}\bar{\eth}\psi_1 + \frac{1}{2k}\partial_z\eth\bar{\eth}\bar{\eth}\psi_2,\\
t_{-2} &= \mathrm{i}\partial_z\bar{\eth}\bar{\eth}\psi_1 + \frac{1}{2k}(\partial_z^2 - k^2)\bar{\eth}\bar{\eth}\psi_2,
\end{aligned} \tag{4.117}$$

where

$$\begin{aligned}
\psi_1 &= (a_1 e^{\gamma z} + b_1 e^{-\gamma z})\,{}_0J_{\alpha m} + (a_2 e^{\gamma z} + b_2 e^{-\gamma z})\,{}_0N_{\alpha m},\\
\psi_2 &= (a_3 e^{\gamma z} + b_3 e^{-\gamma z})\,{}_0J_{\alpha m} + (a_4 e^{\gamma z} + b_4 e^{-\gamma z})\,{}_0N_{\alpha m},
\end{aligned}$$

which are solutions of the scalar Helmholtz equation.

Similarly, one finds that if the field given by (4.114) has vanishing divergence, then its components can be written in the form (4.117) with

$$\begin{aligned}
\psi_1 &= (a_1 + b_1 z)\,{}_0J_{\alpha m} + (a_2 + b_2 z)\,{}_0N_{\alpha m},\\
\psi_2 &= (a_3 + b_3 z)\,{}_0J_{\alpha m} + (a_4 + b_4 z)\,{}_0N_{\alpha m},
\end{aligned}$$

which satisfy the scalar Helmholtz equation, and

$$a_1 \equiv \frac{i}{\alpha^3}(A_1 - A_{-1}), \qquad a_2 \equiv \frac{i}{\alpha^3}(C_1 - C_{-1}),$$
$$b_1 \equiv \frac{i}{\alpha^3}(B_1 - B_{-1}), \qquad b_2 \equiv \frac{i}{\alpha^3}(D_1 - D_{-1}),$$
$$a_3 \equiv -\frac{k}{\alpha^4}(A_2 + A_{-2}), \qquad a_4 \equiv -\frac{k}{\alpha^4}(C_2 + C_{-2}),$$
$$b_3 \equiv \frac{k}{\alpha^3}(A_1 + A_{-1}), \qquad b_4 \equiv \frac{k}{\alpha^3}(C_1 + C_{-1}).$$

Equations (4.117) are equivalent to

$$t_{ij} = W_{ij}(\psi_1) + \frac{1}{k} Z_{ij}(\psi_2), \tag{4.118}$$

where the Cartesian components of the tensor operators W_{ij} and Z_{ij} are defined by

$$W_{ij}(\psi) \equiv iM_i N_j \psi + iM_j N_i \psi, \qquad Z_{ij}(\psi) \equiv \varepsilon_{imn} \partial_m W_{nj}(\psi), \tag{4.119}$$

with

$$\mathbf{M} \equiv -i\mathbf{e}_z \times \nabla, \qquad \mathbf{N} \equiv i\nabla \times \mathbf{M}. \tag{4.120}$$

For any well-behaved function ψ, $W_{ij}(\psi)$ and $Z_{ij}(\psi)$ are symmetric, traceless, divergenceless tensor fields and $\varepsilon_{imn}\partial_m Z_{nj}(\psi) = -W_{ij}(\nabla^2\psi)$.

As shown in Section 3.7, the Einstein field equations linearized about the Minkowski metric imply that the "electric" and "magnetic" parts, E_{ij} and B_{ij}, of the curvature to first order in the metric perturbation (3.176) are divergenceless and satisfy the wave equation [see (3.177) and (3.178)]. Therefore, if E_{ij} and B_{ij} have a harmonic time dependence with frequency ω, they are divergenceless solutions of the spin-2 Helmholtz equation with $k = \omega/c$. Hence, there exist solutions to the scalar Helmholtz equation, ψ_1 and ψ_2, such that

$$\begin{aligned} E_{ij} &= \mathrm{Re}\left[\left(W_{ij}(\psi_1) + \frac{1}{k} Z_{ij}(\psi_2)\right) e^{-i\omega t}\right] \\ &= \frac{1}{c}\partial_t W_{ij}(\chi_M) - Z_{ij}(\chi_E), \end{aligned} \tag{4.121}$$

where $\chi_M = \mathrm{Re}\,(i/k)\psi_1 e^{-i\omega t}$ and $\chi_E = -\mathrm{Re}\,(1/k)\psi_2 e^{-i\omega t}$ are solutions of the scalar wave equation. Then, from (3.178) it follows that

$$B_{ij} = -\frac{1}{c}\partial_t W_{ij}(\chi_E) - Z_{ij}(\chi_M). \tag{4.122}$$

The tensor fields (4.121) and (4.122) are the curvature perturbations produced by the metric perturbations

$$\begin{aligned} h_{00} &= -2\left(\partial_z^2 + \frac{1}{c^2}\partial_t^2\right)\chi_{\rm E}, \\ h_{0j} &= -4\delta_{3j}\partial_z\frac{1}{c}\partial_t\chi_{\rm E} + 2{\rm i}M_j\partial_z\chi_{\rm M}, \\ h_{jk} &= -2\delta_{jk}\left(\partial_z^2 - \frac{1}{c^2}\partial_t^2\right)\chi_{\rm E} - 4\delta_{3j}\delta_{3k}\frac{1}{c^2}\partial_t^2\chi_{\rm E} + 4{\rm i}\delta_{3(j}M_{k)}\frac{1}{c}\partial_t\chi_{\rm M}, \end{aligned} \tag{4.123}$$

where M_i are the Cartesian components of the operator $\mathbf{M}$ [(4.120)], or by any metric perturbation obtained from (4.123) by means of the gauge transformations $h_{\alpha\beta} \mapsto h_{\alpha\beta} - \partial_\alpha\xi_\beta - \partial_\beta\xi_\alpha$ [*cf.* (3.181)]. The spin-weighted components of the complex traceless symmetric tensor field $F_{ij} \equiv E_{ij} + {\rm i}B_{ij}$ are given by

$$\begin{aligned} F_{+2} &= -\frac{1}{2}\left(\frac{1}{c}\partial_t + \partial_z\right)^2 \eth\eth\chi, \\ F_{+1} &= \frac{1}{2}\left(\frac{1}{c}\partial_t + \partial_z\right)\bar{\eth}\eth\eth\chi, \\ F_0 &= -\frac{1}{2}\bar{\eth}\bar{\eth}\eth\eth\chi, \\ F_{-1} &= \frac{1}{2}\left(\frac{1}{c}\partial_t - \partial_z\right)\eth\bar{\eth}\bar{\eth}\chi, \\ F_{-2} &= -\frac{1}{2}\left(\frac{1}{c}\partial_t - \partial_z\right)^2 \bar{\eth}\bar{\eth}\chi, \end{aligned}$$

where $\chi \equiv \chi_{\rm E} + {\rm i}\chi_{\rm M}$ is a solution of the wave equation.

4.4 Parabolic and elliptic coordinates

The method of separation of variables is one of the most useful techniques employed in the solution of partial differential equations; however, the partial differential equations governing vector, tensor, or spinor fields written in noncartesian coordinates usually correspond to systems of partial differential equations that cannot be solved by separation of variables in a straightforward way.

In the case of spherical and circular cylindrical coordinates, the use of spin-weighted quantities and of the corresponding raising and lowering operators allows one to reduce nonscalar partial differential equations to sets of ordinary differential equations, by expressing the fields in terms of spin-weighted harmonics.

In this section we extend the main results of foregoing sections, which deal with circular cylindrical coordinates only, to the parabolic cylindrical and elliptic cylindrical coordinates. Following Section 6.3, the spin weight and the raising and lowering operators are defined for any system of orthogonal cylindrical coordinates; the usual vector operators are expressed in terms of spin-weighted quantities and the spin-weighted harmonics are defined. In Section 4.5, the eigenfunctions of the curl operator, the divergenceless vector fields, the solution of the vector Helmholtz equation and of the Dirac equation in parabolic cylindrical and elliptic cylindrical coordinates are expressed in terms of the corresponding spin-weighted harmonics.

We shall consider cylindrical coordinates (u, v, z), where

$$u = u(x, y), \qquad v = v(x, y)$$

and (x, y, z) are Cartesian coordinates. We shall further assume that (u, v, z) are orthogonal coordinates and that the induced orthonormal basis $\{\mathbf{e}_u, \mathbf{e}_v, \mathbf{e}_z\}$ is right-handed. A quantity η has spin weight s if under the rotation through an angle θ about $\mathbf{e}_z$, given by

$$\mathbf{e}_u + \mathrm{i}\mathbf{e}_v \mapsto \mathrm{e}^{\mathrm{i}\theta}(\mathbf{e}_u + \mathrm{i}\mathbf{e}_v),$$

it transforms according to

$$\eta \mapsto \mathrm{e}^{\mathrm{i}s\theta}\eta.$$

If η has spin weight s then its complex conjugate $\overline{\eta}$ has spin weight $-s$. For an arbitrary vector field $\mathbf{F}$, the scalar fields

$$F_0 \equiv -\frac{1}{\sqrt{2}}\mathbf{F}\cdot\mathbf{e}_z, \qquad F_{\pm 1} \equiv \pm\frac{1}{\sqrt{2}}\mathbf{F}\cdot(\mathbf{e}_u \pm \mathrm{i}\mathbf{e}_v) \tag{4.124}$$

have spin weight 0 and ± 1, and we have

$$\mathbf{F} = -\sqrt{2}\,F_0\,\mathbf{e}_z - \frac{1}{\sqrt{2}}F_{-1}(\mathbf{e}_\rho + \mathrm{i}\mathbf{e}_\phi) + \frac{1}{\sqrt{2}}F_{+1}(\mathbf{e}_\rho - \mathrm{i}\mathbf{e}_\phi). \tag{4.125}$$

For a quantity η with spin weight s, we define

$$\begin{aligned} \eth\eta &\equiv -\left(\frac{1}{h_1}\partial_u + \frac{\mathrm{i}}{h_2}\partial_v\right)\eta + \frac{s}{h_1 h_2}(h_{2,u} + \mathrm{i}h_{1,v})\eta, \\ \overline{\eth}\eta &\equiv -\left(\frac{1}{h_1}\partial_u - \frac{\mathrm{i}}{h_2}\partial_v\right)\eta - \frac{s}{h_1 h_2}(h_{2,u} - \mathrm{i}h_{1,v})\eta, \end{aligned} \tag{4.126}$$

where h_1, h_2 are the scale factors corresponding to the coordinates u and v, respectively (*i.e.*, $\mathrm{d}x^2 + \mathrm{d}y^2 = h_1^2\mathrm{d}u^2 + h_2^2\mathrm{d}v^2$), and the comma indicates partial differentiation. Then, $\eth\eta$ and $\overline{\eth}\eta$ have spin weight $s+1$ and $s-1$, respectively [see (6.53)]. Using the definitions (4.126) one finds that if η has spin weight s, then $\bar{\eth}\eth\eta = \eth\bar{\eth}\eta$ and

$$\begin{aligned}
&\bar{\eth}\eth\eta \\
&= \frac{1}{h_1 h_2}\left[\partial_u\left(\frac{h_2}{h_1}\partial_u\eta\right) + \partial_v\left(\frac{h_1}{h_2}\partial_v\eta\right)\right] - \frac{2\mathrm{i}s}{h_1 h_2}\left(\frac{h_{1,v}}{h_1}\partial_u\eta - \frac{h_{2,u}}{h_2}\partial_v\eta\right) \\
&\quad + s\left[\left(\frac{1}{h_1}\partial_u + \frac{\mathrm{i}}{h_2}\partial_v\right)\frac{h_{2,u} - \mathrm{i}h_{1,v}}{h_1 h_2} - \frac{s-1}{(h_1 h_2)^2}(h_{2,u}^2 + h_{1,v}^2)\right]\eta. \qquad (4.127)
\end{aligned}$$

Similarly, one finds that the gradient of a function f with spin weight 0 is given by

$$\nabla f = (\partial_z f)\mathbf{e}_z - \tfrac{1}{2}(\bar{\eth} f)(\mathbf{e}_u + \mathrm{i}\mathbf{e}_v) - \tfrac{1}{2}(\eth f)(\mathbf{e}_u - \mathrm{i}\mathbf{e}_v), \qquad (4.128)$$

and the divergence and the curl of a vector field $\mathbf{F}$ can be expressed as

$$\nabla \cdot \mathbf{F} = -\sqrt{2}\,\partial_z F_0 + \frac{1}{\sqrt{2}}\eth F_{-1} - \frac{1}{\sqrt{2}}\bar{\eth} F_{+1}, \qquad (4.129)$$

and

$$\begin{aligned}
\nabla \times \mathbf{F} &= \frac{\mathrm{i}}{\sqrt{2}}(\eth F_{-1} + \bar{\eth} F_{+1})\mathbf{e}_z + \frac{\mathrm{i}}{\sqrt{2}}(\partial_z F_{-1} + \bar{\eth} F_0)(\mathbf{e}_u + \mathrm{i}\mathbf{e}_v) \\
&\quad + \frac{\mathrm{i}}{\sqrt{2}}(\partial_z F_{+1} - \eth F_0)(\mathbf{e}_u - \mathrm{i}\mathbf{e}_v), \qquad (4.130)
\end{aligned}$$

in terms of the spin-weighted components F_s defined by (4.124). (Note that (4.128)–(4.130) hold for all the orthogonal cylindrical coordinate systems.)

From (4.128) and (4.129) and the commutativity of $\eth$ and $\bar{\eth}$ it follows that the Laplacian of a function of spin weight 0 is given by

$$\nabla^2 f = \bar{\eth}\eth f + \partial_z{}^2 f. \qquad (4.131)$$

Using the identity $\nabla \times (\nabla \times \mathbf{F}) = \nabla(\nabla \cdot \mathbf{F}) - \nabla^2\mathbf{F}$ and (4.128)–(4.130), one finds that

$$\begin{aligned}
\nabla^2\mathbf{F} &= -\sqrt{2}(\bar{\eth}\eth F_0 + \partial_z{}^2 F_0)\mathbf{e}_z - \frac{1}{\sqrt{2}}(\bar{\eth}\eth F_{-1} + \partial_z{}^2 F_{-1})(\mathbf{e}_u + \mathrm{i}\mathbf{e}_v) \\
&\quad + \frac{1}{\sqrt{2}}(\bar{\eth}\eth F_{+1} + \partial_z{}^2 F_{+1})(\mathbf{e}_u - \mathrm{i}\mathbf{e}_v). \qquad (4.132)
\end{aligned}$$

Let ${}_sF_\alpha$ be a function of u and v with spin weight s such that

$$\bar{\eth}\eth\,({}_sF_\alpha) = -\alpha^2\,{}_sF_\alpha, \qquad (4.133)$$

where α is a (real or complex) constant. Since $\eth$ and $\bar{\eth}$ commute, we can normalize the functions ${}_sF_\alpha$ in such a way that, for $\alpha \neq 0$,

$$\begin{aligned}
\eth_s F_\alpha &= \alpha\,{}_{s+1}F_\alpha, \\
\bar{\eth}_s F_\alpha &= -\alpha\,{}_{s-1}F_\alpha
\end{aligned} \qquad (4.134)$$

[*cf.* (4.16)]. (The solution of (4.133) is not unique; as we shall show below, the solutions of (4.133) can be characterized by an additional label, λ, which takes values in a discrete set. Furthermore, for given values of s, α, and λ, with real α, there is only one linearly independent bounded solution of (4.133).)

The simplest case of (4.133) corresponds to $s = 0$ [see (4.127)], in which case (4.133) reduces to the two-dimensional Helmholtz equation [see (4.127) and (4.131)],

$$\frac{1}{h_1 h_2}\left[\partial_u\left(\frac{h_2}{h_1}\partial_u({}_0F_\alpha)\right) + \partial_v\left(\frac{h_1}{h_2}\partial_v({}_0F_\alpha)\right)\right] + \alpha^2\,{}_0F_\alpha = 0,$$

which admits separable solutions in Cartesian, polar, parabolic and elliptic coordinates (see, *e.g.*, Miller 1977). Since (4.133) has been solved in polar coordinates in Section 4.1, in what follows we shall restrict ourselves to parabolic and elliptic coordinates, which are defined by

$$x = uv, \qquad y = \tfrac{1}{2}(v^2 - u^2) \tag{4.135}$$

and

$$x = a\cosh u \cos v, \qquad y = a \sinh u \sin v, \tag{4.136}$$

where a is a constant scale factor, respectively.

A straightforward computation shows that the coordinate transformations (4.135) and (4.136) satisfy the Cauchy–Riemann conditions

$$\partial_u x = \partial_v y, \qquad \partial_v x = -\partial_u y.$$

Therefore, the scale factors h_1 and h_2 coincide,

$$h_1 = h_2 = \sqrt{(\partial_u x)^2 + (\partial_v x)^2} \equiv h, \tag{4.137}$$

and expressions (4.126) and (4.127) reduce to

$$\begin{aligned} \eth\eta &= -h^{s-1}(\partial_u + \mathrm{i}\partial_v)(h^{-s}\eta), \\ \bar{\eth}\eta &= -h^{-s-1}(\partial_u - \mathrm{i}\partial_v)(h^{s}\eta), \end{aligned} \tag{4.138}$$

and

$$\begin{aligned} \bar{\eth}\eth\eta = {} & \frac{1}{h^2}\left(\partial_u^2\eta + \partial_v^2\eta\right) - \frac{2\mathrm{i}s}{h^3}(h_{,v}\partial_u\eta - h_{,u}\partial_v\eta) \\ & + s\left[\frac{1}{h^3}(h_{,uu} + h_{,vv}) - \frac{s+1}{h^4}(h_{,u}^2 + h_{,v}^2)\right]\eta, \end{aligned} \tag{4.139}$$

respectively. We shall consider now the solutions of (4.133) in parabolic and elliptic coordinates separately.

4.4.1 Spin-weighted parabolic harmonics

The scale factor h for the parabolic coordinates defined by (4.135) is given by

$$h = \sqrt{u^2 + v^2} \tag{4.140}$$

[(4.137)], therefore, using (4.139) and (4.140) one finds that (4.133) amounts to

$$\left[\frac{1}{u^2+v^2}(\partial_u^{\,2} + \partial_v^{\,2}) - \frac{2\mathrm{i}s}{(u^2+v^2)^2}(v\partial_u - u\partial_v) - \frac{s^2}{(u^2+v^2)^2} + \alpha^2\right] {}_sF_\alpha(u, v) = 0. \tag{4.141}$$

This last equation admits separable solutions only if $s = 0$, in which case it has the separable solutions

$${}_0F_{\alpha\lambda}(u, v) \equiv U_{\alpha\lambda}(u) V_{\alpha\lambda}(v),$$

where

$$\frac{\mathrm{d}^2 U_{\alpha\lambda}}{\mathrm{d}u^2} + (\alpha^2 u^2 - \lambda^2) U_{\alpha\lambda} = 0, \qquad \frac{\mathrm{d}^2 V_{\alpha\lambda}}{\mathrm{d}v^2} + (\alpha^2 v^2 + \lambda^2) V_{\alpha\lambda} = 0, \tag{4.142}$$

and λ is a separation constant. Hence, if $\alpha \neq 0$, $U_{\alpha\lambda}$ and $V_{\alpha\lambda}$ can be expressed in terms of the parabolic cylinder functions (Weber functions) (see, *e.g.*, Morse and Feshbach 1953, Lebedev 1965, Miller 1977).

By virtue of (4.134), we can obtain the functions ${}_sF_{\alpha\lambda}$, for integral values of s and $\alpha \neq 0$, in terms of ${}_0F_{\alpha\lambda}$. In fact, using (4.134) and (4.138) one finds that

$${}_sF_{\alpha\lambda} = \begin{cases} \left(\dfrac{1}{\alpha}\right)^s \eth^s\, {}_0F_{\alpha\lambda} = \left(-\dfrac{h}{\alpha}\right)^s \left[\dfrac{1}{h^2}(\partial_u + \mathrm{i}\partial_v)\right]^s {}_0F_{\alpha\lambda}, & s \geqslant 0, \\ \left(-\dfrac{1}{\alpha}\right)^{-s} \bar{\eth}^{-s}\, {}_0F_{\alpha\lambda} = \left(\dfrac{h}{\alpha}\right)^{-s} \left[\dfrac{1}{h^2}(\partial_u - \mathrm{i}\partial_v)\right]^{-s} {}_0F_{\alpha\lambda}, & s \leqslant 0. \end{cases} \tag{4.143}$$

Since the functions ${}_{\pm\frac{1}{2}}F_{\alpha\lambda}$ appear in the solution of the Dirac equation (see Section 4.5.3), one can obtain these functions from the solutions to the Dirac equation given in Villalba (1990); in this manner we get

$${}_{\pm\frac{1}{2}}F_{\alpha\lambda} = (1 \mp \mathrm{i}) h^{-1/2} (\sqrt{h+u} \mp \mathrm{i}\sqrt{h-u}) \big(\widetilde{U}(u) V(v) \mp U(u) \widetilde{V}(v)\big), \tag{4.144}$$

where

$$\begin{aligned} &\frac{\mathrm{d}U}{\mathrm{d}u} + \mathrm{i}\alpha u U = \lambda \widetilde{U}, && \frac{\mathrm{d}\widetilde{U}}{\mathrm{d}u} - \mathrm{i}\alpha u \widetilde{U} = \lambda U, \\ &\frac{\mathrm{d}V}{\mathrm{d}v} + \mathrm{i}\alpha v V = \mathrm{i}\lambda \widetilde{V}, && \frac{\mathrm{d}\widetilde{V}}{\mathrm{d}v} - \mathrm{i}\alpha v \widetilde{V} = \mathrm{i}\lambda V, \end{aligned} \tag{4.145}$$

and λ is a separation constant. Combining (4.145) one obtains the parabolic cylinder equations

$$\frac{\mathrm{d}^2U}{\mathrm{d}u^2} + (\alpha^2u^2 + \mathrm{i}\alpha - \lambda^2)U = 0,$$
$$\frac{\mathrm{d}^2\widetilde{U}}{\mathrm{d}u^2} + (\alpha^2u^2 - \mathrm{i}\alpha - \lambda^2)\widetilde{U} = 0,$$
$$\frac{\mathrm{d}^2V}{\mathrm{d}v^2} + (\alpha^2v^2 + \mathrm{i}\alpha + \lambda^2)V = 0,$$
$$\frac{\mathrm{d}^2\widetilde{V}}{\mathrm{d}v^2} + (\alpha^2v^2 - \mathrm{i}\alpha + \lambda^2)\widetilde{V} = 0,$$

[*cf.* (4.142)]. Using (4.134) and (4.144) one can find ${}_sF_{\alpha\lambda}$ for half-integral values of s and $\alpha \neq 0$.

Finally, using the fact that

$$(\partial_u^2 + \partial_v^2)\ln h = 0, \tag{4.146}$$

it can be verified that the most general solution of (4.133) with $\alpha = 0$ is given by

$${}_sF_0 = h^s f(u + \mathrm{i}v) + h^{-s} g(u - \mathrm{i}v), \tag{4.147}$$

where f and g are arbitrary (differentiable) functions. As in the case of the circular cylindrical coordinates, in some applications the boundary conditions exclude the spin-weighted harmonics with $\alpha = 0$ (note that the functions (4.147) either diverge at the origin or at infinity, or do not vanish at infinity (unless, of course, they are identically zero)).

4.4.2 Spin-weighted elliptic harmonics

In the case of the elliptic coordinates defined by (4.136), the scale factor (4.137) is

$$h = a\sqrt{\sinh^2 u + \sin^2 v} = a\sqrt{\cosh^2 u - \cos^2 v}, \tag{4.148}$$

and, using (4.139) one finds that (4.133) takes the explicit form

$$\begin{aligned}\Bigg[&\frac{1}{\sinh^2 u + \sin^2 v}(\partial_u^2 + \partial_v^2) \\ &- \frac{2\mathrm{i}s}{(\sinh^2 u + \sin^2 v)^2}(\sin v \cos v\, \partial_u - \sinh u \cosh u\, \partial_v) \\ &- \frac{s^2(\cosh^2 u - \sin^2 v)}{(\sinh^2 u + \sin^2 v)^2} + a^2\alpha^2\Bigg]{}_sF_\alpha(u, v) = 0.\end{aligned} \tag{4.149}$$

This partial differential equation admits separable solutions only if $s = 0$. Substituting

$$ {}_0F_{\alpha\lambda}(u, v) \equiv U_{\alpha\lambda}(u)V_{\alpha\lambda}(v) \tag{4.150}$$

into (4.149) with $s = 0$ one finds that

$$\frac{d^2U_{\alpha\lambda}}{du^2} + (a^2\alpha^2 \sinh^2 u + \lambda^2)U_{\alpha\lambda} = 0, \qquad \frac{d^2V_{\alpha\lambda}}{dv^2} + (a^2\alpha^2 \sin^2 v - \lambda^2)V_{\alpha\lambda} = 0, \tag{4.151}$$

where λ is a separation constant. The solutions of (4.151) are linear combinations of Mathieu functions (see, *e.g.*, Morse and Feshbach 1953, Miller 1977).

As in the preceding case, the functions ${}_sF_{\alpha\lambda}$, for integral values of s and $\alpha \neq 0$, are given by (4.143) with h and ${}_0F_{\alpha\lambda}$ given by (4.148)–(4.150) and (4.151), respectively. Using the results of Villalba (1990) one finds that

$$\begin{aligned} {}_{\pm\frac{1}{2}}F_{\alpha\lambda} &= \mp iah^{-3/2}\sqrt{\cosh u + \cos v}\,(\sqrt{h + a\sin v} \mp i\sqrt{h - a\sin v}) \\ &\quad \times \left(\widetilde{V}(v)U(u) \mp V(v)\widetilde{U}(u)\right), \end{aligned} \tag{4.152}$$

where

$$\begin{aligned} \frac{dU}{du} + ia\alpha \sinh u\, U &= i\lambda\widetilde{U}, & \frac{d\widetilde{U}}{du} - ia\alpha \sinh u\, \widetilde{U} &= i\lambda U, \\ \frac{dV}{dv} - ia\alpha \sin v\, V &= -\lambda\widetilde{V}, & \frac{d\widetilde{V}}{dv} + ia\alpha \sin v\, \widetilde{V} &= -\lambda V, \end{aligned} \tag{4.153}$$

and λ is a separation constant. By combining the first-order differential equations (4.153) one gets

$$\begin{aligned} \frac{d^2U}{du^2} + (a^2\alpha^2 \sinh^2 u + ia\alpha \cosh u + \lambda^2)U &= 0, \\ \frac{d^2\widetilde{U}}{du^2} + (a^2\alpha^2 \sinh^2 u - ia\alpha \cosh u + \lambda^2)\widetilde{U} &= 0, \\ \frac{d^2V}{dv^2} + (a^2\alpha^2 \sin^2 v - ia\alpha \cos v - \lambda^2)V &= 0, \\ \frac{d^2\widetilde{V}}{dv^2} + (a^2\alpha^2 \sin^2 v + ia\alpha \cos v - \lambda^2)\widetilde{V} &= 0, \end{aligned}$$

which are Whittaker–Hill equations (see Villalba 1990 and the references cited therein). Then, the functions ${}_sF_{\alpha\lambda}$, with half-integral values of s and $\alpha \neq 0$, can be obtained from (4.134) and (4.152).

Since the scale factor (4.148) also satisfies (4.146), the most general solution of (4.133) with $\alpha = 0$ is also given by (4.147).

By contrast with the spin-weighted harmonics in spherical and circular cylindrical coordinates, in the cases of parabolic cylindrical and elliptic cylindrical coordinates, the functions ${}_sF_{\alpha\lambda}(u, v)$ with $s \neq 0$, are not separable; however, one can find the solutions of (4.141) and (4.149) for $s = \pm 1/2, \pm 1, \ldots$, by means of (4.134).

4.5 Applications

In this section we give some examples of the usefulness of the spin-weighted functions ${}_sF_{\alpha\lambda}(u, v)$ in the solution of nonscalar equations in parabolic cylindrical and elliptic cylindrical coordinates.

4.5.1 Solution of the vector Helmholtz equation

According to (4.125) and (4.132), the vector Helmholtz equation, $\nabla^2\mathbf{F} + k^2\mathbf{F} = 0$, in circular, parabolic, or elliptic cylindrical coordinates, is equivalent to the three uncoupled equations

$$\bar{\eth}\eth F_s + \partial_z{}^2 F_s + k^2 F_s = 0, \qquad s = 0, \pm 1, \tag{4.154}$$

which admit solutions of the form

$$F_s = {}_sF_{\alpha\lambda}(u, v) g_s(z), \tag{4.155}$$

where the $g_s(z)$ are functions of z that, owing only to (4.133) and (4.154) and (4.155), satisfy the differential equations

$$\frac{d^2 g_s}{dz^2} + (k^2 - \alpha^2) g_s = 0, \qquad s = 0, \pm 1.$$

Following the steps given in Section 4.3.1, one finds that any divergenceless solution of the vector Helmholtz equation can be written in the form

$$\mathbf{F} = \mathbf{e}_z \times \nabla\psi_1 + \frac{1}{k}\nabla \times (\mathbf{e}_z \times \nabla\psi_2), \tag{4.156}$$

where ψ_1 and ψ_2 are solutions of the scalar Helmholtz equation, which coincides with (4.53).

As in the case of the circular cylindrical coordinates, from (4.156) it follows that the eigenfunctions of the curl operator with eigenvalue $\lambda \neq 0$ can be expressed in the form

$$\mathbf{u} = \lambda\, \mathbf{e}_z \times \nabla\psi + \nabla \times (\mathbf{e}_z \times \nabla\psi), \tag{4.157}$$

with $\nabla^2\psi + \lambda^2\psi = 0$.

4.5.2 Divergenceless vector fields

Let $\mathbf{F}$ be a vector field with vanishing divergence; then its spin-weighted components satisfy

$$\partial_z F_0 - \tfrac{1}{2}\eth F_{-1} + \tfrac{1}{2}\bar{\eth} F_{+1} = 0 \qquad (4.158)$$

[(4.129)]. Assuming that any function with spin weight s can be expanded in terms of the ${}_sF_{\alpha\lambda}(u, v)$, with $\alpha \neq 0$, we can write

$$F_s = \int d\alpha \sum_{\lambda} g_s(\alpha, \lambda, z)\, {}_sF_{\alpha\lambda}(u, v), \qquad s = 0, \pm 1. \qquad (4.159)$$

Substitution of (4.159) into (4.158), making use of (4.134), yields

$$\tfrac{1}{2}\big(g_1(\alpha, \lambda, z) + g_{-1}(\alpha, \lambda, z)\big) = \frac{1}{\alpha}\partial_z g_0(\alpha, \lambda, z), \qquad (4.160)$$

hence, using (4.133), (4.134) and (4.160), from (4.159) one finds that

$$\begin{aligned} F_{+1} &= -\frac{i}{\sqrt{2}}\eth\psi_1 + \frac{1}{\sqrt{2}}\partial_z\eth\psi_2, \\ F_0 &= -\frac{1}{\sqrt{2}}\bar{\eth}\eth\psi_2, \\ F_{-1} &= -\frac{i}{\sqrt{2}}\bar{\eth}\psi_1 - \frac{1}{\sqrt{2}}\partial_z\bar{\eth}\psi_2, \end{aligned} \qquad (4.161)$$

where

$$\begin{aligned} \psi_1 &\equiv \int d\alpha \sum_{\lambda} \frac{i}{\sqrt{2}\,\alpha}\big(g_1(\alpha, \lambda, z) - g_{-1}(\alpha, \lambda, z)\big)\, {}_0F_{\alpha\lambda}(u, v), \\ \psi_2 &\equiv \int d\alpha \sum_{\lambda} \frac{\sqrt{2}}{\alpha^2} g_0(\alpha, \lambda, z)\, {}_0F_{\alpha\lambda}(u, v). \end{aligned}$$

Owing to (4.128) and (4.130), (4.161) are equivalent to

$$\mathbf{F} = \mathbf{e}_z \times \nabla\psi_1 + \nabla \times (\mathbf{e}_z \times \nabla\psi_2). \qquad (4.162)$$

Thus, any divergenceless vector field can be expressed in the form (4.162), where ψ_1 and ψ_2 are two scalar functions. (Note that (4.156) is a special case of (4.162).)

4.5.3 Solution of the Dirac equation

Following the same steps as in Section 4.3.4 or by means of (6.136) and (6.50)–(6.52), one finds that the Dirac equation written in terms of spin-weighted quantities is given by

$$
\begin{aligned}
\frac{1}{c}\partial_t u_- &= -\partial_z v_- + \bar{\eth} v_+ - \frac{\mathrm{i}Mc}{\hbar} u_-, \\
\frac{1}{c}\partial_t u_+ &= \eth v_- + \partial_z v_+ - \frac{\mathrm{i}Mc}{\hbar} u_+, \\
\frac{1}{c}\partial_t v_- &= -\partial_z u_- + \bar{\eth} u_+ + \frac{\mathrm{i}Mc}{\hbar} v_-, \\
\frac{1}{c}\partial_t v_+ &= \eth u_- + \partial_z u_+ + \frac{\mathrm{i}Mc}{\hbar} v_+,
\end{aligned}
\tag{4.163}
$$

where $u_\pm$, $v_\pm$ are the components of the Dirac spinor with respect to the spin basis induced by the coordinates (u, v, z); u_- and v_- have spin weight $-1/2$, while u_+ and v_+ have spin weight $1/2$. (Alternative derivations of (4.163) are given in Ley-Koo and Wang 1988, Villalba 1990.) Equations (4.163) admit solutions of the form

$$
\begin{aligned}
u_- &= {}_{-\frac{1}{2}}F_{\alpha\lambda}(u, v)\, g(z) \mathrm{e}^{-\mathrm{i}Et/\hbar}, \\
u_+ &= {}_{\frac{1}{2}}F_{\alpha\lambda}(u, v)\, G(z) \mathrm{e}^{-\mathrm{i}Et/\hbar}, \\
v_- &= {}_{-\frac{1}{2}}F_{\alpha\lambda}(u, v)\, f(z) \mathrm{e}^{-\mathrm{i}Et/\hbar}, \\
v_+ &= {}_{\frac{1}{2}}F_{\alpha\lambda}(u, v)\, F(z) \mathrm{e}^{-\mathrm{i}Et/\hbar}.
\end{aligned}
\tag{4.164}
$$

Substituting (4.164) into (4.163) one obtains the set of equations (4.99). Thus, (4.163) admits solutions of the form

$$
\begin{pmatrix} u_- \\ u_+ \\ v_- \\ v_+ \end{pmatrix} = \begin{pmatrix} A(z)X_{\alpha\lambda} \\ \mathrm{i}C(z)X_{-\alpha\lambda} \end{pmatrix} \mathrm{e}^{-\mathrm{i}Et/\hbar} + \begin{pmatrix} B(z)X_{-\alpha\lambda} \\ \mathrm{i}D(z)X_{\alpha\lambda} \end{pmatrix} \mathrm{e}^{-\mathrm{i}Et/\hbar}, \tag{4.165}
$$

where

$$
X_{\alpha\lambda} \equiv \begin{pmatrix} {}_{-\frac{1}{2}}F_{\alpha\lambda} \\ {}_{\frac{1}{2}}F_{\alpha\lambda} \end{pmatrix}, \qquad X_{-\alpha\lambda} \equiv \begin{pmatrix} -({}_{-\frac{1}{2}}F_{\alpha\lambda}) \\ {}_{\frac{1}{2}}F_{\alpha\lambda} \end{pmatrix}, \qquad (\alpha \neq 0). \tag{4.166}
$$

(The bounded solutions with $\alpha = 0$ correspond to plane waves traveling along the z-axis.)

Using the fact that

$$
QX_{\pm\alpha\lambda} = \pm\alpha X_{\pm\alpha\lambda},
$$

where

$$
Q \equiv \begin{pmatrix} 0 & -\bar{\eth} \\ \eth & 0 \end{pmatrix},
$$

one finds that each term on the right-hand side of (4.165) is an eigenfunction of the operator

$$K \equiv \begin{pmatrix} -Q & 0 \\ 0 & Q \end{pmatrix},$$

with eigenvalue $-\alpha$ and α, respectively [*cf.* (4.108)].

Since the equations for fields of any spin, written in terms of spin-weighted quantities and the operators $\eth$ and $\bar{\eth}$, have the same form in circular cylindrical coordinates as in parabolic cylindrical and elliptic cylindrical coordinates, the solutions to such equations, given in terms of the spin-weighted harmonics and the operators $\eth$ and $\bar{\eth}$, have the same form in any of these coordinate systems. Thus, for instance, the divergenceless solutions of the spin-2 Helmholtz equation in parabolic cylindrical or elliptic cylindrical coordinates is given by (4.117).

5
Spinor Algebra

In this chapter a unified treatment of the algebraic properties of the spinors in three-dimensional spaces is given. In Section 5.1 it is shown that every tensor index can be replaced by a pair of spinor indices that take two values only and, using this correspondence, in Section 5.2 all the orthogonal transformations are expressed in terms of 2×2 matrices with unit determinant. In Section 5.3 the conditions satisfied by the spinor equivalent of a real tensor are obtained and it is shown that spinors can be classified according to the repetitions of their principal spinors. In Section 5.4 it is shown that, under certain conditions, a one-index spinor defines a basis for the original three-dimensional space.

5.1 The spinor equivalent of a tensor

Let V be a real vector space of dimension 3 with a metric tensor g, *i.e.*, g is a bilinear, nondegenerate, symmetric form, not necessarily positive definite. In most applications, V will be the tangent space to a three-dimensional Riemannian manifold at some point (see Chapter 6).

One can always find an orthogonal basis of V, $\{\mathbf{e}_1, \mathbf{e}_2, \mathbf{e}_3\}$, such that for $a = 1, 2, 3$, $g(\mathbf{e}_a, \mathbf{e}_a)$ is equal to $+1$ or -1, *e.g.*, making use of the Gram–Schmidt procedure, and such a basis will be referred to as an *orthonormal basis*. That is, if

$$g_{ab} \equiv g(\mathbf{e}_a, \mathbf{e}_b), \tag{5.1}$$

where, as in what follows, $a, b, \ldots = 1, 2, 3$, then (g_{ab}) is a diagonal matrix whose diagonal entries are $+1$ or -1. Thus, the metric tensor is positive definite if $(g_{ab}) = \mathrm{diag}(1, 1, 1)$ and the metric tensor is indefinite if one of the diagonal entries of (g_{ab}) is different from the other two. Without loss of generality, we can assume that $(g_{ab}) = \mathrm{diag}(1, 1, 1)$ or that only one of the entries of (g_{ab}) is equal to -1. (The two remaining possibilities are obtained by reversing the sign of (g_{ab}).)

The results of Sections 1.2 and 1.4 show that the following proposition holds.

Proposition. Let (g_{ab}) be a diagonal matrix whose diagonal entries are $+1$ or -1; then there exist scalars σ_{aAB} such that

$$\sigma_{aAB} = \sigma_{aBA}, \tag{5.2}$$

$$\sigma_{aAB}\sigma_b{}^{AB} = -2g_{ab}. \tag{5.3}$$

The spinor indices are raised and lowered by means of

$$(\varepsilon_{AB}) = \begin{pmatrix} 0 & 1 \\ -1 & 0 \end{pmatrix} = (\varepsilon^{AB}), \tag{5.4}$$

according to the convention

$$\psi_A = \varepsilon_{AB}\psi^B, \qquad \psi^B = \varepsilon^{AB}\psi_A \tag{5.5}$$

(*i.e.*, $\psi_1 = \psi^2$ and $\psi_2 = -\psi^1$), which implies that $\psi_A\phi^A = -\psi^A\phi_A$ and $\varepsilon^A{}_B = \delta^A_B$.

In effect, (1.62) and (1.105), or (1.119), give solutions to the conditions (5.2) and (5.3) in the two cases of interest, relabeling the basis vectors if necessary.

Given g_{ab}, the *connection symbols*, σ_{aAB}, are not uniquely defined by (5.2) and (5.3). If σ_a denotes the 2×2 matrix with entries

$$(\sigma_a)^A{}_B \equiv \varepsilon^{CA}\sigma_{aCB}, \tag{5.6}$$

then (5.2) and (5.3) are equivalent to

$$\operatorname{tr}\sigma_a = 0 \qquad \text{and} \qquad \operatorname{tr}\sigma_a\sigma_b = 2g_{ab}, \tag{5.7}$$

respectively. Given a set of matrices σ_a satisfying (5.7), the matrices $\sigma'_a = M\sigma_a M^{-1}$ also satisfy these conditions for *any* nonsingular 2×2 matrix M [see, *e.g.*, (1.124)]. From (5.2) and (5.3) it follows that

$$g^{ab}\sigma_{aAB}\sigma_{bCD} = -(\varepsilon_{AC}\varepsilon_{BD} + \varepsilon_{AD}\varepsilon_{BC}). \tag{5.8}$$

If $t_{ab\ldots c}$ are the components of an n-index tensor relative to the orthonormal basis $\{\mathbf{e}_1, \mathbf{e}_2, \mathbf{e}_3\}$, the components of its *spinor equivalent* are defined by

$$t_{ABCD\ldots EF} \equiv \left(\frac{1}{\sqrt{2}}\sigma^a{}_{AB}\right)\left(\frac{1}{\sqrt{2}}\sigma^b{}_{CD}\right)\cdots\left(\frac{1}{\sqrt{2}}\sigma^c{}_{EF}\right)t_{ab\ldots c}. \tag{5.9}$$

The indices a, b, ..., are lowered and raised by means of (g_{ab}) and its inverse (g^{ab}), *e.g.*, $\sigma^a{}_{CD} = g^{ab}\sigma_{bCD}$. Since the matrices (g_{ab}) and (g^{ab}) allow us to lower or raise the tensor indices, it suffices to consider tensors with all their indices down. According to (5.3), the tensor components are given in terms of the spinor components by

$$t_{ab...c} = \left(-\frac{1}{\sqrt{2}}\sigma_a{}^{AB}\right)\left(-\frac{1}{\sqrt{2}}\sigma_b{}^{CD}\right)\cdots\left(-\frac{1}{\sqrt{2}}\sigma_c{}^{EF}\right) t_{ABCD...EF}, \quad (5.10)$$

and from (5.8) and (5.10) it follows that

$$t_{...a...}\, s^{...a...} = -t_{...AB...}\, s^{...AB...}. \quad (5.11)$$

Thus, any tensor equation can be written in terms of spinor equivalents, replacing each tensor index by a pair of spinor indices and introducing a factor -1 by each contracted tensor index. It may be noticed that (5.3) and (5.8) mean that the spinor equivalent of g_{ab} is $-\frac{1}{2}(\varepsilon_{AC}\varepsilon_{BD} + \varepsilon_{AD}\varepsilon_{BC}) = -\varepsilon_{(A|C|}\varepsilon_{B)D}$, where the parenthesis denotes symmetrization on the indices enclosed and the indices between bars are excluded from the symmetrization.

As a consequence of the fact that the spinor indices take only two values, any quantity anti-symmetric in three or more indices must vanish identically. Thus, in particular, the antisymmetrization of $\varepsilon_{AB}\varepsilon_{CD}$ on any three indices is equal to zero, *e.g.*,

$$\varepsilon_{AB}\varepsilon_{CD} + \varepsilon_{BC}\varepsilon_{AD} + \varepsilon_{CA}\varepsilon_{BD} = 0. \quad (5.12)$$

This equation is equivalent to the identity

$$\psi_{...A...B...} - \psi_{...B...A...} = \psi_{...}{}^{C}{}_{...C...}\varepsilon_{AB} \quad (5.13)$$

[*cf.* (2.4)].

Owing to (5.2) the components of the spinor equivalent of a tensor are symmetric on each pair of spinor indices corresponding to a tensor index, $t_{ABCD...EF} = t_{(AB)(CD)...(EF)}$. The components $t_{ABCD...EF}$ may have additional symmetries depending on those of $t_{ab...c}$. For instance, if t_{ab} are the components of an anti-symmetric tensor (or *bivector*), $t_{ab} = -t_{ba}$, the corresponding spinor components satisfy $t_{ABCD} = -t_{CDAB}$, therefore, making use of the identity (5.13) one gets

$$\begin{aligned} t_{ABCD} &= \tfrac{1}{2}(t_{ABCD} - t_{CBAD}) + \tfrac{1}{2}(t_{ABCD} - t_{ADCB}) \\ &= \tfrac{1}{2}t^{R}{}_{BRD}\varepsilon_{AC} + \tfrac{1}{2}t_{A}{}^{R}{}_{CR}\varepsilon_{BD} \\ &= \tfrac{1}{2}t^{R}{}_{BRD}\varepsilon_{AC} + \tfrac{1}{2}t^{R}{}_{ARC}\varepsilon_{BD}. \end{aligned}$$

Letting $\tau_{BD} = \frac{1}{2}t^{R}{}_{BRD}$, we find that τ_{AB} is symmetric;

$$\tau_{BD} = \tfrac{1}{2}t^{R}{}_{BRD} = -\tfrac{1}{2}t_{RD}{}^{R}{}_{B} = \tfrac{1}{2}t^{R}{}_{DRB} = \tau_{DB}. \quad (5.14)$$

Thus, the spinor equivalent of an anti-symmetric tensor is given by a symmetric object with two spinor indices

$$t_{ABCD} = \tau_{BD}\varepsilon_{AC} + \tau_{AC}\varepsilon_{BD}. \tag{5.15}$$

In particular, the dual of a vector F_a, ${}^*F_{ab} \equiv \varepsilon_{abc}F^c$, is an anti-symmetric tensor whose spinor components are

$${}^*F_{ABCD} = -\varepsilon_{ABCDEG}F^{EG} \tag{5.16}$$

where, following the rule (5.9),

$$\varepsilon_{ABCDEG} \equiv \frac{1}{2\sqrt{2}}\sigma^a{}_{AB}\sigma^b{}_{CD}\sigma^c{}_{EG}\,\varepsilon_{abc}.$$

The antisymmetry of ε_{abc}, (5.2) and (5.12) imply that ε_{ABCDEG} must be a multiple of $\varepsilon_{AC}\varepsilon_{BE}\varepsilon_{DG} + \varepsilon_{BD}\varepsilon_{AG}\varepsilon_{CE}$, and the proportionality factor is real or pure imaginary depending on the signature of (g_{ab}). Making use of the expressions (1.62), and (1.105) or (1.119), we find that

$$\varepsilon_{ABCDEG} = \begin{cases} \dfrac{\mathrm{i}}{\sqrt{2}}(\varepsilon_{AC}\varepsilon_{BE}\varepsilon_{DG} + \varepsilon_{BD}\varepsilon_{AG}\varepsilon_{CE}) & \text{if } (g_{ab}) = \mathrm{diag}(1,1,1), \\ -\dfrac{1}{\sqrt{2}}(\varepsilon_{AC}\varepsilon_{BE}\varepsilon_{DG} + \varepsilon_{BD}\varepsilon_{AG}\varepsilon_{CE}) & \text{if } (g_{ab}) = \mathrm{diag}(1,1,-1), \end{cases} \tag{5.17}$$

hence,

$${}^*F_{ABCD} = \begin{cases} -\dfrac{\mathrm{i}}{\sqrt{2}}(F_{BD}\varepsilon_{AC} + F_{AC}\varepsilon_{BD}) & \text{if } (g_{ab}) = \mathrm{diag}(1,1,1), \\ \dfrac{1}{\sqrt{2}}(F_{BD}\varepsilon_{AC} + F_{AC}\varepsilon_{BD}) & \text{if } (g_{ab}) = \mathrm{diag}(1,1,-1), \end{cases} \tag{5.18}$$

which are of the form (5.15).

Similarly, if the t_{ab} are the components of a symmetric two-index tensor, then, in addition to the symmetries $t_{ABCD} = t_{(AB)(CD)}$, we have $t_{ABCD} = t_{CDAB}$, but not necessarily t_{ABCD} will coincide with, *e.g.*, t_{ACBD}; in fact, using (5.13) and (5.11),

$$\begin{aligned} t_{ABCD} - t_{ACBD} &= t_A{}^R{}_{RD}\varepsilon_{BC} = \frac{1}{2}(t_A{}^R{}_{RD} + t_{RDA}{}^R)\varepsilon_{BC} \\ &= \frac{1}{2}(t_A{}^R{}_{RD} - t_D{}^R{}_{RA})\varepsilon_{BC} = \frac{1}{2}t^{SR}{}_{RS}\varepsilon_{AD}\varepsilon_{BC} \\ &= -\frac{1}{2}t^a{}_a\varepsilon_{AD}\varepsilon_{BC}. \end{aligned}$$

Hence, the components t_{ABCD} are totally symmetric if and only if t_{ab} is symmetric and trace-free. In an analogous way one finds that $t_{ab...c}$ is symmetric and trace-free if and only if its spinor equivalent $t_{ABCD...EF}$ is totally symmetric (*cf.* Section 2.1).

5.2 The orthogonal and spin groups

If the basis $\{\mathbf{e}_1, \mathbf{e}_2, \mathbf{e}_3\}$ is replaced by a second orthonormal basis $\{\mathbf{e}'_1, \mathbf{e}'_2, \mathbf{e}'_3\}$ such that

$$g(\mathbf{e}'_a, \mathbf{e}'_b) = g(\mathbf{e}_a, \mathbf{e}_b), \tag{5.19}$$

then the components of an n-index tensor with respect to the new basis, $t'_{ab...c}$, are given by

$$t'_{ab...c} = L_a{}^d L_b{}^e \cdots L_c{}^f t_{de...f}, \tag{5.20}$$

where, owing to (5.1) and (5.19), $(L_a{}^b)$ is a real matrix such that

$$g_{ab} = L_a{}^c L_b{}^d g_{cd}. \tag{5.21}$$

The matrices $(L_a{}^b)$ satisfying (5.21) form the group $\mathrm{O}(p, q)$, where p and q are the numbers of positive and negative eigenvalues of the matrix (g_{ab}) or vice versa. Equation (5.21) implies that $\det(L_a{}^b) = \pm 1$. The matrices with unit determinant that satisfy (5.21) form the subgroup $\mathrm{SO}(p, q)$ of $\mathrm{O}(p, q)$.

Taking into account that there are n contracted indices in (5.20), the spinor equivalent of (5.20) is

$$t'_{ABCD...} = (-1)^n L_{AB}{}^{RS} L_{CD}{}^{TV} \cdots t_{RSTV...} \tag{5.22}$$

[see (5.11)], where

$$L_{AB}{}^{CD} \equiv \tfrac{1}{2} \sigma^a{}_{AB} \sigma_b{}^{CD} L_a{}^b. \tag{5.23}$$

Similarly, the spinor equivalent of (5.21) is

$$\varepsilon_{AC}\varepsilon_{BD} + \varepsilon_{AD}\varepsilon_{BC} = L_{AB}{}^{RS} L_{CD}{}^{TV} (\varepsilon_{RT}\varepsilon_{SV} + \varepsilon_{RV}\varepsilon_{ST}). \tag{5.24}$$

Then, using (5.23), (5.13), (5.3), (5.21), and (5.8), we obtain

$$\begin{aligned}
\varepsilon_{AC} L_{11}{}^{AB} L_{11}{}^{CD} &= \tfrac{1}{4} \sigma_b{}^{AB} \sigma_{dA}{}^D \sigma^a{}_{11} \sigma^c{}_{11} L_a{}^b L_c{}^d \\
&= \tfrac{1}{8} (\sigma_b{}^{AB} \sigma_{dA}{}^D + \sigma_d{}^{AB} \sigma_{bA}{}^D) \sigma^a{}_{11} \sigma^c{}_{11} L_a{}^b L_c{}^d \\
&= \tfrac{1}{8} (\sigma_b{}^{AB} \sigma_{dA}{}^D - \sigma_b{}^{AD} \sigma_{dA}{}^B) \sigma^a{}_{11} \sigma^c{}_{11} L_a{}^b L_c{}^d \\
&= \tfrac{1}{8} \varepsilon^{BD} \sigma_b{}^{AR} \sigma_{dAR} \sigma^a{}_{11} \sigma^c{}_{11} L_a{}^b L_c{}^d \\
&= -\tfrac{1}{4} g_{bd} \varepsilon^{BD} \sigma^a{}_{11} \sigma^c{}_{11} L_a{}^b L_c{}^d \\
&= -\tfrac{1}{4} \varepsilon^{BD} g_{ac} \sigma^a{}_{11} \sigma^c{}_{11} = 0.
\end{aligned} \tag{5.25}$$

This last equation is of the form $\varepsilon_{AC} M^{AB} M^{CD} = 0$, and since

$$\varepsilon_{AC} M^{AB} M^{CD} = \det(M^{RS})\, \varepsilon^{BD}, \tag{5.26}$$

it follows that $\det(M^{AB}) = 0$; therefore, the rows of (M^{AB}) are proportional to each other and the columns of (M^{AB}) are proportional to each other, thus M^{AB} is of the form $M^{AB} = \alpha^A \beta^B$. If (M^{AB}) is symmetric, then β^A must be proportional to α^A and, absorbing the proportionality factor into α_A, we find that $M^{AB} = \alpha^A \alpha^B$. Hence, from (5.25) it follows that

$$L_{11}{}^{AB} = \alpha^A \alpha^B, \tag{5.27}$$

for some α^A. In a similar manner, one finds that

$$L_{22}{}^{AB} = \beta^A \beta^B, \tag{5.28}$$

for some β^A. From (5.24) we have $\varepsilon_{AB} \varepsilon_{CD} L_{11}{}^{AC} L_{22}{}^{BD} = 1$, which, by virtue of (5.27) and (5.28), yields

$$(\alpha^A \beta_A)^2 = 1. \tag{5.29}$$

Then, using (5.27)–(5.29), the fact that $\alpha^A \beta^B - \alpha^B \beta^A = (\alpha^C \beta_C) \varepsilon^{AB}$, and

$$\varepsilon_{AC} \varepsilon_{BD} + \varepsilon_{AD} \varepsilon_{BC} = L^{RS}{}_{AB} L^{TV}{}_{CD} (\varepsilon_{RT} \varepsilon_{SV} + \varepsilon_{RV} \varepsilon_{ST}),$$

which is condition (5.24) applied to the inverse of (5.23), one finds that $L_{12}{}^{AB}$ is equal to $\alpha^{(A} \beta^{B)}$ or to $-\alpha^{(A} \beta^{B)}$. In the second case, replacing α^A by $-\alpha^A$, which leaves (5.27) and (5.29) unchanged, we have $L_{12}{}^{AB} = \alpha^{(A} \beta^{B)}$ and therefore we can always write $L_{12}{}^{AB}$ in the form

$$L_{12}{}^{AB} = \alpha^{(A} \beta^{B)}. \tag{5.30}$$

If $\alpha^A \beta_A = 1$, we define $U_1{}^A \equiv \alpha^A$, $U_2{}^A \equiv \beta^A$; then, (5.27), (5.28), and (5.30) are equivalent to

$$L_{CD}{}^{AB} = U_C{}^{(A} U_D{}^{B)} \tag{5.31}$$

with

$$\det(U_B{}^A) = 1, \tag{5.32}$$

while if $\alpha^A \beta_A = -1$, we make $U_1{}^A \equiv \mathrm{i}\alpha^A$, $U_2{}^A \equiv \mathrm{i}\beta^A$ and from (5.27), (5.28), and (5.30) we get

$$L_{CD}{}^{AB} = -U_C{}^{(A} U_D{}^{B)}, \tag{5.33}$$

where $(U_B{}^A)$ again satisfies (5.32).

Thus from (5.23), (5.31), and (5.33) it follows that any matrix $(L_b{}^a)$ belonging to $\mathrm{O}(p, q)$ can be expressed in the form

$$L_a{}^b = \pm \tfrac{1}{2} \sigma_a{}^{AB} \sigma^b{}_{CD} U_A{}^C U_B{}^D$$

or, equivalently,

$$L^a{}_b = \pm\tfrac{1}{2}\sigma^a{}_{AB}\sigma_b{}^{CD}U^A{}_C U^B{}_D, \tag{5.34}$$

where $(U_B{}^A)$ and, hence, $(U^A{}_B)$, has unit determinant. The determinant of the matrix $(L^a{}_b)$ given by (5.34) is equal to -1 or $+1$ according to whether one takes the positive or the negative sign on the right-hand side, respectively. It may be noted that the two matrices $(U_B{}^A)$ and $-(U_B{}^A)$ give rise to the same orthogonal matrix $(L^a{}_b)$.

Hence, the orthogonal transformations with *unit* determinant, *i.e.*, the elements of SO(p, q), can be expressed in the form

$$L^a{}_b = -\tfrac{1}{2}\sigma^a{}_{AB}\sigma_b{}^{CD}U^A{}_C U^B{}_D. \tag{5.35}$$

From the relation

$$\varepsilon_{AC}U^A{}_B U^C{}_D = \det(U^M{}_N)\varepsilon_{BD}, \tag{5.36}$$

it follows that the inverse of a matrix $(U^A{}_B)$ with unit determinant is given by

$$(U^{-1})^A{}_B = -\varepsilon_{BC}U^C{}_D\varepsilon^{DA} = -U_B{}^A, \tag{5.37}$$

thus, using (5.6), one finds that (5.35) can also be written as $L^a{}_b = \frac{1}{2}\operatorname{tr}\sigma^a U\sigma_b U^{-1}$, where $U = (U^A{}_B)$.

Making use of (5.8), (5.36), and (5.3), it can be verified that if $(U^A{}_B)$ is *any* complex matrix with unit determinant, then the matrix $(L^a{}_b)$ given by (5.34) indeed satisfies (5.21); however, $(L^a{}_b)$ may be complex. As shown below, the conditions that $(U^A{}_B)$ has to satisfy in order for $(L^a{}_b)$ to be real, depend on the choice of the connection symbols σ_{aAB}.

5.2.1 Positive definite metric

The connection symbols given by (1.62),

$$(\sigma_{1AB}) = \begin{pmatrix} 1 & 0 \\ 0 & -1 \end{pmatrix}, \quad (\sigma_{2AB}) = \begin{pmatrix} \mathrm{i} & 0 \\ 0 & \mathrm{i} \end{pmatrix}, \quad (\sigma_{3AB}) = \begin{pmatrix} 0 & -1 \\ -1 & 0 \end{pmatrix}, \tag{5.38}$$

satisfy the conditions (5.2) and (5.3) with $g_{ab} = \delta_{ab}$ and, under complex conjugation,

$$\overline{\sigma_{aAB}} = -\sigma_a{}^{AB} \tag{5.39}$$

[see (1.68)]. Therefore, assuming that $(L^a{}_b)$ is real, from (5.23) and (5.39) we obtain

$$\overline{L_{AB}{}^{CD}} = L^{AB}{}_{CD}$$

or, according to (5.31) and (5.33),

$$\overline{U_A{}^{(C}U_B{}^{D)}} = U^A{}_{(C}U^B{}_{D)}$$

which leads to $\overline{U_A{}^C} = \pm U^A{}_C$. The determinant of a matrix such that $\overline{U_A{}^C} = U^A{}_C$ cannot be equal to 1, since $\det(U^A{}_B) = U^1{}_1U^2{}_2 - U^1{}_2U^2{}_1 = \overline{U_1{}^1}U^2{}_2 - \overline{U_1{}^2}U^2{}_1 = -\overline{U^2{}_2}U^2{}_2 - \overline{U^2{}_1}U^2{}_1 < 0$, hence $(U^A{}_B)$ must satisfy the condition

$$\overline{U^A{}_C} = -U_A{}^C, \tag{5.40}$$

which, by virtue of (5.37), means that $(U^A{}_B)$ is unitary; therefore, $(U^A{}_B)$ belongs to SU(2). Thus, *all* the O(3) matrices can be expressed in the form (5.34) with $(U^A{}_B) \in$ SU(2) and (5.35) gives the well-known two-to-one mapping (in fact, homomorphism) of SU(2) onto SO(3) [(5.35) is equivalent to (1.35)].

Substituting (5.33) into (5.22) one finds that under the SO(3) transformation defined by the SU(2) matrix $(U^A{}_B)$, the spinor equivalent of a tensor transforms as

$$t'^{ABCD\dots} = U^A{}_R U^B{}_S U^C{}_T U^D{}_V \cdots t^{RSTV\dots}.$$

By definition, the components of a spinor, $\psi^{ABC\dots}$, where the number of indices can be even or odd, transform as

$$\psi'^{ABC\dots} = U^A{}_R U^B{}_S U^C{}_T \cdots \psi^{RST\dots} \tag{5.41}$$

or, equivalently [see (5.37)],

$$\psi'_{ABC\dots} = (U^{-1})^R{}_A (U^{-1})^S{}_B (U^{-1})^T{}_C \cdots \psi_{RST\dots}. \tag{5.42}$$

5.2.2 Indefinite metric

The matrices given in (1.119),

$$(\sigma_{1AB}) \equiv \begin{pmatrix} 1 & 0 \\ 0 & -1 \end{pmatrix}, \quad (\sigma_{2AB}) \equiv \begin{pmatrix} 0 & 1 \\ 1 & 0 \end{pmatrix}, \quad (\sigma_{3AB}) \equiv \begin{pmatrix} 1 & 0 \\ 0 & 1 \end{pmatrix}, \tag{5.43}$$

satisfy (5.2) and (5.3) with

$$(g_{ab}) = \operatorname{diag}(1, 1, -1). \tag{5.44}$$

Since in this case the σ_{aAB} are all real, the components of the spinor equivalent of $L^a{}_b$ are also real, hence $U_A{}^{(C}U_B{}^{D)}$ are real, which means that $(U^A{}_B)$ is real or pure imaginary; in the first case $(U^A{}_B)$ belongs to SL(2,ℝ). An explicit calculation shows that if $(U^A{}_B) \in$ SL(2, ℝ), then the SO(2,1) matrix given by (5.35) satisfies $L^3{}_3 > 0$ and that $L^3{}_3 < 0$ if $(U_B{}^A)$ is pure imaginary. Equation (5.35) establishes in this case a two-to-one homomorphism of SL(2,$\mathbb{R}$) onto $SO_0(2,1)$ — the connected component of the identity in SO(2,1).

Since SL(2,$\mathbb{R}$) is connected, the preceding results show that the group O(2,1) has four connected components and that SL(2,$\mathbb{R}$) is a double covering group of the connected component of the identity, $SO_0(2,1)$.

Alternatively, if the connection symbols are chosen as in (1.105),

$$(\sigma_{1AB}) \equiv \begin{pmatrix} \mathrm{i} & 0 \\ 0 & \mathrm{i} \end{pmatrix}, \quad (\sigma_{2AB}) \equiv \begin{pmatrix} 1 & 0 \\ 0 & -1 \end{pmatrix}, \quad (\sigma_{3AB}) \equiv \begin{pmatrix} 0 & -\mathrm{i} \\ -\mathrm{i} & 0 \end{pmatrix}, \tag{5.45}$$

then (5.2) and (5.3) are satisfied with (g_{ab}) given by (5.44) and

$$\overline{\sigma_{aAB}} = -\eta_{AR}\eta_{BS}\sigma_a{}^{RS}, \tag{5.46}$$

where

$$(\eta_{AB}) \equiv \begin{pmatrix} 1 & 0 \\ 0 & -1 \end{pmatrix} \tag{5.47}$$

[see (1.107) and (1.88)].

Proceeding as in the previous subsection, from (5.23) and (5.46) one finds that if $(L^a{}_b)$ is real, then

$$\overline{L_{AB}{}^{CD}} = \eta_{AR}\eta_{BS}\eta^{CP}\eta^{DQ}L^{RS}{}_{PQ},$$

hence

$$\eta_{AR}\eta^{CP}\overline{U^A{}_C} = \pm U_R{}^P \tag{5.48}$$

or, equivalently,

$$U^\dagger \eta U = \pm\eta, \tag{5.49}$$

where $U \equiv (U^A{}_B)$ and $\eta \equiv (\eta_{AB})$. The matrices $U = (U^A{}_B)$ that satisfy (5.48) or (5.49) with the positive sign form the group SU(1,1) (which is isomorphic to SL(2,$\mathbb{R}$), see (5.53) and Section 1.4).

Furthermore, $L^3{}_3$ is positive if U satisfies (5.49) with the positive sign and $L^3{}_3$ is negative if U satisfies (5.49) with the negative sign. Therefore, with the σ_{aAB} given by (5.45), (5.35) defines a two-to-one correspondence between SU(1,1) and $SO_0(2,1)$, which is a group homomorphism.

As pointed out in Section 5.1, the fact that the connection symbols (5.43) and (5.45) satisfy (5.2) and (5.3), with the same metric tensor (g_{ab}), implies the existence of a matrix $(M^A{}_B)$ with unit determinant, defined up to sign, such that

$$\sigma_{aAB} = M^C{}_A M^D{}_B \overset{(\mathrm{r})}{\sigma}_{aCD}, \tag{5.50}$$

where $\sigma^{(r)}_{aCD}$ and σ_{aAB} denote the connection symbols given by (5.43) and (5.45), respectively. One can verify that (5.50) is satisfied by

$$(M^A{}_B) = -\tfrac{1}{2}\begin{pmatrix} 1+\mathrm{i} & -1-\mathrm{i} \\ 1-\mathrm{i} & 1-\mathrm{i} \end{pmatrix} \tag{5.51}$$

and that

$$\overline{M^{AB}} = \mathrm{i} M^A{}_C \eta^{CB}. \tag{5.52}$$

The matrix $(M^A{}_B)$ represents a change of basis in spin space. By virtue of (5.52), if $(U^A{}_B)$ belongs to SL(2,$\mathbb{R}$) then

$$W^A{}_B = (M^{-1})^A{}_C U^C{}_D M^D{}_B = -M_C{}^A M^D{}_B U^C{}_D \tag{5.53}$$

belongs to SU(1,1).

When the metric tensor is indefinite, the components of a spinor transform according to (5.42), where $(U^A{}_B)$ is a matrix belonging to SL(2,$\mathbb{R}$) or to SU(1,1), depending on the choice of the connection symbols.

5.3 Algebraic classification

As shown in Chapter 1, we can associate to each spinor a second spinor, called its mate or conjugate. The mate of a one-index spinor ψ_A, denoted by $\widehat{\psi}_A$, is defined by

$$\widehat{\psi}_A \equiv \begin{cases} \overline{\psi^A} & \text{if the } \sigma_{aAB} \text{ are given by (5.38),} \\ \overline{\psi_A} & \text{if the } \sigma_{aAB} \text{ are given by (5.43),} \\ \mathrm{i}\eta_{AB}\overline{\psi^B} & \text{if the } \sigma_{aAB} \text{ are given by (5.45)} \end{cases} \tag{5.54}$$

or, equivalently,

$$\widehat{\psi}^A \equiv \begin{cases} -\overline{\psi_A} & \text{if the } \sigma_{aAB} \text{ are given by (5.38),} \\ \overline{\psi^A} & \text{if the } \sigma_{aAB} \text{ are given by (5.43),} \\ -\mathrm{i}\eta^{AB}\overline{\psi_B} & \text{if the } \sigma_{aAB} \text{ are given by (5.45),} \end{cases} \tag{5.55}$$

where, in accordance with the rules (5.5), $\eta^{AB} = \varepsilon^{CA}\varepsilon^{DB}\eta_{CD}$, i.e.,

$$(\eta^{AB}) = \begin{pmatrix} -1 & 0 \\ 0 & 1 \end{pmatrix}. \tag{5.56}$$

Then $\eta_{AB}\eta^{BC} = -\delta^C_A$.

In each case, one can verify that $\widehat{\psi}_A$ transforms in the same manner as ψ_A under the corresponding spin transformations (see also Chapter 1). If the metric is positive definite, from (5.54), (5.42), and (5.40) it follows that

$$\widehat{\psi}'_A = \overline{\psi'^A} = \overline{U^A{}_R \psi^R} = (U^{-1})^R{}_A \widehat{\psi}_R.$$

Similarly, if the connection symbols are given by (5.43), $(U^A{}_B)$ is real and

$$\widehat{\psi}'_A = \overline{\psi'_A} = \overline{(U^{-1})^R{}_A \psi^R} = (U^{-1})^R{}_A \widehat{\psi}_R,$$

and if the connection symbols are given by (5.45), using (5.54) and (5.48) one finds that

$$\widehat{\psi}'_A = \mathrm{i}\eta_{AB}\overline{\psi'^B} = \mathrm{i}\eta_{AB}\overline{U^B{}_S \psi^S} = \mathrm{i}\eta_{SB}(U^{-1})^B{}_A \overline{\psi^S} = (U^{-1})^B{}_A \widehat{\psi}_B.$$

One can verify, making use of (5.52), that the two definitions of the mate of a spinor given in (5.54) or (5.55) when the metric is indefinite are equivalent and they are just two expressions of the same mapping with respect to two different bases of the spin space.

From (5.54) and (5.55) it follows that

$$\widehat{\widehat{\psi}}_A = \begin{cases} -\psi_A & \text{if } (g_{ab}) = \operatorname{diag}(1,1,1), \\ \psi_A & \text{if } (g_{ab}) = \operatorname{diag}(1,1,-1). \end{cases} \tag{5.57}$$

(Hence, the map $\psi_A \mapsto \widehat{\psi}_A$ is a quaternionic structure if the metric is positive definite and a real structure if the metric is indefinite (see, *e.g.*, Friedrich 2000, Chap. 1).) Furthermore

$$\overline{\alpha^A \beta_A} = \widehat{\alpha}^A \widehat{\beta}_A, \tag{5.58}$$

in all cases. The map $\psi_A \mapsto \widehat{\psi}_A$ is antilinear and in the case where $(g_{ab}) = \operatorname{diag}(1,1,1)$ it is, except for a factor, the time reversal operation for spin-1/2 particles in quantum mechanics (see, *e.g.*, Schiff 1968, Sakurai 1994).

If $\widehat{\psi}_A = \lambda\psi_A$, then $\widehat{\widehat{\psi}}_A = \overline{\lambda}\widehat{\psi}_A = |\lambda|^2\psi_A$ and, comparing with (5.57), we see that only in the case where the metric is indefinite do there exist nontrivial solutions of

$$\widehat{\psi}_A = \lambda\psi_A \tag{5.59}$$

and necessarily $|\lambda| = 1$.

The mate of a spinor with any number of indices can be defined by requiring that the mate of the tensor product of two spinors be the tensor product of the mates of the spinors. Thus, according to (5.54) and (5.55), the mate of an m-index spinor $\psi_{AB...L}$ will be given by

$$\widehat{\psi}_{AB...L} \equiv \begin{cases} \overline{\psi^{AB...L}} & \text{if the } \sigma_{aAB} \text{ are given by (5.38)}, \\ \overline{\psi_{AB...L}} & \text{if the } \sigma_{aAB} \text{ are given by (5.43)}, \\ \mathrm{i}^m \eta_{AR}\eta_{BS}\cdots\eta_{LW}\overline{\psi^{RS...W}} & \text{if the } \sigma_{aAB} \text{ are given by (5.45)} \end{cases} \tag{5.60}$$

or,

$$\widehat{\psi}^{AB...L} \equiv \begin{cases} (-1)^m\overline{\psi_{AB...L}} & \text{if the } \sigma_{aAB} \text{ are given by (5.38)}, \\ \overline{\psi^{AB...L}} & \text{if the } \sigma_{aAB} \text{ are given by (5.43)}, \\ (-\mathrm{i})^m \eta^{AR}\eta^{BS}\cdots\eta^{LW}\overline{\psi_{RS...W}} & \text{if the } \sigma_{aAB} \text{ are given by (5.45)}, \end{cases} \tag{5.61}$$

therefore, for an m-index spinor

$$\widehat{\widehat{\psi}}_{AB...L} = \begin{cases} (-1)^m\psi_{AB...L} & \text{if } (g_{ab}) = \mathrm{diag}(1,1,1), \\ \psi_{AB...L} & \text{if } (g_{ab}) = \mathrm{diag}(1,1,-1). \end{cases} \tag{5.62}$$

It may be noticed that $\widehat{\varepsilon}_{AB} = \varepsilon_{AB}$ in all cases, which implies (5.58).

According to the definitions (5.60) and (5.61), conditions (5.40) and (5.48) can be expressed as $\widehat{U}_A{}^C = U_A{}^C$ and $\widehat{U}_R{}^P = \pm U_R{}^P$, respectively; therefore the spin transformations, which represent the orthogonal transformations belonging to the connected component of the identity, correspond in all cases to the 2×2 matrices with unit determinant such that $\widehat{U}_A{}^B = U_A{}^B$. This is equivalent to the fact that a spinor and its mate transform in the same way under the spin transformations.

Since the connection symbols can be complex, the components of the spinor equivalent of a real tensor may be complex [see (5.9)]. In the case where the metric is positive definite, with the connection symbols given by (5.38), using (5.9) and (5.39) one finds that the spinor equivalent of an n-index tensor $t_{ab...c}$ satisfies

$$\overline{t_{ABCD...EF}} = (-1)^n t^{ABCD...EF} \tag{5.63}$$

if and only if the tensor components $t_{ab...c}$ are real.

On the other hand, when the metric is indefinite and the connection symbols are real, the spinor components of a tensor are real if and only if the tensor is real. If the connection symbols are given by (5.45), then the spinor equivalent of an n-index tensor $t_{ab...c}$ satisfies the conditions

$$\overline{t_{AB...EF}} = (-1)^n \eta_{AR}\eta_{BS}\cdots\eta_{EW}\eta_{FX}t^{RS...WX} \tag{5.64}$$

if and only if the tensor components $t_{ab...c}$ are real.

Comparison of (5.61) with (5.63) and (5.64) shows that an n-index tensor $t_{ab...c}$ is real if and only if its spinor equivalent satisfies

$$\widehat{t}_{AB...EF} = \begin{cases} (-1)^n t_{AB...EF} & \text{if } (g_{ab}) = \mathrm{diag}(1, 1, 1), \\ t_{AB...EF} & \text{if } (g_{ab}) = \mathrm{diag}(1, 1, -1). \end{cases} \tag{5.65}$$

Making use of the of the mate of a spinor we can define an inner product. When the metric of V is positive definite, the expression

$$\langle \alpha, \beta \rangle \equiv \widehat{\alpha}_A \beta^A \tag{5.66}$$

gives a positive definite Hermitian inner product for the complex two-dimensional space of one-index spinors, as can be seen by noting that from (5.54) we have $\langle \alpha, \beta \rangle = \overline{\alpha^1}\beta^1 + \overline{\alpha^2}\beta^2$. This inner product can be extended in a natural manner to spinors of higher rank; if $\phi_{AB...L}$ and $\psi_{AB...L}$ are two m-index spinors we define

$$\langle \phi, \psi \rangle \equiv \widehat{\phi}_{AB...L} \psi^{AB...L}. \tag{5.67}$$

Then, if v_{AB} and w_{AB} are the spinor equivalents of two real vectors v_a and w_a, respectively, the inner product $\langle v, w \rangle$ coincides with the inner product of the vectors v_a and w_a, $\langle v, w \rangle = \widehat{v}_{AB} w^{AB} = -v_{AB} w^{AB} = v_a w^a$ [see (5.65) and (5.11)].

When the metric of V is indefinite, with the definitions (5.66) and (5.54) one finds that $\langle \alpha, \alpha \rangle$ is pure imaginary for any one-index spinor α_A; in fact, $\langle \alpha, \beta \rangle = \mathrm{i}\eta_{AB}\overline{\alpha^B}\beta^A = \mathrm{i}(\overline{\alpha^1}\beta^1 - \overline{\alpha^2}\beta^2)$, which shows that, apart from a factor i, the inner product is indefinite.

Principal spinors

As in the case of four-dimensional spaces, the fact that each spinor index can take only two values and that the spin transformations are given by unimodular matrices imply that the irreducible parts of an arbitrary spinor correspond to totally symmetric spinors and each totally symmetric n-index spinor $\phi_{AB...L}$ can be expressed as the symmetrized tensor product of n one-index spinors (Penrose 1960, Penrose and Rindler 1984, Stewart 1990)

$$\phi_{AB...L} = \alpha_{(A}\beta_B \cdots \delta_{L)}. \tag{5.68}$$

This decomposition is unique except for scale factors. The existence and uniqueness of the expression (5.68) is a consequence of the fundamental theorem of algebra. If ξ^A is an arbitrary spinor, then assuming, *e.g.*, $\xi^2 \neq 0$,

$$\begin{aligned} &\phi_{AB...L}\xi^A\xi^B \cdots \xi^L \\ &= \phi_{11...1}(\xi^1)^n + n\phi_{21...1}(\xi^1)^{n-1}\xi^2 + \cdots + \phi_{22...2}(\xi^2)^n \\ &= (\xi^2)^n\{\phi_{11...1}(\xi^1/\xi^2)^n + n\phi_{21...1}(\xi^1/\xi^2)^{n-1} + \cdots + \phi_{22...2}\}, \end{aligned}$$

hence, $(\xi^2)^{-n}\phi_{AB\ldots L}\xi^A\xi^B\cdots\xi^L$ is an nth degree polynomial in (ξ^1/ξ^2) which can be factorized as $\phi_{11\ldots1}(\xi^1/\xi^2 - z_1)(\xi^1/\xi^2 - z_2)\cdots(\xi^1/\xi^2 - z_n)$; therefore, $\phi_{AB\ldots L}\xi^A\xi^B\cdots\xi^L = \phi_{11\ldots1}(\xi^1 - z_1\xi^2)(\xi^1 - z_2\xi^2)\cdots(\xi^1 - z_n\xi^2)$, which is the product of n homogeneous first degree polynomials in ξ^A, *i.e.*,

$$\phi_{AB\ldots L}\xi^A\xi^B\cdots\xi^L = (\alpha_A\xi^A)(\beta_B\xi^B)\cdots(\delta_L\xi^L), \tag{5.69}$$

which implies (5.68). The spinors α_A, β_A, ..., δ_A, appearing in (5.68) are called principal spinors of $\phi_{AB\ldots L}$. Equation (5.69) shows that ξ_A is a principal spinor of $\phi_{AB\ldots L}$ if and only if $\phi_{AB\ldots L}\xi^A\xi^B\cdots\xi^L = 0$.

Since the tensor $t_{ab\ldots c}$ is trace-free and totally symmetric if and only if its spinor equivalent $t_{AB\ldots EF}$ is totally symmetric, according to (5.68), if $t_{ab\ldots c}$ is an n-index trace-free, totally symmetric tensor, $t_{AB\ldots EF}$ can be expressed in the form

$$t_{AB\ldots EF} = \alpha_{(A}\beta_B\cdots\gamma_E\delta_{F)}, \tag{5.70}$$

and making use of (5.65) it follows that $t_{ab\ldots c}$ is real if and only if

$$\alpha_{(A}\beta_B\cdots\gamma_E\delta_{F)} = \begin{cases} (-1)^n\widehat{\alpha}_{(A}\widehat{\beta}_B\cdots\widehat{\gamma}_E\widehat{\delta}_{F)} & \text{if } (g_{ab}) = \mathrm{diag}(1,1,1), \\ \widehat{\alpha}_{(A}\widehat{\beta}_B\cdots\widehat{\gamma}_E\widehat{\delta}_{F)} & \text{if } (g_{ab}) = \mathrm{diag}(1,1,-1). \end{cases} \tag{5.71}$$

As in the case of the spinor formalism employed in the four-dimensional space-time of general relativity, the totally symmetric spinors of a given rank can be classified according to the multiplicity of their principal spinors. In the case of three-dimensional spaces, when two principal spinors are not proportional to each other, a further subclassification can be obtained according to whether one of them is proportional to the mate of the other or not.

The simplest nontrivial case of this algebraic classification corresponds to a two-index symmetric spinor, v_{AB}, which is equivalent to a (possibly complex) vector v_a. The components v_{AB} can be expressed as

$$v_{AB} = \alpha_{(A}\beta_{B)}, \tag{5.72}$$

hence

$$v^a v_a = -v^{AB}v_{AB} = \tfrac{1}{2}(\alpha^A\beta_A)^2 \tag{5.73}$$

[see (5.11)].

Considering the case where the metric is positive definite, the vector v_a is real if and only if $\alpha_{(A}\beta_{B)} = -\widehat{\alpha}_{(A}\widehat{\beta}_{B)}$ [see (5.71)]. Since in this case a one-index spinor cannot be proportional to its mate, the last equation implies that

$$\widehat{\alpha}_A = \lambda\beta_A, \qquad \widehat{\beta}_A = -\lambda^{-1}\alpha_A, \tag{5.74}$$

for some scalar λ. By combining (5.74), using the fact that the map $\psi_A \mapsto \widehat{\psi}_A$ is antilinear, we obtain

$$\widehat{\widehat{\alpha}}_A = \overline{\lambda}\widehat{\beta}_A = -\overline{\lambda}\lambda^{-1}\alpha_A, \tag{5.75}$$

which must coincide with $-\alpha_A$ [see (5.57)]; hence, λ must be real. If λ is positive, substituting the first equation (5.74) into (5.72) we have $v_{AB} = \alpha_{(A}\lambda^{-1}\widehat{\alpha}_{B)} = \lambda^{-1/2}\alpha_{(A}\lambda^{-1/2}\widehat{\alpha}_{B)}$; hence, absorbing the (real) factor $\lambda^{-1/2}$ into α_A we find that

$$v_{AB} = \alpha_{(A}\widehat{\alpha}_{B)}. \tag{5.76}$$

In a similar way, if λ is negative, (5.72) and (5.74) give $v_{AB} = -\lambda\widehat{\beta}_{(A}\beta_{B)} = (-\lambda)^{1/2}\widehat{\beta}_{(A}(-\lambda)^{1/2}\beta_{B)}$, which is also of the form (5.76).

On the other hand, when $(g_{ab}) = \mathrm{diag}(1, 1, -1)$, v_a is real if and only if $\alpha_{(A}\beta_{B)} = \widehat{\alpha}_{(A}\widehat{\beta}_{B)}$, which leads to the two possibilities

$$\text{(i)} \quad \widehat{\alpha}_A = \lambda\alpha_A, \qquad \widehat{\beta}_A = \lambda^{-1}\beta_A, \tag{5.77}$$

with $|\lambda| = 1$, and

$$\text{(ii)} \quad \widehat{\alpha}_A = \lambda\beta_A, \qquad \widehat{\beta}_A = \lambda^{-1}\alpha_A. \tag{5.78}$$

In the case (i), λ must be of the form $e^{i\theta}$; then, from (5.77) we have $(e^{i\theta/2}\alpha_A)\widehat{\ } = e^{-i\theta/2}e^{i\theta}\alpha_A = e^{i\theta/2}\alpha_A$ and $(e^{-i\theta/2}\beta_A)\widehat{\ } = e^{-i\theta/2}\beta_A$. Therefore, by rewriting (5.72) in the form $v_{AB} = e^{i\theta/2}\alpha_{(A}e^{-i\theta/2}\beta_{B)}$ and absorbing the factors $e^{\pm i\theta/2}$ into α_A and β_A, we find that in the case (i) v_{AB} can be expressed as

$$\text{(i)} \quad v_{AB} = \alpha_{(A}\beta_{B)} \qquad \text{with } \widehat{\alpha}_A = \alpha_A, \ \ \widehat{\beta}_A = \beta_A. \tag{5.79}$$

Using (5.58) and (5.73) one finds that the vectors of the form (5.79) are such that $v^a v_a \geqslant 0$.

In the case (ii), from (5.78) we obtain $\widehat{\widehat{\alpha}}_A = \overline{\lambda}\widehat{\beta}_A = \overline{\lambda}\lambda^{-1}\alpha_A$, which means that λ is real. Hence, $v_{AB} = \lambda^{-1}\alpha_{(A}\widehat{\alpha}_{B)} = \pm|\lambda|^{-1/2}\alpha_{(A}|\lambda|^{-1/2}\widehat{\alpha}_{B)}$ and, absorbing the factor $|\lambda|^{-1/2}$ into α_A, we conclude that in the case (ii) v_{AB} can be expressed as

$$\text{(ii)} \quad v_{AB} = \pm\alpha_{(A}\widehat{\alpha}_{B)}. \tag{5.80}$$

From (5.58) and (5.73) it follows that (5.80) corresponds to a real vector such that $v^a v_a \leqslant 0$.

In the special case where $v^a v_a = 0$, (5.73) implies that v_{AB} must be of the form $v_{AB} = \alpha_A\alpha_B$ and v_{AB} corresponds to a real vector if and only if $\alpha_A\alpha_B = \widehat{\alpha}_A\widehat{\alpha}_B$, which amounts to $\widehat{\alpha}_A = \pm\alpha_A$; therefore, the spinor equivalent of a real null vector is of the form

$$v_{AB} = \pm\alpha_A\alpha_B \qquad \text{with } \widehat{\alpha}_A = \alpha_A. \tag{5.81}$$

Thus, when the metric is indefinite, the spinor equivalent of a real vector is of the form

$$v_{AB} = \begin{cases} \alpha_{(A}\beta_{B)} & \text{with } \widehat{\alpha}_A = \alpha_A,\ \widehat{\beta}_A = \beta_A,\ \alpha^A\beta_A \neq 0 & \text{if } v^a v_a > 0, \\ \pm\alpha_A\alpha_B & \text{with } \widehat{\alpha}_A = \alpha_A & \text{if } v^a v_a = 0, \\ \pm\alpha_{(A}\widehat{\alpha}_{B)} & \text{with } \widehat{\alpha}^A\alpha_A \neq 0 & \text{if } v^a v_a < 0. \end{cases} \tag{5.82}$$

As a second example we consider a four-index totally symmetric spinor Φ_{ABCD} which is equivalent to a trace-free symmetric tensor Φ_{ab} and, according to (5.68), can be written as

$$\Phi_{ABCD} = \alpha_{(A}\beta_B\gamma_C\delta_{D)}. \tag{5.83}$$

Making use of (5.71) one finds that Φ_{ab} is real if and only if

$$\alpha_{(A}\beta_B\gamma_C\delta_{D)} = \widehat{\alpha}_{(A}\widehat{\beta}_B\widehat{\gamma}_C\widehat{\delta}_{D)}. \tag{5.84}$$

In the case with signature $(+++)$, condition (5.84) severely restricts the possible multiplicities in the principal spinors of Φ_{ABCD}. In fact, it can be verified, with the help of (5.57), that the only possible algebraic types are

$$\begin{aligned} \Phi_{ABCD} &= \pm\alpha_{(A}\widehat{\alpha}_B\alpha_C\widehat{\alpha}_{D)}, \\ \Phi_{ABCD} &= \alpha_{(A}\widehat{\alpha}_B\beta_C\widehat{\beta}_{D)}. \end{aligned} \tag{5.85}$$

By contrast, when the metric is indefinite, the solutions of (5.84) are of the form

$$\begin{aligned}
\Phi_{ABCD} &= \alpha_{(A}\beta_B\gamma_C\delta_{D)} && \text{with } \widehat{\alpha}_A = \alpha_A,\ \widehat{\beta}_A = \beta_A,\ \widehat{\gamma}_A = \gamma_A,\ \widehat{\delta}_A = \delta_A, \\
\Phi_{ABCD} &= \alpha_{(A}\beta_B\gamma_C\widehat{\gamma}_{D)} && \text{with } \widehat{\alpha}_A = \alpha_A,\ \widehat{\beta}_A = \beta_A, \\
\Phi_{ABCD} &= \pm\alpha_{(A}\widehat{\alpha}_B\beta_C\widehat{\beta}_{D)}, \\
\Phi_{ABCD} &= \alpha_{(A}\alpha_B\beta_C\gamma_{D)} && \text{with } \widehat{\alpha}_A = \alpha_A,\ \widehat{\beta}_A = \beta_A,\ \widehat{\gamma}_A = \gamma_A, \\
\Phi_{ABCD} &= \pm\alpha_{(A}\alpha_B\beta_C\widehat{\beta}_{D)} && \text{with } \widehat{\alpha}_A = \alpha_A, \\
\Phi_{ABCD} &= \pm\alpha_{(A}\alpha_B\beta_C\beta_{D)} && \text{with } \widehat{\alpha}_A = \alpha_A,\ \widehat{\beta}_A = \beta_A, \\
\Phi_{ABCD} &= \pm\alpha_{(A}\alpha_B\widehat{\alpha}_C\widehat{\alpha}_{D)}, \\
\Phi_{ABCD} &= \alpha_{(A}\alpha_B\alpha_C\beta_{D)} && \text{with } \widehat{\alpha}_A = \alpha_A,\ \widehat{\beta}_A = \beta_A, \\
\Phi_{ABCD} &= \pm\alpha_A\alpha_B\alpha_C\alpha_D && \text{with } \widehat{\alpha}_A = \alpha_A.
\end{aligned} \tag{5.86}$$

Using the (real) connection symbols (5.43), the components of the spinor equivalent of a real symmetric tensor are real and the nine algebraic types (5.86) correspond to the character and multiplicities of the roots of the polynomial

$$\Phi_{1111}\zeta^4 + 4\Phi_{1112}\zeta^3 + 6\Phi_{1122}\zeta^2 + 4\Phi_{1222}\zeta + \Phi_{2222}.$$

This follows from the fact that, when the connection symbols are real, the components $\widehat{\alpha}_A$ are the complex conjugates of the components α_A. In a similar manner, one can obtain the classification (5.82) by considering the polynomial

$$v_{11}\zeta^2 + 2v_{12}\zeta + v_{22} \tag{5.87}$$

whose roots are $-v_{12} \pm \sqrt{-\frac{1}{2}v^{AB}v_{AB}} = -v_{12} \pm \sqrt{\frac{1}{2}v^a v_a}$; hence, the roots of (5.87) are real and different, repeated, or one is the complex conjugate of the other, depending on whether $v^a v_a$ is positive, equal to zero or negative, respectively.

A trace-free symmetric tensor Φ_{ab} in a three-dimensional space with indefinite metric can also be classified according to the character and the coincidences of its eigenvectors (Hall and Capocci 1999); in that way only four different algebraic types are obtained, by contrast with the nine algebraic types given by (5.86).

Making use of the identity

$$\alpha_A\beta_B - \alpha_B\beta_A = \alpha^R\beta_R\,\varepsilon_{AB} \tag{5.88}$$

and (5.8), a straightforward computation shows that if $v_{AB} \equiv \alpha_{(A}\beta_{B)}$ and $w_{AB} \equiv \gamma_{(A}\delta_{B)}$, then (5.83) amounts to

$$\Phi_{ABCD} = \tfrac{1}{2}(v_{AB}w_{CD} + w_{AB}v_{CD}) - \tfrac{1}{6}v^{EF}w_{EF}(\varepsilon_{AC}\varepsilon_{BD} + \varepsilon_{AD}\varepsilon_{BC}) \tag{5.89}$$

or, equivalently,

$$\Phi_{ab} = v_{(a}w_{b)} - \tfrac{1}{3}v^c w_c\, g_{ab}, \tag{5.90}$$

where v_a and w_a are the vector equivalents of v_{AB} and w_{AB}, respectively. In other words, Φ_{ab} is the trace-free symmetric part of the tensor product of two vectors.

More generally, if $t_{AB\ldots D}$ is the spinor equivalent of a real n-index trace-free totally symmetric tensor, $t_{ab\ldots d}$, then α^A is a principal spinor of $t_{AB\ldots D}$ if and only if $t_{AB\ldots D}\alpha^A\alpha^B\cdots\alpha^D = 0$ and since $\widehat{t}_{AB\ldots D}$ is proportional to $t_{AB\ldots D}$, it follows that $t_{AB\ldots D}\widehat{\alpha}^A\widehat{\alpha}^B\cdots\widehat{\alpha}^D = 0$, which means that $\widehat{\alpha}^A$ is also a principal spinor of $t_{AB\ldots D}$. Therefore, when the metric is positive definite, $t_{AB\ldots L}$ must be of the form $\alpha_{(A}\widehat{\alpha}_B\beta_C\widehat{\beta}_D\cdots\eta_K\widehat{\eta}_{L)}$ and this, in turn, is equivalent to the existence of n (real) vectors $u_a, v_a, \ldots, w_a$ (the vector equivalents of $\alpha_{(A}\widehat{\alpha}_{B)}$, $\beta_{(A}\widehat{\beta}_{B)}$, …, $\eta_{(A}\widehat{\eta}_{B)}$), such that $t_{ab\ldots d}$ is the trace-free symmetric part of $u_a v_b\cdots w_d$. The directions of $u_a, v_a, \ldots, w_a$, are uniquely defined by $t_{ab\ldots d}$.

By contrast, when the metric is indefinite, a one-index spinor can be proportional to its mate and, therefore, a real, trace-free, totally symmetric n-index tensor can be expressed as the trace-free totally symmetric part of the product of n real vectors, the directions of which may not be uniquely defined. For instance, the tensor equivalent of the first row of (5.86) is given by (5.90), with v_a being the vector equivalent of $\alpha_{(A}\beta_{B)}$, $\alpha_{(A}\gamma_{B)}$, or $\alpha_{(A}\delta_{B)}$ and w_a being the vector equivalent of $\gamma_{(A}\delta_{B)}$, $\beta_{(A}\delta_{B)}$, or $\beta_{(A}\gamma_{B)}$, respectively.

Bivectors

Any antisymmetric 2-index tensor, or bivector, t_{ab}, is the dual of some vector,

$$t_{ab} = \varepsilon_{abc} t^c. \tag{5.91}$$

In effect, given t_{ab}, we can define $t_c \equiv \frac{1}{2} \det(g_{rs})\, \varepsilon_{abc} t^{ab}$, then with the aid of

$$\det(g_{rs})\, g^{ac} \varepsilon_{abe} \varepsilon_{cdf} = g_{bd} g_{ef} - g_{bf} g_{ed}, \tag{5.92}$$

one can verify that (5.91) holds. The bivector t_{ab} is real if and only if t_a is real. Therefore, in the case where the metric is positive definite, t_{ab} is real if and only if the spinor equivalent of t_a is of the form $\alpha_{(A}\widehat{\alpha}_{B)}$ [see (5.76)] and from (5.18) and (5.88) we find that the spinor equivalent of t_{ab} is given by

$$\begin{aligned} t_{ABCD} &= -\frac{\mathrm{i}}{\sqrt{2}} \frac{1}{\alpha^R \widehat{\alpha}_R} \left\{ \alpha_{(B}\widehat{\alpha}_{D)} (\alpha_A \widehat{\alpha}_C - \alpha_C \widehat{\alpha}_A) + \alpha_{(A}\widehat{\alpha}_{C)} (\alpha_B \widehat{\alpha}_D - \alpha_D \widehat{\alpha}_B) \right\} \\ &= -\frac{\mathrm{i}}{\sqrt{2}} \frac{1}{\alpha^R \widehat{\alpha}_R} (\alpha_A \alpha_B \widehat{\alpha}_C \widehat{\alpha}_D - \widehat{\alpha}_A \widehat{\alpha}_B \alpha_C \alpha_D), \end{aligned}$$

i.e.,

$$t_{ab} = \frac{\mathrm{i}}{\sqrt{2}} (m_a \overline{m}_b - \overline{m}_a m_b), \tag{5.93}$$

where we have denoted by m_a the tensor equivalent of $\alpha_A \alpha_B / \sqrt{\alpha^R \widehat{\alpha}_R}$ and, hence, $-\widehat{\alpha}_A \widehat{\alpha}_B / \sqrt{\alpha^R \widehat{\alpha}_R}$ is the spinor equivalent of $\overline{m}_a$.

When the metric is indefinite, there are three different cases distinguished by the value of $t^a t_a$. If $t^a t_a > 0$, the spinor equivalent of t_a is of the form $\alpha_{(A}\beta_{B)}$ with $\widehat{\alpha}_A = \alpha_A$, $\widehat{\beta}_A = \beta_A$ and $\alpha^A \beta_A \neq 0$, hence, proceeding as in the previous case, using (5.18), we find that

$$t_{ABCD} = \frac{1}{\sqrt{2}} \frac{1}{\alpha^R \beta_R} (\alpha_A \alpha_B \beta_C \beta_D - \beta_A \beta_B \alpha_C \alpha_D). \tag{5.94}$$

By virtue of (5.58), $\alpha^R \beta_R$ is real; hence, assuming, *e.g.*, that $\alpha^R \beta_R$ is greater than 0, it follows that (5.94) is equivalent to

$$t_{ab} = \frac{1}{\sqrt{2}} (v_a w_b - w_a v_b), \tag{5.95}$$

where v_a and w_a are two real null vectors which are the tensor equivalents of $\alpha_A \alpha_B / \sqrt{\alpha^R \beta_R}$ and $\beta_A \beta_B / \sqrt{\alpha^R \beta_R}$, respectively [see (5.81)].

Similarly if $t^a t_a < 0$, the spinor equivalent of t_a is of the form $\pm \alpha_{(A}\widehat{\alpha}_{B)}$, with $\alpha^A \widehat{\alpha}_A \neq 0$, hence,

$$t_{ABCD} = \pm \frac{1}{\sqrt{2}} \frac{1}{\alpha^R \widehat{\alpha}_R} (\alpha_A \alpha_B \widehat{\alpha}_C \widehat{\alpha}_D - \widehat{\alpha}_A \widehat{\alpha}_B \alpha_C \alpha_D). \tag{5.96}$$

Now $\alpha^R \widehat{\alpha}_R$ is pure imaginary and assuming $\mathrm{i}\alpha^R \widehat{\alpha}_R > 0$, we find that

$$t_{ab} = \pm \frac{\mathrm{i}}{\sqrt{2}} (m_a \overline{m}_b - \overline{m}_a m_b), \tag{5.97}$$

where m_a is the tensor equivalent of $\alpha_A \alpha_B / \sqrt{\mathrm{i}\alpha^R \widehat{\alpha}_R}$.

Finally, if $t^a t_a = 0$, the spinor equivalent of t_a is of the form $\pm\alpha_A \alpha_B$, with $\widehat{\alpha}_A = \alpha_A$. We can always find a spinor β_A such that $\alpha^A \beta_A = 1$ and $\widehat{\beta}_A = \beta_A$, then, using (5.18) and (5.86), we obtain

$$t_{ABCD} = \pm\sqrt{2}(\alpha_A \alpha_B \alpha_{(C} \beta_{D)} - \alpha_{(A} \beta_{B)} \alpha_C \alpha_D), \tag{5.98}$$

which amounts to

$$t_{ab} = \pm\sqrt{2}(t_a s_b - s_a t_b), \tag{5.99}$$

where s_a is the tensor equivalent of $\alpha_{(A} \beta_{B)}$ and therefore s_a is real and $s^a s_a = \frac{1}{2}$. The preceding results show that in three dimensions every bivector is simple, *i.e.*, it is the antisymmetrized tensor product of two vectors. (It must be noticed that this conclusion applies only to bivectors at a point and not to tensor fields, see Penrose and Rindler 1984, §3.5.)

Among the differences between the spinor formalism employed in general relativity and the spinor formalism of three-dimensional spaces is the fact that, in the latter case, any vector can be expressed in terms of two one-index spinors [(5.72)]. As we have shown, when the metric has signature $(+ + -)$, the algebraic classification of the spinor equivalents of vectors amounts to classifying the vectors according to whether $v^a v_a$ is positive, negative, or equal to zero.

Spin-s particles

The spin states of a particle with nonvanishing rest-mass and spin s are given by totally symmetric spinors with $2s$ indices, $\psi_{AB...L}$, in a space with positive definite metric. If α_A is a principal spinor of $\psi_{AB...L}$, then $\alpha_{(A} \widehat{\alpha}_{B)}$ is the spinor equivalent of a real vector which defines a direction or, equivalently, a point of the unit sphere. In this manner, the $2s$ principal spinors of $\psi_{AB...L}$ correspond to $2s$ (not necessarily distinct) points of the unit sphere (see Penrose 1994 and the references cited therein); since $\widehat{\alpha}_{(A} \widehat{\widehat{\alpha}}_{B)} = -\alpha_{(A} \widehat{\alpha}_{B)}$, α_A and $\widehat{\alpha}_A$ correspond to antipodal points of the sphere. (Note that, for $s > 1/2$, each principal spinor, α_A, of $\psi_{AB...L}$ is defined up to a complex factor, but the *direction* of the vector equivalent of $\alpha_{(A} \widehat{\alpha}_{B)}$ is uniquely defined.) Conversely, taking into account that the state vector of any quantum system is defined up to a complex factor, a set of $2s$ points of the unit sphere defines a state of a spin-s particle with nonvanishing rest-mass.

If α_A is normalized in such a way that $\alpha^A\widehat{\alpha}_A = 1$, then the $2s+1$ totally symmetric spinors

$$_{(m)}\chi_{AB\ldots L} \equiv \sqrt{\frac{(2s)!}{(s+m)!(s-m)!}}\,\underbrace{\alpha_{(A}\alpha_B\cdots\alpha_D}_{s+m}\underbrace{\widehat{\alpha}_E\widehat{\alpha}_F\cdots\widehat{\alpha}_{L)}}_{s-m}$$

$(m = -s, -s+1, \ldots, s-1, s)$ satisfy $_{(m)}\chi^{AB\ldots L}\,{}_{(m')}\widehat{\chi}_{AB\ldots L} = \delta_{mm'}$ [*i.e.*, are orthonormal with respect to the inner product (5.67)] and $_{(m)}\chi_{AB\ldots L}$ is an eigenspinor of the operator corresponding to the spin along the vector equivalent of $\alpha_{(A}\widehat{\alpha}_{B)}$, with eigenvalue $m\hbar$. Any spin state, $\psi_{AB\ldots L}$, can be expressed as a linear combination of the basis states $_{(m)}\chi_{AB\ldots L}$,

$$\psi_{AB\ldots L} = \sum_{m=-s}^{s} c_{(m)}\,{}_{(m)}\chi_{AB\ldots L},$$

with

$$\begin{aligned} c_{(m)} &= (-1)^{2s}\,{}_{(m)}\widehat{\chi}^{AB\ldots L}\psi_{AB\ldots L} \\ &= (-1)^{s+m}\sqrt{\frac{(2s)!}{(s+m)!(s-m)!}}\,\underbrace{\widehat{\alpha}^A\widehat{\alpha}^B\cdots\widehat{\alpha}^D}_{s+m}\underbrace{\alpha^E\alpha^F\cdots\alpha^L}_{s-m}\psi_{AB\ldots L}. \end{aligned}$$

According to the standard interpretation, if $\psi^{AB\ldots L}\widehat{\psi}_{AB\ldots L} = 1$, $|c_{(m)}|^2$ is the probability of obtaining the value $m\hbar$ when the projection of the spin along the vector equivalent of $\alpha_{(A}\widehat{\alpha}_{B)}$ is measured. Hence, if α_A is a p-fold repeated principal spinor of $\psi_{AB\ldots L}$, then $c_{(-s)} = c_{(-s+1)} = \cdots = c_{(-s+p-1)} = 0$; *i.e.*, the probabilities of obtaining the values $-s\hbar$, $(-s+1)\hbar$, ..., $(-s+p-1)\hbar$ when the projection of the spin along the vector equivalent of $\alpha_{(A}\widehat{\alpha}_{B)}$ is measured are equal to zero.

Spin transformations

A spin transformation, (U_{AB}), can be decomposed as the sum of its anti-symmetric and symmetric parts

$$U_{AB} = a\,\varepsilon_{AB} + w_{AB}.$$

Since $\widehat{U}_{AB} = U_{AB}$ and $\widehat{\varepsilon}_{AB} = \varepsilon_{AB}$, it follows that a is real and $\widehat{w}_{AB} = w_{AB}$. Then w_{AB} is the spinor equivalent of a real or pure imaginary eigenvector of $(L^a{}_b)$, the SO(p,q) transformation corresponding to $(U^A{}_B)$, since, according to (5.33) and making use of the symmetry of w_{AB},

$$\begin{aligned} -L^{AB}{}_{CD}w^{CD} &= U^{(A}{}_C U^{B)}{}_D w^{CD} = U^A{}_C U^B{}_D w^{CD} \\ &= (a\delta^A_C + w^A{}_C)(a\delta^B_D + w^B{}_D)w^{CD} \\ &= (a^2 + \tfrac{1}{2}w^{CD}w_{CD})w^{AB}. \end{aligned}$$

In the case where the metric is positive definite w_{AB} is of the form $\mathrm{i}\beta_{(A}\widehat{\beta}_{B)}$, for some spinor β_A. Then, β^A and $\widehat{\beta}^A$ are eigenspinors of $(U^A{}_B)$,

$$U^A{}_B\beta^B = (a + \tfrac{1}{2}\mathrm{i}\beta^B\widehat{\beta}_B)\beta^A, \qquad U^A{}_B\widehat{\beta}^B = (a - \tfrac{1}{2}\mathrm{i}\beta^B\widehat{\beta}_B)\widehat{\beta}^A.$$

The determinant of $(U^A{}_B)$ is the product of its eigenvalues; therefore, $\det(U^A{}_B) = a^2 + \frac{1}{4}(\beta^A\widehat{\beta}_A)^2 = 1$ and we can write $a = \cos(\theta/2)$ and $\beta^A\widehat{\beta}_A = 2\sin(\theta/2)$, for $0 \leqslant \theta < 2\pi$ (taking into account that $\beta^A\widehat{\beta}_A \geqslant 0$). Making $\alpha_A \equiv [\sqrt{2}\,\sin(\theta/2)]^{-1/2}\beta_A$ (excluding the trivial case $\theta = 0$) we have

$$U_{AB} = \cos(\theta/2)\,\varepsilon_{AB} + \mathrm{i}\sqrt{2}\,\sin(\theta/2)\,\alpha_{(A}\widehat{\alpha}_{B)}, \tag{5.100}$$

with $\alpha^A\widehat{\alpha}_A = \sqrt{2}$, so that $\alpha_{(A}\widehat{\alpha}_{B)}$ is the spinor equivalent of a real unit vector, which lies along the axis of the rotation corresponding to $(U^A{}_B)$ [*cf.* (1.15)]. Thus, $U^A{}_B\alpha^B = \mathrm{e}^{\mathrm{i}\theta/2}\alpha^A$ and $U^A{}_B\widehat{\alpha}^B = \mathrm{e}^{-\mathrm{i}\theta/2}\widehat{\alpha}^A$, which implies that $\alpha_{(A}\widehat{\alpha}_{B)}$, $\alpha_A\alpha_B$, and $\widehat{\alpha}_A\widehat{\alpha}_B$ are the spinor equivalents of a real and two complex eigenvectors of the orthogonal transformation $(L^a{}_b)$ defined by $(U^A{}_B)$, with eigenvalues 1, $\mathrm{e}^{\mathrm{i}\theta}$, and $\mathrm{e}^{-\mathrm{i}\theta}$, respectively.

When the metric is indefinite, w_{AB} is the spinor equivalent of a real vector [see (5.65)] and, according to (5.82), any spin transformation is of the form

$$U_{AB} = \begin{cases} a\,\varepsilon_{AB} + \alpha_{(A}\gamma_{B)} & \text{with } \widehat{\alpha}_A = \alpha_A,\ \widehat{\gamma}_A = \gamma_A,\ \alpha^A\gamma_A \neq 0, \\ a\,\varepsilon_{AB} \pm \beta_A\beta_B & \text{with } \widehat{\beta}_A = \beta_A, \\ a\,\varepsilon_{AB} \pm \beta_{(A}\widehat{\beta}_{B)} & \text{with } \beta^A\widehat{\beta}_A \neq 0. \end{cases}$$

In the first of these cases, α^A and γ^A are eigenspinors of $(U^A{}_B)$, with the real eigenvalues $a \pm \frac{1}{2}\alpha^A\gamma_A$; hence, $a^2 - \frac{1}{4}(\alpha^A\gamma_A)^2 = 1$ and we can write $a = \pm\cosh(\theta/2)$, $\alpha^A\gamma_A = \pm 2\sinh(\theta/2)$, for some $\theta \in \mathbb{R}$. Taking $\beta_A \equiv \pm[\sqrt{2}\,\sinh(\theta/2)]^{-1}\gamma_A$, we have $\alpha^A\beta_A = \sqrt{2}$ and

$$U_{AB} = \pm[\cosh(\theta/2)\,\varepsilon_{AB} + \sqrt{2}\,\sinh(\theta/2)\,\alpha_{(A}\beta_{B)}]. \tag{5.101}$$

Hence, $U^A{}_B\alpha^B = \pm\mathrm{e}^{\theta/2}\alpha^A$, $U^A{}_B\beta^B = \pm\mathrm{e}^{-\theta/2}\beta^A$ and, as a consequence, $\alpha_A\alpha_B$ and $\beta_A\beta_B$ are the spinor equivalents of two real null eigenvectors of the orthogonal transformation defined by $(U^A{}_B)$, with eigenvalues e^{θ} and $\mathrm{e}^{-\theta}$, respectively, while $\alpha_{(A}\beta_{B)}$ is the spinor equivalent of a real unit eigenvector with eigenvalue 1 of that orthogonal transformation.

In the second case, $\det(U^A{}_B) = a^2 = 1$ and we can write

$$U_{AB} = \pm[\varepsilon_{AB} + \sqrt{2}\,(\theta/2)\,\alpha_A\alpha_B], \tag{5.102}$$

where α_A is a multiple of β_A and θ is some real number. The eigenspinors of $(U^A{}_B)$ are proportional to α^A; therefore, the eigenvectors of the orthogonal transformation corresponding to $(U^A{}_B)$ are multiples of the vector equivalent of $\alpha_A\alpha_B$, with eigenvalue 1. In the third case, β^A and $\widehat{\beta}^A$ are eigenspinors of $(U^A{}_B)$ with the complex-conjugate eigenvalues $a \pm \frac{1}{2}\beta^A\widehat{\beta}_A$, with $a^2 - \frac{1}{4}(\beta^A\widehat{\beta}_A)^2 = 1$. Then, $a = \cos(\theta/2)$ and $\beta^A\widehat{\beta}_A = 2\mathrm{i}\sin(\theta/2)$, for $0 \leqslant \theta < 2\pi$ (by interchanging β_A and $\widehat{\beta}_A$ if necessary). Making $\alpha_A \equiv [\sqrt{2}\,\sin(\theta/2)]^{-1/2}\beta_A$ we obtain $\alpha^A\widehat{\alpha}_A = \mathrm{i}\sqrt{2}$ and $U_{AB} = \cos(\theta/2)\,\varepsilon_{AB} \pm \sqrt{2}\,\sin(\theta/2)\,\alpha_A\widehat{\alpha}_B$ or, allowing θ to take any real value,

$$U_{AB} = \cos(\theta/2)\,\varepsilon_{AB} + \sqrt{2}\,\sin(\theta/2)\,\alpha_{(A}\widehat{\alpha}_{B)}. \tag{5.103}$$

In this case, $\alpha_{(A}\widehat{\alpha}_{B)}$ is the spinor equivalent of a real unit eigenvector of the orthogonal transformation defined by $(U^A{}_B)$ [*cf.* (1.86)].

Reflections

If the metric of the vector space V is positive definite, the reflection on a plane passing through the origin with unit normal n_a is represented by the orthogonal matrix

$$L^a{}_b = \delta^a_b - 2n^a n_b,$$

whose spinor equivalent is [see (5.23) and (5.8)]

$$L^{AB}{}_{CD} = \tfrac{1}{2}\sigma_a{}^{AB}\sigma^b{}_{CD}(\delta^a_b - 2n^a n_b) = -\delta^{(A}_C\delta^{B)}_D - 2n^{AB}n_{CD},$$

where n_{AB} is the spinor equivalent of n_a. Since $n^a n_a = 1$, we have $n^{AB}n_{BC} = -\frac{1}{2}\delta^A_C$, and, using (5.13),

$$\begin{aligned} n^{AB}n_{CD} &= n^A{}_C n^B{}_D + n^{AB}n_{CD} - n^A{}_C n^B{}_D = n^A{}_C n^B{}_D + \delta^B_C n^{AR}n_{RD} \\ &= n^A{}_C n^B{}_D - \tfrac{1}{2}\delta^B_C\delta^A_D, \end{aligned}$$

hence,

$$L^{AB}{}_{CD} = -2n^{(A}{}_C n^{B)}{}_D, \tag{5.104}$$

which is of the form (5.31) with $U^A{}_B = \mathrm{i}\sqrt{2}\,n^A{}_B \in \mathrm{SU}(2)$.

Then, the composition of reflections on planes passing through the origin with unit normals n_a and l_a is represented by the SU(2) matrix $U^A{}_B = -2l^A{}_R n^R{}_B$, which can be written in the form

$$\begin{aligned} U_{AB} &= 2l_{[A}{}^R n_{B]R} + 2l_{(A}{}^R n_{B)R} \\ &= l^{RS}n_{RS}\varepsilon_{AB} + 2l_{(A}{}^R n_{B)R} \\ &= -\cos(\theta/2)\,\varepsilon_{AB} - \mathrm{i}\sqrt{2}\,v_{AB}, \end{aligned}$$

where $\theta/2$ is the angle between l_a and n_a and v_{AB} is the spinor equivalent of the cross product of l_a by n_a, $v_{AB} = \varepsilon_{ABCDEF}l^{CD}n^{EF} = \mathrm{i}\sqrt{2}\,l_{(A}{}^R n_{B)R}$. Hence,

$v_{AB} = \sin(\theta/2)\, u_{AB}$, where u_{AB} is the spinor equivalent of a real unit vector and we obtain

$$U_{AB} = -\cos(\theta/2)\, \varepsilon_{AB} - \mathrm{i}\sqrt{2}\, \sin(\theta/2)\, u_{AB},$$

which is of the form (5.100).

It can be readily seen that any rotation can be obtained through the composition of two reflections (see also Cartan 1966, Misner, Thorne and Wheeler 1973). In fact, given the spin transformation (5.100), defining

$$\begin{aligned} n_{AB} &= \tfrac{1}{2}(\mathrm{e}^{-\mathrm{i}\theta/4}\alpha_A\alpha_B - \mathrm{e}^{\mathrm{i}\theta/4}\widehat{\alpha}_A\widehat{\alpha}_B), \\ l_{AB} &= \tfrac{1}{2}(\mathrm{e}^{\mathrm{i}\theta/4}\alpha_A\alpha_B - \mathrm{e}^{-\mathrm{i}\theta/4}\widehat{\alpha}_A\widehat{\alpha}_B), \end{aligned}$$

which are the spinor equivalents of two real unit vectors, using the fact that $\alpha^A\widehat{\alpha}_A = \sqrt{2}$, we obtain $U^A{}_B = -2l^A{}_R n^R{}_B$.

When the metric is indefinite, the foregoing derivation applies with slight modifications. If n_a is a vector such that $n^a n_a = \pm 1$, the reflection on the plane normal to n_a is represented by

$$L^{AB}{}_{CD} = \mp 2n^{(A}{}_C n^{B)}{}_D, \tag{5.105}$$

which is of the form (5.31) with

$$U^A{}_B = \begin{cases} \mathrm{i}\sqrt{2}\, n^A{}_B & \text{if } n^a n_a = 1, \\ \sqrt{2}\, n^A{}_B & \text{if } n^a n_a = -1. \end{cases} \tag{5.106}$$

Note that, in all cases, $\det(n^A{}_B) = \frac{1}{2} n^{AB} n_{AB} = -\frac{1}{2} n^a n_a$ and, therefore, the determinants of the matrices $(U^A{}_B)$, given by (5.106), are equal to 1.

The composition of any two reflections on planes passing through the origin is an orthogonal transformation; however, from (5.106) one concludes that the composition of two reflections on planes through the origin with normal vectors n_a and l_a yields an $SO_0(2,1)$ transformation if and only if $n^a n_a$ and $l^a l_a$ are both positive or negative.

When the metric is indefinite, any spin transformation is also the composition of two reflections. The spin transformation (5.101) is the composition of the reflections on the planes passing through the origin with normal vectors n_a and l_a whose spinor equivalents are

$$\begin{aligned} n_{AB} &= \tfrac{1}{2}(\mathrm{e}^{-\theta/4}\alpha_A\alpha_B + \mathrm{e}^{\theta/4}\beta_A\beta_B), \\ l_{AB} &= \mp\tfrac{1}{2}(\mathrm{e}^{\theta/4}\alpha_A\alpha_B + \mathrm{e}^{-\theta/4}\beta_A\beta_B). \end{aligned}$$

Similarly, the spin transformation (5.102) can be expressed in the form $U^A{}_B = -2l^A{}_R n^R{}_B$ with

$$\begin{aligned} n_{AB} &= \alpha_{(A}\beta_{B)} + (\theta/4)\,\alpha_A\alpha_B, \\ l_{AB} &= \mp[\alpha_{(A}\beta_{B)} - (\theta/4)\,\alpha_A\alpha_B], \end{aligned}$$

where β_A is such that $\widehat{\beta_A} = \beta_A$ and $\alpha^A \beta_A = \sqrt{2}$. Finally, the spin transformation (5.103) can be expressed as $U^A{}_B = -2l^A{}_R n^R{}_B$ with

$$\begin{aligned} n_{AB} &= \tfrac{1}{2}(\mathrm{e}^{-\mathrm{i}\theta/4}\alpha_A\alpha_B + \mathrm{e}^{\mathrm{i}\theta/4}\widehat{\alpha}_A\widehat{\alpha}_B), \\ l_{AB} &= -\tfrac{1}{2}(\mathrm{e}^{\mathrm{i}\theta/4}\alpha_A\alpha_B + \mathrm{e}^{-\mathrm{i}\theta/4}\widehat{\alpha}_A\widehat{\alpha}_B). \end{aligned}$$

5.4 The triad defined by a spinor

In the spinor calculus employed in general relativity, two linearly independent one-index spinors define a tetrad of vectors. In the case of three-dimensional spaces, a single one-index spinor determines a basis. When the metric is positive definite, this relationship is well known and allows the representation of a spinor by means of an ax or a flag (see Section 1.2).

If $(g_{ab}) = \mathrm{diag}(1, 1, 1)$, given a one-index spinor, ψ_A, different from zero, one can define the vectors **R** and **M** with components

$$R_a \equiv -\sigma_{aAB}\widehat{\psi}^A\psi^B, \qquad M_a \equiv \sigma_{aAB}\psi^A\psi^B \tag{5.107}$$

[*cf.* (1.59) and (1.60)]. The components R_a are real [the spinor equivalent of R_a is $R_{AB} = \sqrt{2}\,\widehat{\psi}_{(A}\psi_{B)}$, which is of the form (5.76)] and the components M_a are complex ($\overline{M_a} = -\sigma_{aAB}\widehat{\psi}^A\widehat{\psi}^B$). Furthermore, $R_a M^a = 0$, $M_a M^a = 0$ and $R_a R^a = (\mathrm{Re}\ M_a)(\mathrm{Re}\ M^a) = (\mathrm{Im}\ M_a)(\mathrm{Im}\ M^a) = (\psi^A\widehat{\psi}_A)^2$. Therefore, if ψ^A is a normalized spinor, in the sense that $\psi^A\widehat{\psi}_A = 1$ (*i.e.*, $|\psi^1|^2 + |\psi^2|^2 = 1$), then $\{\mathrm{Re}\ \mathbf{M}, \mathrm{Im}\ \mathbf{M}, \mathbf{R}\}$ is an orthonormal basis with the same orientation as $\{\mathbf{e}_1, \mathbf{e}_2, \mathbf{e}_3\}$. The spinors ψ^A and $-\psi^A$ define the same triad.

Conversely, given an orthonormal basis with the same orientation as $\{\mathbf{e}_1, \mathbf{e}_2, \mathbf{e}_3\}$, there is a normalized spinor, defined up to sign, such that the triad $\{\mathrm{Re}\ \mathbf{M}, \mathrm{Im}\ \mathbf{M}, \mathbf{R}\}$ coincides with the given basis.

When the metric is indefinite, any nonvanishing one-index spinor, ψ_A, such that $\widehat{\psi}_A$ is not proportional to ψ_A, defines the vectors **R** and **M** with components

$$R_a \equiv -\sigma_{aAB}\widehat{\psi}^A\psi^B, \qquad M_a \equiv \sigma_{aAB}\psi^A\psi^B. \tag{5.108}$$

R is real and **M** is complex (with $\overline{M_a} = \sigma_{aAB}\widehat{\psi}^A\widehat{\psi}^B$). Then $R^3 = -R_3 \geqslant 0$, $R_a M^a = 0$, and $M_a M^a = 0$, which means that the real and imaginary parts of M_a are orthogonal to R_a and to each other and that they have the same magnitude. Furthermore

$$R_a R^a = -(\mathrm{Re}\ M_a)(\mathrm{Re}\ M^a) = -(\mathrm{Im}\ M_a)(\mathrm{Im}\ M^a) = (\psi^A\widehat{\psi}_A)^2. \tag{5.109}$$

Thus, if ψ^A satisfies the condition $\psi^A\widehat{\psi}_A = \pm\mathrm{i}$ (which is possible since, according to (5.58), $\psi^A\widehat{\psi}_A$ is pure imaginary), then $\{\mathrm{Re}\ \mathbf{M}, \mathrm{Im}\ \mathbf{M}, \mathbf{R}\}$ is an orthonormal basis with the same orientation as $\{\mathbf{e}_1, \mathbf{e}_2, \mathbf{e}_3\}$. The spinors ψ^A and $-\psi^A$ give rise to the same triad. (If $\widehat{\psi}^A$ is proportional to ψ^A, then $\psi^A\widehat{\psi}_A = 0$ and R_a and M_a are proportional to each other.)

When the metric is indefinite, one can consider a "null basis", $\{\mathbf{l}, \mathbf{n}, \mathbf{s}\}$, formed by the three real vectors such that

$$l^a l_a = n^a n_a = l^a s_a = n^a s_a = 0, \qquad l^a n_a = -1, \qquad s^a s_a = 1.$$

The basis $\{\mathbf{l}, \mathbf{n}, \mathbf{s}\}$ can be related to the orthonormal basis $\{\mathrm{Re}\,\mathbf{M}, \mathrm{Im}\,\mathbf{M}, \mathbf{R}\}$ defined by a one-index spinor by

$$\mathbf{l} = \frac{1}{\sqrt{2}}(\mathbf{R} - \mathrm{Re}\,\mathbf{M}), \qquad \mathbf{n} = \frac{1}{\sqrt{2}}(\mathbf{R} + \mathrm{Re}\,\mathbf{M}), \qquad \mathbf{s} = \mathrm{Im}\,\mathbf{M}. \tag{5.110}$$

By introducing the spinors

$$\alpha_A = \frac{1}{\sqrt{2}}(\psi_A + \widehat{\psi}_A), \qquad \beta_A = \frac{1}{\sqrt{2}\,\mathrm{i}}(\psi_A - \widehat{\psi}_A),$$

which satisfy $\widehat{\alpha}_A = \alpha_A$, $\widehat{\beta}_A = \beta_A$ and $\alpha^A\beta_A = \mp 1$, one finds that

$$l_a = -\frac{1}{\sqrt{2}}\sigma_{aAB}\alpha^A\alpha^B, \qquad n_a = -\frac{1}{\sqrt{2}}\sigma_{aAB}\beta^A\beta^B, \qquad s_a = \sigma_{aAB}\alpha^A\beta^B. \tag{5.111}$$

The direction of the real null vector l_a is preserved under the replacement of α_A by $C\alpha_A$, where C is a real number different from zero; then, the conditions $\widehat{\beta}_A = \beta_A$ and $\alpha^A\beta_A = \mp 1$ are preserved if β_A is replaced by $C^{-1}(\beta_A + b\alpha_A)$, with $b \in \mathbb{R}$. Thus, from (5.111) we obtain

$$l_a \mapsto C^2 l_a, \qquad n_a \mapsto C^{-2}(n_a - \sqrt{2}\,b s_a + b^2 l_a), \qquad s_a \mapsto s_a - \sqrt{2}\,b l_a. \tag{5.112}$$

The $SO_0(2,1)$ transformations given by (5.112) are called null rotations about l_a.

6
Spinor Analysis

6.1 Covariant differentiation

Let M be a differentiable manifold of dimension 3 with a Riemannian metric, not necessarily positive definite. In an open neighborhood of each point of M we can find three (real, differentiable) vector fields, ∂_a, which form an *orthonormal rigid triad*, that is, at each point of their domain of definition, the vector fields ∂_a form an orthonormal basis of the tangent space to M at that point. In order to make use of the results of the preceding chapter, we shall assume that the (constant) components of the metric with respect to the basis $\{\partial_1, \partial_2, \partial_3\}$ are given by $(g_{ab}) = \mathrm{diag}(1, 1, 1)$ or by $(g_{ab}) = \mathrm{diag}(1, 1, -1)$.

As is well known, in a Riemannian manifold there exists a unique connection such that the metric tensor is covariantly constant and the torsion vanishes (the Levi-Civita, or Riemannian, connection). If we denote by ∇_a the covariant derivative with respect to ∂_a, the components of the Levi-Civita connection relative to the basis $\{\partial_1, \partial_2, \partial_3\}$ are the real-valued functions $\Gamma^c{}_{ba}$ (the Ricci rotation coefficients) given by

$$\nabla_a \partial_b = \Gamma^c{}_{ba} \partial_c. \tag{6.1}$$

Then, following a notation similar to that employed in the tensor calculus, the components of the covariant derivative of a tensor field $t^{ab\ldots}_{cd\ldots}$ are given by

$$\begin{aligned}\nabla_a t^{bc\ldots}_{de\ldots} &= \partial_a t^{bc\ldots}_{de\ldots} + \Gamma^b{}_{ma} t^{mc\ldots}_{de\ldots} + \Gamma^c{}_{ma} t^{bm\ldots}_{de\ldots} + \cdots \\ &\quad - \Gamma^m{}_{da} t^{bc\ldots}_{me\ldots} - \Gamma^m{}_{ea} t^{bc\ldots}_{dm\ldots} - \cdots .\end{aligned} \tag{6.2}$$

Since the torsion of the connection vanishes, the Lie bracket, or commutator, $[X, Y]$, of any pair of vector fields X, Y on M, is given by

$$[X, Y] = \nabla_X Y - \nabla_Y X.$$

Hence, the functions $\Gamma^c{}_{ba}$ satisfy

$$[\partial_a, \partial_b] = (\Gamma^c{}_{ba} - \Gamma^c{}_{ab})\, \partial_c. \qquad (6.3)$$

Using the fact that the metric is covariantly constant ($\nabla_a g_{bc} = 0$) and that the basis $\{\partial_1, \partial_2, \partial_3\}$ is rigid ($\partial_a g_{bc} = 0$), from (6.2) we find that $\Gamma^d{}_{ba} g_{dc} + \Gamma^d{}_{ca} g_{bd} = 0$; therefore, the functions

$$\Gamma_{abc} \equiv g_{ad} \Gamma^d{}_{bc}$$

are anti-symmetric in the first pair of indices,

$$\Gamma_{abc} = -\Gamma_{bac}. \qquad (6.4)$$

Given the vector fields ∂_a, the relations (6.3) and (6.4) allow us to find the Ricci rotation coefficients; by virtue of (6.4), we have $\Gamma_{abc} = \Gamma_{a[bc]} - \Gamma_{b[ac]} - \Gamma_{c[ab]}$, with the functions $\Gamma_{c[ba]} \equiv \frac{1}{2}(\Gamma_{cba} - \Gamma_{cab})$ being determined by (6.3).

The antisymmetry of Γ_{abc} in the first two indices implies that the spinor equivalent of Γ_{abc}, Γ_{ABCDEF}, can be written as

$$\Gamma_{ABCDEF} = -\Gamma_{ACEF}\varepsilon_{BD} - \Gamma_{BDEF}\varepsilon_{AC}, \qquad (6.5)$$

where

$$\Gamma_{ABCD} = -\tfrac{1}{2}\varepsilon^{RS}\Gamma_{RASBCD} = -\tfrac{1}{2}\Gamma^R{}_{ARBCD}$$

(*cf.* (5.14), the minus signs are introduced for later convenience). (Even though the Γ_{abc} are not the components of a tensor field, one can employ the decomposition (5.15), which applies for any anti-symmetric object, regardless of its transformation properties.) The components Γ_{ABCD} are symmetric in the first and second pairs of indices

$$\Gamma_{ABCD} = \Gamma_{(AB)(CD)}, \qquad (6.6)$$

and from (5.63) and (5.64), using the fact that $\det(\eta_{AB}) = -1$, one obtains

$$\overline{\Gamma_{ABCD}} = \begin{cases} -\Gamma^{ABCD} & \text{if the } \sigma_{aAB} \text{ are given by (5.38),} \\ \Gamma_{ABCD} & \text{if the } \sigma_{aAB} \text{ are given by (5.43),} \\ \eta_{AR}\eta_{BS}\eta_{CT}\eta_{DV}\Gamma^{RSTV} & \text{if the } \sigma_{aAB} \text{ are given by (5.45).} \end{cases} \qquad (6.7)$$

These relations imply that, when the connection symbols σ_{aAB} are complex, there are four independent complex components Γ_{ABCD} and one real or pure imaginary (Γ_{1212}); by contrast, when the connection symbols are all real, there are nine independent real components Γ_{ABCD}. Following the terminology used in Penrose and Rindler (1984), the functions Γ_{ABCD} will be called spin-coefficients.

Denoting by ∂_{AB} the differential operators (or vector fields)

$$\partial_{AB} \equiv \frac{1}{\sqrt{2}}\sigma^a{}_{AB}\partial_a, \qquad (6.8)$$

and making use of (5.11) and (6.5), one finds that the spinor equivalent of (6.1) is

$$\nabla_{AB}\partial_{CD} = \Gamma^R{}_{CAB}\partial_{RD} + \Gamma^R{}_{DAB}\partial_{CR}, \tag{6.9}$$

where ∇_{AB} denotes the covariant derivative with respect to ∂_{AB}. Hence, the commutators of the vector fields ∂_{AB} are given by

$$[\partial_{AB}, \partial_{CD}] = \Gamma^R{}_{CAB}\partial_{RD} + \Gamma^R{}_{DAB}\partial_{CR} - \Gamma^R{}_{ACD}\partial_{RB} - \Gamma^R{}_{BCD}\partial_{AR} \tag{6.10}$$

or, in a more explicit form,

$$\begin{aligned}
[\partial_{11}, \partial_{12}] &= (2\Gamma_{1212} - \Gamma_{2211})\,\partial_{11} - 2\Gamma_{1112}\,\partial_{12} + \Gamma_{1111}\,\partial_{22}, \\
[\partial_{11}, \partial_{22}] &= 2\Gamma_{1222}\,\partial_{11} - 2(\Gamma_{1122} + \Gamma_{2211})\,\partial_{12} + 2\Gamma_{1211}\,\partial_{22}, \\
[\partial_{22}, \partial_{12}] &= -\Gamma_{2222}\,\partial_{11} + 2\Gamma_{2212}\,\partial_{12} - (2\Gamma_{1212} - \Gamma_{1122})\,\partial_{22}.
\end{aligned} \tag{6.11}$$

When the connection symbols are complex, the last of these equations is the complex conjugate of the first one.

From (5.11) and (6.5) it follows that the spinor equivalent of (6.2) is

$$\begin{aligned}
\nabla_{AB} t^{CD\ldots}_{FG\ldots} &= \partial_{AB} t^{CD\ldots}_{FG\ldots} + \Gamma^C{}_{RAB} t^{RD\ldots}_{FG\ldots} + \Gamma^D{}_{RAB} t^{CR\ldots}_{FG\ldots} + \cdots \\
&\quad - \Gamma^R{}_{FAB} t^{CD\ldots}_{RG\ldots} - \Gamma^R{}_{GAB} t^{CD\ldots}_{FR\ldots} - \cdots.
\end{aligned}$$

Then, the covariant derivative of a spinor field $\psi^{CD\ldots}_{FG\ldots}$ with respect to ∂_{AB} is defined by

$$\begin{aligned}
\nabla_{AB} \psi^{CD\ldots}_{FG\ldots} &= \partial_{AB} \psi^{CD\ldots}_{FG\ldots} + \Gamma^C{}_{RAB} \psi^{RD\ldots}_{FG\ldots} + \Gamma^D{}_{RAB} \psi^{CR\ldots}_{FG\ldots} + \cdots \\
&\quad - \Gamma^R{}_{FAB} \psi^{CD\ldots}_{RG\ldots} - \Gamma^R{}_{GAB} \psi^{CD\ldots}_{FR\ldots} - \cdots.
\end{aligned} \tag{6.12}$$

In the case of a function, f, $\nabla_{AB} f = \partial_{AB} f$. The symmetry of Γ_{ABCD} in the first pair of indices [see (6.6)] implies that the covariant derivatives of ε_{AB} vanish and, therefore, the covariant derivative commutes with the raising and lowering of spinor indices. From (6.9) or (6.12), making use of (5.37), it follows that under the spin transformation (5.42), where now the entries $U^A{}_B$ are functions defined on M, the components Γ_{ABCD} transform according to

$$\Gamma'_{ABCD} = U_C{}^T U_D{}^V (U_A{}^R U_B{}^S \Gamma_{RSTV} + U_A{}^M\, \partial_{TV} U_{BM}). \tag{6.13}$$

By extending the definition (5.60) to ∂_{AB} and Γ_{ABCD}, conditions (5.39), (5.46), and (6.7) are equivalent to

$$\widehat{\partial_{AB}} = \begin{cases} -\partial_{AB} & \text{if } (g_{ab}) = \text{diag}(1, 1, 1), \\ \partial_{AB} & \text{if } (g_{ab}) = \text{diag}(1, 1, -1), \end{cases}$$

and

$$\widehat{\Gamma}_{ABCD} = \begin{cases} -\Gamma_{ABCD} & \text{if } (g_{ab}) = \mathrm{diag}(1,1,1), \\ \Gamma_{ABCD} & \text{if } (g_{ab}) = \mathrm{diag}(1,1,-1). \end{cases}$$

Then, from (6.12) it follows that

$$(\nabla_{AB}\psi^{CD\ldots}_{FG\ldots})^{\widehat{}} = \begin{cases} -\nabla_{AB}\widehat{\psi}^{CD\ldots}_{FG\ldots} & \text{if } (g_{ab}) = \mathrm{diag}(1,1,1), \\ \nabla_{AB}\widehat{\psi}^{CD\ldots}_{FG\ldots} & \text{if } (g_{ab}) = \mathrm{diag}(1,1,-1), \end{cases} \tag{6.14}$$

for any spinor field.

As a consequence of (6.14), the covariant derivative of spinors (6.12) is compatible with the inner product (5.66), $\langle\phi, \psi\rangle \equiv \widehat{\phi}_A \psi^A$, in the sense that, for any real tangent vector v^a, $v^{AB}\nabla_{AB}\langle\phi, \psi\rangle = \langle v^{AB}\nabla_{AB}\phi, \psi\rangle + \langle\phi, v^{AB}\nabla_{AB}\psi\rangle$. It can be readily seen that (6.12) is characterized by the conditions that the covariant derivatives of ε_{AB} and $\langle\ ,\ \rangle$ are equal to zero and, acting on vector fields, the torsion is also equal to zero.

Killing vector fields and the Lie derivative of a spinor field

A Killing vector field is the infinitesimal generator of a one-parameter group of isometries; that is, K is a Killing vector field if the Lie derivative of the metric tensor with respect to K vanishes. This last condition can also be expressed as

$$\nabla_a K_b + \nabla_b K_a = 0, \tag{6.15}$$

where ∇ denotes the Levi-Civita connection. Equations (6.15) are known as the Killing equations. Since the components of the covariant derivative of a Killing vector field are antisymmetric in their two indices, the spinor equivalent of $\nabla_a K_b$ is of the form

$$\nabla_{AB}K_{CD} = \varepsilon_{AC}L_{BD} + \varepsilon_{BD}L_{AC}, \tag{6.16}$$

with L_{AB} being symmetric,

$$L_{AB} = \tfrac{1}{2}\nabla^R{}_A K_{RB}.$$

The Lie derivative of a vector field Y with respect to a vector field X, denoted by $\pounds_X Y$, coincides with their Lie bracket $[X, Y]$; hence, if the torsion vanishes, $\pounds_X Y = \nabla_X Y - \nabla_Y X$. If $K = K^a\partial_a = -K^{AB}\partial_{AB}$ is a Killing vector field and $X = -X^{AB}\partial_{AB}$ is an arbitrary vector field, according to (6.16), the spinor components of the Lie derivative of X with respect to K, $\pounds_K X = (K^{CD}\nabla_{CD}X^{AB} - X^{CD}\nabla_{CD}K^{AB})\partial_{AB}$, are given by

$$\pounds_K X^{AB} = -K^{CD}\nabla_{CD}X^{AB} - L^A{}_C X^{CB} - L^B{}_D X^{AD}.$$

This formula suggests the following definition. The components of the Lie derivative of a spinor field, $\psi_{FG\ldots}^{CD\ldots}$ with respect to a Killing vector field K are given by

$$\begin{aligned}\pounds_K \psi_{FG\ldots}^{CD\ldots} &= -K^{AB}\nabla_{AB}\psi_{FG\ldots}^{CD\ldots} - L^C{}_R\psi_{FG\ldots}^{RD\ldots} - L^D{}_R\psi_{FG\ldots}^{CR\ldots} - \cdots \\ &\quad + L^R{}_F\psi_{RG\ldots}^{CD\ldots} + L^R{}_G\psi_{FR\ldots}^{CD\ldots} + \cdots \end{aligned} \tag{6.17}$$

[*cf.* (6.12)]. Assuming that K is real, (6.17) yields

$$(\pounds_K \psi_{FG\ldots}^{CD\ldots})\widehat{\ } = \pounds_K \widehat{\psi}_{FG\ldots}^{CD\ldots}.$$

The Lie derivative of a spinor field can also be defined along conformal Killing vector fields, that is, vector fields obeying the condition $\nabla_a K_b + \nabla_b K_a = 2\chi g_{ab}$, for some real-valued function χ. Now we have

$$\nabla_{AB} K_{CD} = \varepsilon_{AC} L_{BD} + \varepsilon_{BD} L_{AC} - \tfrac{1}{2}\chi(\varepsilon_{AC}\varepsilon_{BD} + \varepsilon_{AD}\varepsilon_{BC}),$$

with L_{AB} symmetric and the components of the Lie derivative of a spinor field $\psi_{FG\ldots}^{CD\ldots}$ with r superscripts and s subscripts are

$$\begin{aligned}\pounds_K \psi_{FG\ldots}^{CD\ldots} &= -K^{AB}\nabla_{AB}\psi_{FG\ldots}^{CD\ldots} - L^C{}_R\psi_{FG\ldots}^{RD\ldots} - L^D{}_R\psi_{FG\ldots}^{CR\ldots} - \cdots \\ &\quad + L^R{}_F\psi_{RG\ldots}^{CD\ldots} + L^R{}_G\psi_{FR\ldots}^{CD\ldots} + \cdots - \tfrac{1}{2}(r-s)\chi\psi_{FG\ldots}^{CD\ldots}. \end{aligned}$$

Thus, $\pounds_K \varepsilon_{AB} = \chi\varepsilon_{AB}$ and $\pounds_K \varepsilon^{AB} = -\chi\varepsilon^{AB}$.

6.2 Curvature

The Riemann, or curvature, tensor of the connection ∇, defined by

$$(\nabla_a\nabla_b - \nabla_b\nabla_a)t_c = -R^d{}_{cab}t_d, \tag{6.18}$$

possesses the symmetries $R_{abcd} = -R_{bacd} = -R_{abdc}$, when the torsion vanishes. Hence, the components of the curvature tensor of the Levi-Civita connection of a three-dimensional manifold can be expressed in the form

$$R_{abcd} = -\det(g_{rs})\,\varepsilon_{abe}\varepsilon_{cdf}G^{ef}, \tag{6.19}$$

where G_{ab} are the components of a tensor (the factor $-\det(g_{rs})$ is introduced for later convenience). Using (5.92), from (6.19) it follows that the components of the Ricci tensor $R_{ab} \equiv R^c{}_{acb}$ are given by $R_{ab} = -g_{ab}G^c{}_c + G_{ba}$, and, therefore, the scalar curvature, $R \equiv R^a{}_a$, is given by $R = -2G^c{}_c$. Thus

$$G_{ab} = R_{ab} - \tfrac{1}{2}R\,g_{ab}, \tag{6.20}$$

which shows that G_{ab} is symmetric.

If the torsion vanishes, the Riemann tensor also satisfies the identity $R_{abcd} + R_{acdb} + R_{adbc} = 0$ or, equivalently, $\varepsilon^{bcd} R_{abcd} = 0$; substituting (6.19) into this last equation, with the aid of (5.92), we find that $\varepsilon_{abc} G^{bc} = 0$, which is equivalent to the symmetry of G_{ab}. Denoting by Φ_{ab} the components of the trace-free part of the Ricci tensor, $\Phi_{ab} \equiv R_{ab} - \frac{1}{3} R\, g_{ab}$, (6.20) gives

$$G_{ab} = \Phi_{ab} - \tfrac{1}{6} R\, g_{ab}. \tag{6.21}$$

Making use of the spinor equivalent of the alternating symbol ε_{abc} given by (5.17), we find that the spinor equivalent of (6.19) is

$$\begin{aligned} R_{ABCDEFHI} = {}& \tfrac{1}{2}(\varepsilon_{AC}\varepsilon_{EH} G_{BDFI} + \varepsilon_{AC}\varepsilon_{FI} G_{BDEH} \\ & + \varepsilon_{BD}\varepsilon_{EH} G_{ACFI} + \varepsilon_{BD}\varepsilon_{FI} G_{ACEH}), \end{aligned} \tag{6.22}$$

where G_{ABCD} are the spinor components of G_{ab}, and from (5.8) and (6.21) we have

$$G_{ABCD} = \Phi_{ABCD} + \tfrac{1}{12} R(\varepsilon_{AC}\varepsilon_{BD} + \varepsilon_{AD}\varepsilon_{BC}), \tag{6.23}$$

where Φ_{ABCD} are the spinor components of Φ_{ab}, which are totally symmetric.

Applying the decomposition (5.15), the spinor equivalent of (6.18) can be written in the form

$$(\varepsilon_{AC} \Box_{BD} + \varepsilon_{BD} \Box_{AC})\, t_{EF} = R^{HI}{}_{EFABCD} t_{HI}, \tag{6.24}$$

where

$$\Box_{AB} \equiv \nabla^R{}_{(A} \nabla_{B)R}. \tag{6.25}$$

Then, from (6.22) and (6.24) it follows that $\Box_{AB} t_{CD} = -\frac{1}{2} G_{ABCE} t^E{}_D - \frac{1}{2} G_{ABDE} t_C{}^E$; therefore, in the case of a one-index spinor field ψ_A,

$$\begin{aligned} \Box_{AB}\psi_C &= -\tfrac{1}{2} G_{ABCD}\psi^D \\ &= -\tfrac{1}{2}\Phi_{ABCD}\psi^D - \tfrac{1}{24} R(\varepsilon_{AC}\psi_B + \varepsilon_{BC}\psi_A). \end{aligned} \tag{6.26}$$

This formula and the relation $\Box_{AB}(\psi_{CD\ldots}\phi_{RS\ldots}) = \psi_{CD\ldots}\Box_{AB}\phi_{RS\ldots} + \phi_{RS\ldots}\Box_{AB}\psi_{CD\ldots}$ allow us to compute the commutator of covariant derivatives of any spinor field. By expanding the left-hand side of (6.26), making use of (6.25) and (6.12), we obtain

$$\begin{aligned} -\tfrac{1}{2}\Phi_{ABCD} - \tfrac{1}{24} R(\varepsilon_{AC}\varepsilon_{BD} + \varepsilon_{AD}\varepsilon_{BC}) = {}& \partial^R{}_{(A}\Gamma_{|DC|B)R} \\ & - \Gamma^S{}_R{}^R{}_{(A}\Gamma_{|DC|B)S} - \Gamma^S{}_{(A}{}^R{}_{B)}\Gamma_{DCSR} - \Gamma^S{}_C{}^R{}_{(A}\Gamma_{|DS|B)R}. \end{aligned} \tag{6.27}$$

(Note that, for a scalar function f, from (6.10) it follows that $\partial^R{}_{(A}\partial_{B)R} f = \Gamma^M{}_R{}^R{}_{(A}\partial_{B)M} f + \Gamma^M{}_{(AB)}{}^R \partial_{MR} f$ and, therefore, $\nabla^R{}_{(A}\nabla_{B)R} f = 0$.)

Since Φ_{ab} is real, from (5.63) and (5.64) we find that

$$\overline{\Phi_{ABCD}} = \begin{cases} \Phi^{ABCD} & \text{if the } \sigma_{aAB} \text{ are given by (5.38)}, \\ \Phi_{ABCD} & \text{if the } \sigma_{aAB} \text{ are given by (5.43)}, \\ \eta_{AR}\eta_{BS}\eta_{CT}\eta_{DV}\Phi^{RSTV} & \text{if the } \sigma_{aAB} \text{ are given by (5.45)}, \end{cases}$$

i.e., $\widehat{\Phi}_{ABCD} = \Phi_{ABCD}$. Owing to (6.19), in a three-dimensional manifold, the Bianchi identities, $\nabla_a R_{bcde} + \nabla_b R_{cade} + \nabla_c R_{abde} = 0$, are equivalent to the contracted Bianchi identities, $\nabla^a G_{ab} = 0$, which amount to

$$\nabla^{AB}\Phi_{ABCD} + \tfrac{1}{6}\partial_{CD}R = 0. \tag{6.28}$$

Making use of Killing's equations (6.15), of (6.18) and the algebraic properties of the Riemann tensor one finds that if K_a is a Killing vector field, then $\nabla_a\nabla_b K_c = R^d{}_{abc}K_d$, which is equivalent to [see (6.16)]

$$\begin{aligned} \nabla_{AB}L_{CD} &= G_{E(A|CD|}K_{B)}{}^E \\ &= \Phi_{CDE(A}K_{B)}{}^E - \tfrac{1}{12}R(\varepsilon_{AC}K_{BD} + \varepsilon_{BD}K_{AC}). \end{aligned}$$

Conformal rescalings

Two metrics of M, $\mathrm{d}s^2$ and $\mathrm{d}s'^2$, are conformally related if there exists a positive function, Ω, such that $\mathrm{d}s'^2 = \Omega^{-2}\mathrm{d}s^2$. If ∂_{AB} is a spinorial triad for the metric $\mathrm{d}s^2$, then

$$\partial'_{AB} \equiv \Omega\,\partial_{AB} \tag{6.29}$$

is a spinorial triad for $\mathrm{d}s'^2$. The components of the connections compatible with $\mathrm{d}s^2$ and $\mathrm{d}s'^2$ are related in a simple way if one makes use of bases related as in (6.29). Indeed, from (6.10) and (6.29) one obtains

$$[\partial'_{AB}, \partial'_{CD}] = \gamma^R{}_{CAB}\,\partial'_{RD} + \gamma^R{}_{DAB}\,\partial'_{CR} - \gamma^R{}_{ACD}\,\partial'_{RB} - \gamma^R{}_{BCD}\,\partial'_{AR}, \tag{6.30}$$

where

$$\gamma^M{}_{CAB} = \Omega\,\Gamma^M{}_{CAB} + \tfrac{1}{2}\delta^M_C\,\partial_{AB}\Omega.$$

If Ω is not a constant, the coefficients γ_{ABCD} are not symmetric in the first pair of indices [see (6.6)] and, therefore, are not the components of the connection for the triad ∂'_{AB}. However, noting that the right-hand side of (6.30) is unchanged if $\gamma^M{}_{CAB}$ is replaced by $\gamma^M{}_{CAB} + \delta^M_{(A}\beta_{B)C}$, provided that $\beta_{AB} = \beta_{BA}$, one finds that taking $\beta_{AB} = -\partial_{AB}\Omega$, $\Gamma'^M{}_{CAB} \equiv \gamma^M{}_{CAB} + \delta^M_{(A}\beta_{B)C}$ has the symmetries (6.6); thus the components of the connection compatible with $\mathrm{d}s'^2$ with respect to ∂'_{AB} are given by

$$\begin{aligned} \Gamma'_{ABCD} &= \Omega\,\Gamma_{ABCD} + \tfrac{1}{2}\varepsilon_{C(A}\partial_{B)D}\Omega + \tfrac{1}{2}\varepsilon_{D(A}\partial_{B)C}\Omega \\ &= \Omega\,\Gamma_{ABCD} - \tfrac{1}{2}\varepsilon_{AD}\partial_{BC}\Omega - \tfrac{1}{2}\varepsilon_{BC}\partial_{AD}\Omega. \end{aligned} \tag{6.31}$$

Note that (6.31) relates the *components* of two different connections with respect to two different bases. Note also that the spinor equivalent of the components of the metric ds'^2 with respect to ∂'_{AB} is the same as the spinor equivalent of the components of the metric ds^2 with respect to ∂_{AB} (namely, $-(\varepsilon_{AC}\varepsilon_{BD}+\varepsilon_{AD}\varepsilon_{BC})$) and in *all cases* the spinor indices are raised or lowered by means of ε^{AB} and ε_{AB}.

The components of the curvature of ds'^2 with respect to ∂'_{AB} can be obtained substituting (6.31) into (6.27); in this manner, we find that

$$\Phi'_{ABCD} = \Omega^2\Phi_{ABCD} + \Omega\nabla_{(AB}\nabla_{CD)}\Omega \tag{6.32}$$

and

$$R' = \Omega^2 R - 4\Omega\nabla^{AB}\nabla_{AB}\Omega + 6(\partial^{AB}\Omega)(\partial_{AB}\Omega). \tag{6.33}$$

The Cotton–York tensor (Schouten 1921, York 1971, Hall and Capocci 1999) is defined by

$$Y_{ab} \equiv \varepsilon_{acd}\big(\nabla^c R_b{}^d - \tfrac{1}{4}\delta_b^d\partial^c R\big) = \varepsilon_{acd}\big(\nabla^c \Phi_b{}^d + \tfrac{1}{12}\delta_b^d\partial^c R\big),$$

hence, the spinor equivalent of the Cotton–York tensor is given by

$$Y_{ABCD} = \sqrt{2}\,\mathrm{i}\big\{\nabla^S{}_{(A}\Phi_{B)SCD} - \tfrac{1}{24}\varepsilon_{C(A}\partial_{B)D}R - \tfrac{1}{24}\varepsilon_{D(A}\partial_{B)C}R\big\}, \tag{6.34}$$

if the metric is positive definite, and

$$Y_{ABCD} = -\sqrt{2}\,\big\{\nabla^S{}_{(A}\Phi_{B)SCD} - \tfrac{1}{24}\varepsilon_{C(A}\partial_{B)D}R - \tfrac{1}{24}\varepsilon_{D(A}\partial_{B)C}R\big\}, \tag{6.35}$$

if the metric has signature $(+\,+\,-)$. Then, by virtue of the Bianchi identities (6.28), $Y^A{}_{BAD} = 0$, which means that Y_{ABCD} is totally symmetric; therefore, Y_{ab} is symmetric and trace-free and

$$Y_{ABCD} = \begin{cases} \sqrt{2}\,\mathrm{i}\nabla^S{}_{(A}\Phi_{BCD)S} & \text{if } (g_{ab}) = \mathrm{diag}(1,1,1), \\ -\sqrt{2}\,\nabla^S{}_{(A}\Phi_{BCD)S} & \text{if } (g_{ab}) = \mathrm{diag}(1,1,-1). \end{cases} \tag{6.36}$$

Making use of (6.31) and (6.32), from (6.36) one finds that under a conformal rescaling, the spinor components of the Cotton–York tensor with respect to the triads ∂'_{AB} and ∂_{AB} are related by

$$Y'_{ABCD} = \Omega^3 Y_{ABCD}.$$

Thus, if M is conformally flat, *i.e.*, if the metric of M is conformally equivalent to a flat metric, then its Cotton–York tensor vanishes. It can be shown that if the Cotton–York tensor vanishes, then M is locally conformally flat.

Since the Cotton–York tensor is real, $\widehat{Y}_{ABCD} = Y_{ABCD}$ [see (5.65)]. The totally symmetric spinors Φ_{ABCD} and Y_{ABCD} can be expressed as symmetrized outer products of their principal spinors. A k-fold repeated principal spinor of Φ_{ABCD}, with $k \geqslant 3$, is, at least, a $(k-2)$-fold repeated principal spinor of Y_{ABCD}.

6.3 Spin weight and priming operation

6.3.1 Positive definite metric

A quantity η has spin weight s if under the spin transformation given by the matrix

$$(U^A{}_B) = \begin{pmatrix} e^{-i\theta/2} & 0 \\ 0 & e^{i\theta/2} \end{pmatrix} \tag{6.37}$$

(which corresponds to the rotation through an angle θ given by $\partial_1 + i\partial_2 \mapsto e^{i\theta}(\partial_1 + i\partial_2)$), it transforms according to

$$\eta \mapsto e^{is\theta}\eta. \tag{6.38}$$

From (5.42), (6.37), and (6.38) it follows that each component $\psi_{AB...D}$ of a spinor has a spin weight equal to one half of the difference between the number of the indices $A, B, \ldots, D$ taking the value 1 and the number of indices taking the value 2. Hence, the $2n + 1$ independent components of a totally symmetric $2n$-index spinor can be labeled by their spin weight:

$$\psi_n \equiv \psi_{11...1}, \qquad \psi_{n-1} \equiv \psi_{21...1}, \qquad \ldots, \qquad \psi_{-n} \equiv \psi_{22...2}. \tag{6.39}$$

Equations (5.39) and (5.9) imply that if $t_{ab...c}$ is a *real* trace-free totally symmetric n-index tensor, then the spinor components $t_n \equiv t_{11...1}, t_{n-1} \equiv t_{21...1}, \ldots, t_{-n} \equiv t_{22...2}$, satisfy the relation

$$\overline{t_s} = (-1)^s t_{-s}. \tag{6.40}$$

Owing to (6.7), the components Γ_{ABCD} are given by the four complex functions

$$\kappa \equiv \Gamma_{1111}, \qquad \beta \equiv \Gamma_{1211}, \qquad \rho \equiv \Gamma_{2211}, \qquad \alpha \equiv \Gamma_{1112}, \tag{6.41}$$

together with the pure imaginary function

$$\varepsilon \equiv \Gamma_{1212}. \tag{6.42}$$

Equations (6.7) and (6.41) give

$$\Gamma_{2222} = -\overline{\kappa}, \qquad \Gamma_{1222} = \overline{\beta}, \qquad \Gamma_{1122} = -\overline{\rho}, \qquad \Gamma_{2212} = \overline{\alpha}. \tag{6.43}$$

Introducing now the definitions

$$D \equiv -\partial_{12}, \qquad \delta \equiv \partial_{11}, \qquad \overline{\delta} \equiv -\partial_{22}, \tag{6.44}$$

or, equivalently,

$$D = \frac{1}{\sqrt{2}}\partial_3, \qquad \delta = \frac{1}{\sqrt{2}}(\partial_1 + i\partial_2), \qquad \overline{\delta} = \frac{1}{\sqrt{2}}(\partial_1 - i\partial_2), \tag{6.45}$$

from (6.11) and (6.41)–(6.44) one gets

$$[D, \delta] = 2\alpha D + (2\varepsilon - \rho)\delta - \kappa\overline{\delta},$$
$$[\delta, \overline{\delta}] = 2(\overline{\rho} - \rho)D - 2\overline{\beta}\delta + 2\beta\overline{\delta}. \tag{6.46}$$

In view of (6.44), under the transformation (6.37) the operators D, δ and $\overline{\delta}$ transform according to

$$D \mapsto D, \qquad \delta \mapsto e^{i\theta}\delta, \qquad \overline{\delta} \mapsto e^{-i\theta}\overline{\delta}, \tag{6.47}$$

and using (6.46) or (6.13) one readily obtains that

$$\kappa \mapsto e^{2i\theta}\kappa, \qquad \rho \mapsto \rho, \qquad \alpha \mapsto e^{i\theta}\alpha, \tag{6.48}$$

and

$$\beta \mapsto e^{i\theta}\left(\beta - \tfrac{1}{2}i\delta\theta\right), \qquad \varepsilon \mapsto \varepsilon + \tfrac{1}{2}iD\theta, \tag{6.49}$$

which means that κ, ρ and α (together with their complex conjugates) have a well-defined spin weight. On the other hand, from (6.38), (6.47), and (6.49) it follows that if η has spin weight s, then $(D - 2s\varepsilon)\eta$, $(\delta + 2s\beta)\eta$, and $(\overline{\delta} - 2s\overline{\beta})\eta$ have spin weight s, $s+1$ and $s-1$, respectively. The operators $D - 2s\varepsilon$, $\delta + 2s\beta$, and $\overline{\delta} - 2s\overline{\beta}$ are the analogs of the Geroch–Held–Penrose operators "thorn", "eth", and "eth-bar" (Geroch, Held and Penrose 1973, Penrose and Rindler 1984). Borrowing the Geroch–Held–Penrose notation, for a quantity η with spin weight s, we define the operators Þ, ð, and $\overline{\eth}$ by

$$Þ\eta \equiv (D - 2s\varepsilon)\eta, \qquad \eth\eta \equiv (\delta + 2s\beta)\eta, \qquad \overline{\eth}\eta \equiv (\overline{\delta} - 2s\overline{\beta})\eta.$$

The operators ð and $\overline{\eth}$ defined in Chapters 2 and 4 differ by a factor from the operators defined here (see below).

For instance, if (u, v, z) are orthogonal cylindrical coordinates in Euclidean three-dimensional space, then, denoting by h_1 and h_2 the scale factors corresponding to u and v (*i.e.*, $dx^2 + dy^2 = h_1^2 du^2 + h_2^2 dv^2$),

$$\partial_1 \equiv \frac{1}{h_1}\partial_u, \qquad \partial_2 \equiv \frac{1}{h_2}\partial_v, \qquad \partial_3 \equiv \partial_z, \tag{6.50}$$

form an orthonormal triad. A straightforward computation gives

$$[D, \delta] = 0, \qquad [\delta, \overline{\delta}] = \frac{1}{\sqrt{2}h_1h_2}\{(h_{2,u} - ih_{1,v})\delta - (h_{2,u} + ih_{1,v})\overline{\delta}\}, \tag{6.51}$$

where the comma indicates partial differentiation. Comparing (6.51) with (6.46) one finds that the only nonvanishing spin-coefficient is given by

$$\beta = -\frac{1}{2\sqrt{2}h_1h_2}(h_{2,u} + ih_{1,v}). \tag{6.52}$$

Therefore, the spin weight raising and lowering operators $\delta + 2s\beta$ and $\overline{\delta} - 2s\overline{\beta}$ are

$$\begin{aligned} \delta + 2s\beta &= \frac{1}{\sqrt{2}}\left(\frac{1}{h_1}\partial_u + \frac{\mathrm{i}}{h_2}\partial_v\right) - \frac{s}{\sqrt{2}\,h_1h_2}(h_{2,u} + \mathrm{i}h_{1,v}), \\ \overline{\delta} - 2s\overline{\beta} &= \frac{1}{\sqrt{2}}\left(\frac{1}{h_1}\partial_u - \frac{\mathrm{i}}{h_2}\partial_v\right) + \frac{s}{\sqrt{2}\,h_1h_2}(h_{2,u} - \mathrm{i}h_{1,v}). \end{aligned} \tag{6.53}$$

Apart from a factor $-\sqrt{2}$, (6.53) reduces to the definitions of the operators $\eth$ and $\overline{\eth}$ given in Chapter 4.

In the case of spherical coordinates (r, θ, ϕ) in Euclidean three-dimensional space, one finds that the triad

$$D = \frac{1}{\sqrt{2}}\partial_r, \quad \delta = \frac{1}{\sqrt{2}\,r}\left(\partial_\theta + \frac{\mathrm{i}}{\sin\theta}\partial_\phi\right), \quad \overline{\delta} = \frac{1}{\sqrt{2}\,r}\left(\partial_\theta - \frac{\mathrm{i}}{\sin\theta}\partial_\phi\right), \tag{6.54}$$

has spin-coefficients

$$\kappa = \alpha = \varepsilon = 0, \qquad \beta = -\frac{\cot\theta}{2\sqrt{2}\,r}, \qquad \rho = \frac{1}{\sqrt{2}\,r}, \tag{6.55}$$

hence, the spin weight raising and lowering operators are

$$\begin{aligned} \delta + 2s\beta &= \frac{1}{\sqrt{2}\,r}\left(\partial_\theta + \frac{\mathrm{i}}{\sin\theta}\partial_\phi - s\cot\theta\right), \\ \overline{\delta} - 2s\overline{\beta} &= \frac{1}{\sqrt{2}\,r}\left(\partial_\theta - \frac{\mathrm{i}}{\sin\theta}\partial_\phi + s\cot\theta\right), \end{aligned} \tag{6.56}$$

which are, apart from a factor $-\sqrt{2}\,r$, the operators $\eth$ and $\overline{\eth}$ defined in Chapter 2.

Owing to (6.31), (6.41), and (6.42), under the conformal rescaling given by

$$D \mapsto \Omega D, \qquad \delta \mapsto \Omega\delta, \qquad \overline{\delta} \mapsto \Omega\overline{\delta},$$

the spin-coefficients are replaced according to

$$\begin{aligned} &\kappa \mapsto \Omega\kappa, \qquad \beta \mapsto \Omega\beta + \tfrac{1}{2}\delta\Omega, \qquad \rho \mapsto \Omega\rho - D\Omega, \\ &\qquad \alpha \mapsto \Omega\alpha - \tfrac{1}{2}\delta\Omega, \qquad \varepsilon \mapsto \Omega\varepsilon. \end{aligned}$$

Therefore, for the spin-weighted operators we have

$$\eth \mapsto \Omega^{1-s}\eth\Omega^{s}, \qquad \overline{\eth} \mapsto \Omega^{1+s}\overline{\eth}\Omega^{-s}, \qquad \text{Þ} \mapsto \Omega\text{Þ}. \tag{6.57}$$

Note that the standard metric of the sphere, written in terms of $\zeta = \mathrm{e}^{\mathrm{i}\phi}\cot\frac{1}{2}\theta$ and its complex conjugate, is given by

$$\mathrm{d}\theta^2 + \sin^2\theta\,\mathrm{d}\phi^2 = \frac{4\mathrm{d}\zeta\,\mathrm{d}\overline{\zeta}}{(1+\zeta\overline{\zeta})^2} = \Omega^{-2}\mathrm{d}\zeta\,\mathrm{d}\overline{\zeta},$$

with $\Omega = \frac{1}{2}(1 + \zeta\bar{\zeta})$; hence, the operators ð and $\bar{\text{ð}}$ adapted to the spherical coordinates, defined in Chapter 2, can be obtained from those corresponding to the plane by means of (6.57) [see (2.40)].

As in the case of the spinor formalism applied in general relativity (Geroch, Held and Penrose 1973, Penrose and Rindler 1984), we can introduce the map $'$ defined by the matrix

$$(U^A{}_B) = \begin{pmatrix} 0 & -\mathrm{i} \\ -\mathrm{i} & 0 \end{pmatrix}, \tag{6.58}$$

which belongs to SU(2) and corresponds essentially to an interchange of the basis spinors. The operators D, δ and $\bar{\delta}$ transform according to

$$D' = -D, \qquad \delta' = \bar{\delta}, \qquad \bar{\delta}' = \delta \tag{6.59}$$

[*cf.* (6.44) and (6.45)], which shows that the matrix (6.58) represents a rotation through π about ∂_1. Then, from (6.46) and (6.59) it is easy to see that

$$\kappa' = -\bar{\kappa}, \qquad \beta' = \bar{\beta}, \qquad \rho' = -\bar{\rho}, \qquad \alpha' = \bar{\alpha}, \qquad \varepsilon' = \varepsilon \ (= -\bar{\varepsilon}).$$

The spin-weighted components of a totally symmetric $2n$-index spinor [see (6.39)] transform according to

$$\psi'_s = \mathrm{i}^{2n}\psi_{-s}.$$

Note that under the priming operation (6.60), and (6.62)–(6.65) below are mapped into themselves.

Following the notation (6.39), the spinor components of the gradient of a scalar function f, $(\text{grad } f)_{AB} \equiv \frac{1}{\sqrt{2}}\sigma^a{}_{AB}\partial_a f = \partial_{AB} f$, are given explicitly by [*cf.* (6.8), (6.44), and (6.45)]

$$(\text{grad } f)_{+1} = \delta f, \qquad (\text{grad } f)_0 = -Df, \qquad (\text{grad } f)_{-1} = -\bar{\delta} f. \tag{6.60}$$

Similarly, the spinor components of any vector field $\mathbf{F}$, $F_{AB} = \frac{1}{\sqrt{2}}\sigma^a{}_{AB}F_a$, are

$$\begin{aligned} F_{+1} &= F_{11} = \frac{1}{\sqrt{2}}(F_1 + \mathrm{i}F_2), \\ F_0 &= F_{12} = -\frac{1}{\sqrt{2}}F_3, \\ F_{-1} &= F_{22} = -\frac{1}{\sqrt{2}}(F_1 - \mathrm{i}F_2), \end{aligned} \tag{6.61}$$

where F_1, F_2, F_3 are the components of $\mathbf{F}$ with respect to the orthonormal basis $\{\partial_1, \partial_2, \partial_3\}$. According to (6.12), and (6.41)–(6.44), the divergence of $\mathbf{F}$ is given by

$$\begin{aligned} \operatorname{div} \mathbf{F} &= \nabla_a F^a = -\nabla_{AB} F^{AB} \\ &= (\bar{\delta} - 2\bar{\beta} + 2\bar{\alpha})F_{+1} - 2(D + \rho + \bar{\rho})F_0 - (\delta - 2\beta + 2\alpha)F_{-1}. \end{aligned} \tag{6.62}$$

Using (5.17) one finds that the spinor components of curl **F** are

$$(\text{curl}\,\mathbf{F})_{AB} = \varepsilon_{ABCDEG}\nabla^{CD}F^{EG} = \sqrt{2}\,\mathrm{i}\nabla^{R}{}_{(A}F_{B)R}, \tag{6.63}$$

therefore

$$\begin{aligned}
(\text{curl}\,\mathbf{F})_{+1} &= \sqrt{2}\,\mathrm{i}\{(D-2\varepsilon+\rho)F_{+1}+(\delta+2\alpha)F_0-\kappa F_{-1}\},\\
(\text{curl}\,\mathbf{F})_{0} &= \sqrt{2}\,\mathrm{i}\{\tfrac{1}{2}(\overline{\delta}-2\overline{\beta})F_{+1}+(\rho-\overline{\rho})F_0+\tfrac{1}{2}(\delta-2\beta)F_{-1}\},\\
(\text{curl}\,\mathbf{F})_{-1} &= \sqrt{2}\,\mathrm{i}\{\overline{\kappa}F_{+1}+(\overline{\delta}+2\overline{\alpha})F_0-(D+2\varepsilon+\overline{\rho})F_{-1}\}.
\end{aligned} \tag{6.64}$$

The divergence of the curl of a vector field vanishes even if the curvature is different from zero. In fact, $\text{div}\,\text{curl}\,\mathbf{F} = -\sqrt{2}\,\mathrm{i}\nabla^{A(B}\nabla^{R)}{}_{A}F_{BR} = -\sqrt{2}\,\mathrm{i}\Box_{BR}F^{BR} = \frac{1}{\sqrt{2}}\mathrm{i}(G_{BR}{}^{B}{}_{D}F^{DR}+G_{BR}{}^{R}{}_{D}F^{BD}) = 0$. Similarly, one finds that curl grad $f = 0$. Substituting (6.60) into (6.64) one readily obtains that curl grad $= 0$ amounts to the commutation relations (6.46).

In the case of a trace-free symmetric 2-index tensor field, t_{ab}, the spinor components of the vector field $(\text{div}\,t)_a \equiv \nabla^b t_{ab}$ are given explicitly by

$$\begin{aligned}
(\text{div}\,t)_{+1} &= (\overline{\delta}-4\overline{\beta}+2\overline{\alpha})t_{+2}-2(D-2\varepsilon+\rho+2\overline{\rho})t_{+1}\\
&\quad -(\delta+6\alpha)t_0+2\kappa t_{-1},\\
(\text{div}\,t)_{0} &= \overline{\kappa}t_{+2}+(\overline{\delta}-2\overline{\beta}+4\overline{\alpha})t_{+1}-(2D+3\rho+3\overline{\rho})t_0\\
&\quad -(\delta-2\beta+4\alpha)t_{-1}+\kappa t_{-2},\\
(\text{div}\,t)_{-1} &= 2\overline{\kappa}t_{+1}+(\overline{\delta}+6\overline{\alpha})t_0-2(D+2\varepsilon+2\rho+\overline{\rho})t_{-1}\\
&\quad -(\delta-4\beta+2\alpha)t_{-2}.
\end{aligned} \tag{6.65}$$

Equation (6.27) leads to the explicit expressions

$$-\tfrac{1}{2}\Phi_{+2} = (D-4\varepsilon+\rho+\overline{\rho})\kappa+(\delta+2\beta+2\alpha)\alpha, \tag{6.66}$$

$$-\tfrac{1}{2}\Phi_{+1} = (D-2\varepsilon+\rho)\beta+(\delta+2\alpha)\varepsilon-(\overline{\alpha}+\overline{\beta})\kappa+\alpha\rho, \tag{6.67}$$

$$-\Phi_{+1} = (\overline{\delta}-4\overline{\beta})\kappa-(\delta+2\alpha)\overline{\rho}+2\alpha\rho, \tag{6.68}$$

$$-\tfrac{1}{2}\Phi_0-\tfrac{1}{12}R = (D+\rho)\rho+(\delta-2\beta+2\alpha)\overline{\alpha}+\kappa\overline{\kappa}, \tag{6.69}$$

$$-\Phi_0+\tfrac{1}{12}R = \overline{\delta}\beta+\delta\overline{\beta}-4\beta\overline{\beta}+2\varepsilon(\rho-\overline{\rho})+\kappa\overline{\kappa}-\rho\overline{\rho}, \tag{6.70}$$

together with the complex conjugates of (6.66)–(6.69), taking into account that $\overline{D} = D$, $\overline{\varepsilon} = -\varepsilon$ and $\overline{\Phi_s} = (-1)^s\Phi_{-s}$.

If η has spin weight s, from (6.46), (6.67), and (6.70) it follows that the commutators of the spin-weighted operators Þ, ð, and $\overline{\text{ð}}$ are given by

$$\begin{aligned}
[\text{Þ},\text{ð}]\eta &= 2\alpha\text{Þ}\eta-\rho\text{ð}\eta-\kappa\overline{\text{ð}}\eta+2s(-\tfrac{1}{2}\Phi_{+1}+\overline{\alpha}\kappa-\alpha\rho)\eta,\\
[\text{ð},\overline{\text{ð}}]\eta &= 2(\overline{\rho}-\rho)\text{Þ}\eta+2s\big(\Phi_0-\tfrac{1}{12}R+\kappa\overline{\kappa}-\rho\overline{\rho}\big)\eta.
\end{aligned} \tag{6.71}$$

It may be noticed that the equations (6.67) and (6.70), employed above, involve the spin-coefficients that do not have a well-defined spin weight (β and ε).

According to the second equation in (6.46), $\rho = \overline{\rho}$ if and only if D is locally surface-orthogonal (*i.e.*, there exists locally a family of two-dimensional surfaces such that, at each point, D is orthogonal to the tangent space to the surface passing through that point). Making use of (6.9) and (6.44) one finds that the shape operator, S, of these surfaces is given by $S(\delta) \equiv -\nabla_\delta(\sqrt{2}D) = \sqrt{2}\,\nabla_{11}\partial_{12} = -\sqrt{2}(\rho\delta + \kappa\overline{\delta})$ and, therefore, $S(\overline{\delta}) = -\sqrt{2}(\overline{\kappa}\delta + \rho\overline{\delta})$. Thus, the Gaussian and the mean curvatures of the surfaces orthogonal to D are $K = 2(\rho^2 - \kappa\overline{\kappa})$ and $H = -\sqrt{2}\rho$, respectively. Furthermore, if κ is real, then $\partial_1 = (\delta + \overline{\delta})/\sqrt{2}$ and $\partial_2 = \mathrm{i}(\overline{\delta} - \delta)/\sqrt{2}$ are eigenvectors of the shape operator with eigenvalues $-\sqrt{2}(\rho + \kappa)$ and $-\sqrt{2}(\rho - \kappa)$, respectively, which means that ∂_1 and ∂_2 are principal vectors and $-\sqrt{2}(\rho \pm \kappa)$ are the principal curvatures. When κ is complex, κ is of the form $\kappa = |\kappa|\mathrm{e}^{\mathrm{i}\chi}$ and under the rotation (6.47) with $\theta = -\chi/2$, $\kappa \mapsto |\kappa|$; thus, the principal vectors form an angle $-(\arg\kappa)/2$ with respect to ∂_1 and ∂_2.

On the other hand, when ρ is real, the scalar curvature of the metric induced on the surfaces orthogonal to D is found to be

$$^{(2)}R = 4(\overline{\delta}\beta + \delta\overline{\beta} - 4\beta\overline{\beta}),$$

hence, the second equation in (6.71) reduces to

$$[\eth, \overline{\eth}]\eta = 2s\big(\Phi_0 - \tfrac{1}{12}R - \tfrac{1}{2}K\big)\eta = -\frac{s}{2}\,{}^{(2)}R\,\eta, \qquad (6.72)$$

furthermore, the Gaussian curvature (defined as the determinant of the shape operator) is equal to $\frac{1}{2}{}^{(2)}R$ if and only if $\Phi_0 - \frac{1}{12}R = 0$. For instance, in the case of the triad (6.54), ρ is real and D is orthogonal to the spheres centered at the origin. The Gaussian curvature of these spheres is $K = 1/r^2$ [see (6.55)] and since the curvature of the Euclidean space is equal to zero, the commutation relation (6.72) reduces to (2.24). Similarly, for the triad adapted to the cylindrical coordinates (6.50), ρ is real ($\rho = 0$) and D is orthogonal to planes, hence, in this case, $\eth$ and $\overline{\eth}$ commute [see, *e.g.*, (4.5)].

In terms of the notation of (6.41)–(6.44), and (6.39), the Bianchi identities are [*cf.* (6.60) and (6.65)]

$$\begin{aligned}(\overline{\delta} - 4\overline{\beta} + 2\overline{\alpha})\Phi_{+2} &- 2(D - 2\varepsilon + \rho + 2\overline{\rho})\Phi_{+1} \\ &- (\delta + 6\alpha)\Phi_0 + 2\kappa\Phi_{-1} - \tfrac{1}{6}\delta R = 0, \qquad (6.73)\\ \overline{\kappa}\Phi_{+2} + (\overline{\delta} - 2\overline{\beta} + 4\overline{\alpha})\Phi_{+1} &- (2D + 3\rho + 3\overline{\rho})\Phi_0 \\ &- (\delta - 2\beta + 4\alpha)\Phi_{-1} + \kappa\Phi_{-2} + \tfrac{1}{6}DR = 0, \qquad (6.74)\\ 2\overline{\kappa}\Phi_{+1} + (\overline{\delta} + 6\overline{\alpha})\Phi_0 &- 2(D + 2\varepsilon + 2\rho + \overline{\rho})\Phi_{-1} \\ &- (\delta - 4\beta + 2\alpha)\Phi_{-2} + \tfrac{1}{6}\overline{\delta}R = 0. \qquad (6.75)\end{aligned}$$

Note that (6.75) can be obtained from (6.73) by complex conjugation or through the priming operation.

In the case where M has a positive definite metric, the spin-weighted components of the Cotton–York tensor are explicitly given by

$$\begin{aligned}
Y_{+2} &= \sqrt{2}\,\mathrm{i}[(D-4\varepsilon+\rho)\Phi_{+2}+(\delta+2\beta+4\alpha)\Phi_{+1}-3\kappa\Phi_0],\\
Y_{+1} &= \sqrt{2}\,\mathrm{i}[-\overline{\alpha}\Phi_{+2}+(D-2\varepsilon+2\rho)\Phi_{+1}+(\delta+3\alpha)\Phi_0-2\kappa\Phi_{-1}-\tfrac{1}{6}\delta R],\\
Y_{+1} &= \frac{\mathrm{i}}{\sqrt{2}}[(\overline{\delta}-4\overline{\beta})\Phi_{+2}+(2\rho-4\overline{\rho})\Phi_{+1}+\delta\Phi_0-2\kappa\Phi_{-1}-\tfrac{1}{12}\delta R], \qquad (6.76)\\
Y_0 &= \sqrt{2}\,\mathrm{i}[-2\overline{\alpha}\Phi_{+1}+(D+3\rho)\Phi_0+(\delta-2\beta+2\alpha)\Phi_{-1}-\kappa\Phi_{-2}+\tfrac{1}{12}DR],\\
Y_0 &= \frac{\mathrm{i}}{\sqrt{2}}[\overline{\kappa}\Phi_{+2}+(\overline{\delta}-2\overline{\beta})\Phi_{+1}+(3\rho-3\overline{\rho})\Phi_0+(\delta-2\beta)\Phi_{-1}-\kappa\Phi_{-2}].
\end{aligned}$$

Eliminating the derivatives of the scalar curvature by means of the Bianchi identities [(6.73)–(6.75)] one obtains the explicit form of (6.36).

EXAMPLE. Curvature of a spherically symmetric metric.

We shall consider the positive definite metric given by

$$[f(r)]^{-2}\mathrm{d}r^2+r^2(\mathrm{d}\theta^2+\sin^2\theta\,\mathrm{d}\phi^2), \qquad (6.77)$$

where f is some real-valued function. The triad

$$\partial_1 \equiv \frac{1}{r}\partial_\theta, \qquad \partial_2 \equiv \frac{1}{r\sin\theta}\partial_\phi, \qquad \partial_3 \equiv f(r)\,\partial_r$$

is orthonormal and the vector fields D, δ, and $\overline{\delta}$ are

$$D=\frac{1}{\sqrt{2}}f(r)\partial_r, \quad \delta=\frac{1}{\sqrt{2}r}\left(\partial_\theta+\frac{\mathrm{i}}{\sin\theta}\partial_\phi\right), \quad \overline{\delta}=\frac{1}{\sqrt{2}r}\left(\partial_\theta-\frac{\mathrm{i}}{\sin\theta}\partial_\phi\right). \qquad (6.78)$$

A straightforward computation gives

$$[D,\delta]=-\frac{f(r)}{\sqrt{2}r}\delta, \qquad [\delta,\overline{\delta}]=\frac{\cot\theta}{\sqrt{2}r}(\delta-\overline{\delta});$$

hence, comparing with (6.46) we find that $\alpha=0$, $\kappa=0$, $\rho=\overline{\rho}$, $2\varepsilon-\rho=-f(r)/(\sqrt{2}r)$, $\beta=-\cot\theta/(2\sqrt{2}r)=\overline{\beta}$. Since ε is pure imaginary [see (6.42)] and in this case ρ is real, we conclude that the only nonvanishing spin-coefficients for the triad (6.78) are

$$\rho=\frac{f(r)}{\sqrt{2}r}, \qquad \beta=-\frac{\cot\theta}{2\sqrt{2}r} \qquad (6.79)$$

[*cf.* (6.55)]. Substituting (6.78) and (6.79) into (6.66)–(6.70) one readily obtains $\Phi_{+2} = \Phi_{+1} = 0$ and

$$\Phi_0 = -\frac{r}{6}\frac{\mathrm{d}}{\mathrm{d}r}\left[\frac{f^2-1}{r^2}\right], \qquad R = -\frac{2}{r^2}\frac{\mathrm{d}}{\mathrm{d}r}\left[r(f^2-1)\right]. \tag{6.80}$$

If we impose the condition $\Phi_{ab} = 0$ (*i.e.*, R_{ab} proportional to g_{ab}, which corresponds to assuming that the metric (6.77) is isotropic), from the first equation in (6.80) we find that

$$f^2 = 1 - kr^2, \tag{6.81}$$

where k is some constant. Then, the second equation (6.80) gives $R = 6k$. The metric (6.77) with the function f given by (6.81) is the metric of a maximally symmetric space; when $k = 0$, the curvature vanishes and (6.77) is the metric of three-dimensional Euclidean space in spherical coordinates. For $k = 1$, (6.77) corresponds to the standard metric of the sphere S^3.

On the other hand, if we require $R = 0$, from (6.80) we find

$$f^2 = 1 - \frac{2m}{r}, \tag{6.82}$$

where m is some constant, and therefore $\Phi_0 = -m/r^3$. The metric given by (6.77) and (6.82) turns out to be the constant-time hypersurfaces of the Schwarzschild metric. If, instead, we require $R = 2\lambda$, where λ is any real constant, we obtain

$$f^2 = 1 - \frac{2m}{r} - \frac{\lambda r^2}{3},$$

which corresponds to the metric of the constant-time hypersurfaces of the Schwarzschild metric with cosmological constant λ.

The only nonvanishing component of the curvature of the metric (6.77) are given by Φ_0 and R, and these functions depend only on r [(6.80)], hence, making use of (6.78), (6.79), and (6.76) one finds that the Cotton–York tensor of the metric (6.77) vanishes identically, which implies that this metric is locally conformally flat. In fact, defining a new variable u by $\mathrm{d}u/u = \mathrm{d}r/[rf(r)]$, one finds that the metric (6.77) can also be written as

$$\left(\frac{r}{u}\right)^2 \left[\mathrm{d}u^2 + u^2(\mathrm{d}\theta^2 + \sin^2\theta\, \mathrm{d}\phi^2)\right].$$

Cartan's structural equations

If the 1-forms θ^1, θ^2, and θ^3 form the dual basis to $\{\partial_1, \partial_2, \partial_3\}$ (hence, $\mathrm{d}s^2 = \theta^1 \otimes \theta^1 + \theta^2 \otimes \theta^2 + \theta^3 \otimes \theta^3$) then the spin-coefficients can be computed by means of

$$\begin{aligned} \mathrm{d}(\theta^1 + \mathrm{i}\theta^2) &= \sqrt{2}\,[\kappa\,\theta^3 \wedge (\theta^1 - \mathrm{i}\theta^2) + \beta(\theta^1 + \mathrm{i}\theta^2) \wedge (\theta^1 - \mathrm{i}\theta^2) \\ &\quad + (2\varepsilon + \overline{\rho})\,\theta^3 \wedge (\theta^1 + \mathrm{i}\theta^2)], \\ \mathrm{d}\theta^3 &= \sqrt{2}\,[-\alpha\,\theta^3 \wedge (\theta^1 - \mathrm{i}\theta^2) + \tfrac{1}{2}(\overline{\rho} - \rho)(\theta^1 + \mathrm{i}\theta^2) \wedge (\theta^1 - \mathrm{i}\theta^2) \\ &\quad - \overline{\alpha}\,\theta^3 \wedge (\theta^1 + \mathrm{i}\theta^2)]. \end{aligned} \tag{6.83}$$

These relations are equivalent to

$$\mathrm{d}\theta^{AB} = 2\Gamma^{C(A} \wedge \theta^{B)}{}_C, \tag{6.84}$$

where

$$\theta^{AB} \equiv \frac{1}{\sqrt{2}} \sigma^{AB}{}_a\,\theta^a \tag{6.85}$$

(*i.e.*, $\theta^{11} = -(\theta^1 - \mathrm{i}\theta^2)/\sqrt{2}$, $\theta^{12} = \theta^3/\sqrt{2}$, $\theta^{22} = (\theta^1 + \mathrm{i}\theta^2)/\sqrt{2}$) and the connection 1-forms, $\Gamma_{AB} = \Gamma_{(AB)}$, are defined by

$$\Gamma_{AB} = -\Gamma_{ABCD}\,\theta^{CD}. \tag{6.86}$$

The spin-weighted components of the curvature can also be computed with the aid of differential forms. We have

$$\mathrm{d}\Gamma^A{}_B + \Gamma^A{}_C \wedge \Gamma^C{}_B = -G^A{}_{BCD} S^{CD}, \tag{6.87}$$

where

$$S^{AB} \equiv \tfrac{1}{2}\theta^{C(A} \wedge \theta^{B)}{}_C = \tfrac{1}{2}\theta^{CA} \wedge \theta^B{}_C \tag{6.88}$$

(hence, $S^{11} = \theta^{11} \wedge \theta^{12}$, $S^{12} = \frac{1}{2}\theta^{11} \wedge \theta^{22}$, $S^{22} = \theta^{12} \wedge \theta^{22}$). The connection 1-forms Γ_{AB} and the curvature 2-forms,

$$\mathcal{R}_{AB} \equiv -G_{ABCD} S^{CD}, \tag{6.89}$$

are given explicitly by

$$\begin{aligned} \Gamma_{11} &= \frac{1}{\sqrt{2}}[\kappa(\theta^1 - \mathrm{i}\theta^2) - 2\alpha\,\theta^3 + \overline{\rho}(\theta^1 + \mathrm{i}\theta^2)], \\ \Gamma_{12} &= \frac{1}{\sqrt{2}}[\beta(\theta^1 - \mathrm{i}\theta^2) - 2\varepsilon\,\theta^3 - \overline{\beta}(\theta^1 + \mathrm{i}\theta^2)], \end{aligned} \tag{6.90}$$

with $\Gamma_{22} = \overline{\Gamma_{11}}$, and

$$\begin{aligned} \mathrm{d}\Gamma_{11} + 2\Gamma_{12} \wedge \Gamma_{11} &= -\tfrac{1}{2}[\Phi_{+2}\,\theta^3 \wedge (\theta^1 - \mathrm{i}\theta^2) + \Phi_{+1}(\theta^1 + \mathrm{i}\theta^2) \wedge (\theta^1 - \mathrm{i}\theta^2) \\ &\quad + \left(\Phi_0 + \tfrac{1}{6}R\right)\theta^3 \wedge (\theta^1 + \mathrm{i}\theta^2)], \\ \mathrm{d}\Gamma_{12} - \Gamma_{11} \wedge \Gamma_{22} &= -\tfrac{1}{2}[\Phi_{+1}\,\theta^3 \wedge (\theta^1 - \mathrm{i}\theta^2) \\ &\quad + \left(\Phi_0 - \tfrac{1}{12}R\right)(\theta^1 + \mathrm{i}\theta^2) \wedge (\theta^1 - \mathrm{i}\theta^2) + \Phi_{-1}\,\theta^3 \wedge (\theta^1 + \mathrm{i}\theta^2)]. \end{aligned} \tag{6.91}$$

Equations (6.84)–(6.88) also apply when the metric is indefinite, with $ds^2 = \theta^1 \otimes \theta^1 + \theta^2 \otimes \theta^2 - \theta^3 \otimes \theta^3$.

6.3.2 Indefinite metric

In this subsection we shall consider the case where the metric has signature $(+ + -)$, with the σ_{aAB} given by (5.45).

A quantity η has spin weight s if under the spin transformation defined by

$$(U^A{}_B) = \begin{pmatrix} e^{-i\theta/2} & 0 \\ 0 & e^{i\theta/2} \end{pmatrix} \tag{6.92}$$

(which belongs to SU(1,1) and corresponds to a rotation through an angle θ about ∂_3), it transforms according to

$$\eta \mapsto e^{is\theta} \eta. \tag{6.93}$$

Making use of (5.42), (6.92), and (6.93) one finds that each component $\psi_{AB...D}$ of a spinor has spin weight equal to one half of the difference between the number of indices $A, B, \ldots, D$ taking the value 1 and those taking the value 2. The $2n + 1$ independent components of a totally symmetric $2n$-index spinor can be labeled by their spin weight

$$\psi_s \equiv \psi_{\underbrace{1\ldots1}_{n+s}\underbrace{2\ldots2}_{n-s}}, \qquad (s = 0, \pm1, \ldots, \pm n). \tag{6.94}$$

From (6.93) it is clear that if η has spin weight s, then $\overline{\eta}$ has spin weight $-s$. The spin-weighted components of a *real* trace-free totally symmetric n-index tensor $t_{ab...c}$, defined by

$$t_s \equiv t_{\underbrace{1\ldots1}_{n+s}\underbrace{2\ldots2}_{n-s}},$$

where $t_{AB...D}$ are the (totally symmetric) spinor components of $t_{ab...c}$, satisfy the relations

$$\overline{t_s} = (-1)^n t_{-s}, \tag{6.95}$$

where we have made use of (5.64) [*cf.* (6.40)].

As in the case where the metric is positive definite, the general expressions (6.11), (6.27), (6.28), and (6.35) can be written in a more compact and convenient form. If we make use of the definitions (6.45),

$$D = \frac{1}{\sqrt{2}}\partial_3, \qquad \delta = \frac{1}{\sqrt{2}}(\partial_1 + i\partial_2), \qquad \overline{\delta} = \frac{1}{\sqrt{2}}(\partial_1 - i\partial_2),$$

then, from (6.8) and (5.45),

$$\partial_{11} = \mathrm{i}\overline{\delta}, \qquad \partial_{12} = \mathrm{i}D, \qquad \partial_{22} = \mathrm{i}\delta.$$

In the present case there are four complex independent components of Γ_{ABCD}, which will be denoted as

$$\Gamma_{1111} = -\mathrm{i}\kappa, \qquad \Gamma_{1211} = -\mathrm{i}\beta, \qquad \Gamma_{2211} = -\mathrm{i}\rho, \qquad \Gamma_{1112} = -\mathrm{i}\alpha,$$

and one real,

$$\Gamma_{1212} = -\mathrm{i}\varepsilon$$

(thus, ε is pure imaginary); then, from (6.7), we have

$$\Gamma_{2222} = \mathrm{i}\overline{\kappa}, \qquad \Gamma_{1222} = \mathrm{i}\overline{\beta}, \qquad \Gamma_{1122} = \mathrm{i}\overline{\rho}, \qquad \Gamma_{2212} = \mathrm{i}\overline{\alpha}.$$

Substituting these definitions into (6.11) one obtains the two independent relations

$$\begin{aligned} [D, \delta] &= -2\overline{\alpha} D - (2\varepsilon + \overline{\rho})\delta + \overline{\kappa}\overline{\delta}, \\ [\delta, \overline{\delta}] &= 2(\overline{\rho} - \rho)D + 2\beta\delta - 2\overline{\beta}\,\overline{\delta}. \end{aligned} \tag{6.96}$$

The spinor components of the connection Γ_{11AB} and Γ_{22AB} have a well-defined spin weight; in fact, from (6.13) one finds that under the spin transformation (6.92)

$$\kappa \mapsto \mathrm{e}^{2\mathrm{i}\theta}\kappa, \qquad \alpha \mapsto \mathrm{e}^{\mathrm{i}\theta}\alpha, \qquad \rho \mapsto \rho,$$
$$\beta \mapsto \mathrm{e}^{\mathrm{i}\theta}(\beta + \tfrac{1}{2}\mathrm{i}\overline{\delta}\theta), \qquad \varepsilon \mapsto \varepsilon + \tfrac{1}{2}\mathrm{i}D\theta,$$

therefore, if η has spin weight s, then $\eth\eta \equiv (\overline{\delta} - 2s\beta)\eta$, $\text{Þ}\eta \equiv (D - 2s\varepsilon)\eta$ and $\overline{\eth}\eta \equiv (\delta + 2s\overline{\beta})\eta$ have spin weight $s+1$, s and $s-1$, respectively.

The unimodular matrix

$$(U^A{}_B) = \begin{pmatrix} 0 & \mathrm{i} \\ \mathrm{i} & 0 \end{pmatrix}, \tag{6.97}$$

which satisfies (5.49) with the negative sign and represents a rotation through π about ∂_2, defines a spin transformation which will be called a priming operation. Under this transformation, the components of a totally symmetric $2n$-index spinor defined by (6.94), transform as

$$\psi'_s = \mathrm{i}^{2n}\psi_{-s}.$$

Using (5.42), (5.46), (6.13), and (6.97) one finds that if $t_{AB\ldots D}$ are the spinor components of a *real* tensor, then

$$t'_{AB\ldots D} = \overline{t_{AB\ldots D}}, \tag{6.98}$$

and

$$\eth'_{AB} = \overline{\eth_{AB}}, \qquad \Gamma'_{ABCD} = \overline{\Gamma_{ABCD}}, \tag{6.99}$$

i.e.,

$$D' = -D, \qquad \delta' = -\overline{\delta}, \qquad \overline{\delta}' = -\delta,$$
$$\kappa' = -\overline{\kappa}, \qquad \beta' = -\overline{\beta}, \qquad \rho' = -\overline{\rho}, \qquad \alpha' = -\overline{\alpha}, \qquad \varepsilon' = -\overline{\varepsilon}.$$

Analogously, from (6.27) we find that the spin-weighted components of the traceless part of the Ricci tensor and the scalar curvature are given by

$$-\tfrac{1}{2}\Phi_{+2} = -(D - 4\varepsilon + \rho + \overline{\rho})\kappa + (\overline{\delta} - 2\beta - 2\alpha)\alpha, \tag{6.100}$$
$$-\tfrac{1}{2}\Phi_{+1} = -(D - 2\varepsilon + \rho)\beta + (\overline{\delta} - 2\alpha)\varepsilon - (\overline{\alpha} + \overline{\beta})\kappa - \alpha\rho, \tag{6.101}$$
$$-\Phi_{+1} = -(\delta + 4\overline{\beta})\kappa - (\overline{\delta} - 2\alpha)\overline{\rho} - 2\alpha\rho, \tag{6.102}$$
$$-\tfrac{1}{2}\Phi_0 - \tfrac{1}{12}R = -(D + \rho)\rho - (\overline{\delta} + 2\beta - 2\alpha)\overline{\alpha} - \kappa\overline{\kappa}, \tag{6.103}$$
$$-\Phi_0 + \tfrac{1}{12}R = -\delta\beta - \overline{\delta\beta} - 4\beta\overline{\beta} - 2\varepsilon(\rho - \overline{\rho}) - \kappa\overline{\kappa} + \rho\overline{\rho}, \tag{6.104}$$

and the Bianchi identities take the form [*cf.* (6.60) and (6.65)]

$$(\delta + 4\overline{\beta} - 2\overline{\alpha})\Phi_{+2} - 2(D - 2\varepsilon + \rho + 2\overline{\rho})\Phi_{+1} + (\overline{\delta} - 6\alpha)\Phi_0 + 2\kappa\Phi_{-1} + \tfrac{1}{6}\overline{\delta}R = 0, \tag{6.105}$$
$$\overline{\kappa}\Phi_{+2} + (\delta + 2\overline{\beta} - 4\overline{\alpha})\Phi_{+1} - (2D + 3\rho + 3\overline{\rho})\Phi_0 + (\overline{\delta} + 2\beta - 4\alpha)\Phi_{-1} + \kappa\Phi_{-2} + \tfrac{1}{6}DR = 0, \tag{6.106}$$
$$2\overline{\kappa}\Phi_{+1} + (\delta - 6\overline{\alpha})\Phi_0 - 2(D + 2\varepsilon + 2\rho + \overline{\rho})\Phi_{-1} + (\overline{\delta} + 4\beta - 2\alpha)\Phi_{-2} + \tfrac{1}{6}\delta R = 0. \tag{6.107}$$

Note that (6.107) can be obtained from (6.105) by complex conjugation or through the priming operation.

If η has spin weight s, the commutators of the spin-weighted operators Þ, $\eth$, and $\overline{\eth}$ are given by

$$[\text{Þ}, \eth]\eta = -2\alpha\text{Þ}\eta - \rho\eth\eta + \kappa\overline{\eth}\eta + 2s(-\tfrac{1}{2}\Phi_{+1} + \overline{\alpha}\kappa + \alpha\rho)\eta,$$
$$[\eth, \overline{\eth}]\eta = 2(\rho - \overline{\rho})\text{Þ}\eta + 2s\big(\Phi_0 - \tfrac{1}{12}R - \kappa\overline{\kappa} + \rho\overline{\rho}\big)\eta.$$

Under the conformal rescaling

$$D \mapsto \Omega D, \qquad \delta \mapsto \Omega\delta, \qquad \overline{\delta} \mapsto \Omega\overline{\delta},$$

the spin-coefficients transform according to

$$\kappa \mapsto \Omega\kappa, \qquad \beta \mapsto \Omega\beta - \tfrac{1}{2}\overline{\delta}\Omega, \qquad \rho \mapsto \Omega\rho - D\Omega,$$
$$\alpha \mapsto \Omega\alpha + \tfrac{1}{2}\overline{\delta}\Omega, \qquad \varepsilon \mapsto \Omega\varepsilon$$

and, therefore, the spin-weighted operators þ, $\eth$, and $\bar{\eth}$ transform in the same manner as in the case of a positive definite metric

$$\eth \mapsto \Omega^{1-s}\eth\Omega^{s}, \qquad \bar{\eth} \mapsto \Omega^{1+s}\bar{\eth}\Omega^{-s}, \qquad \text{þ} \mapsto \Omega\text{þ}.$$

The components of the Cotton–York tensor are explicitly given by

$$\begin{aligned}
Y_{+2} &= \sqrt{2}\,\mathrm{i}[(D - 4\varepsilon + \rho)\Phi_{+2} - (\bar{\delta} - 2\beta - 4\alpha)\Phi_{+1} - 3\kappa\Phi_0],\\
Y_{+1} &= \sqrt{2}\,\mathrm{i}[\bar{\alpha}\Phi_{+2} + (D - 2\varepsilon + 2\rho)\Phi_{+1} - (\bar{\delta} - 3\alpha)\Phi_0 - 2\kappa\Phi_{-1} + \tfrac{1}{6}\bar{\delta}R],\\
Y_{+1} &= \frac{\mathrm{i}}{\sqrt{2}}[(\delta + 4\bar{\beta})\Phi_{+2} + (2\rho - 4\bar{\rho})\Phi_{+1} - \bar{\delta}\Phi_0 - 2\kappa\Phi_{-1} + \tfrac{1}{12}\bar{\delta}R],\\
Y_0 &= \sqrt{2}\,\mathrm{i}[2\bar{\alpha}\Phi_{+1} + (D + 3\rho)\Phi_0 - (\bar{\delta} + 2\beta - 2\alpha)\Phi_{-1} - \kappa\Phi_{-2} + \tfrac{1}{12}DR],\\
Y_0 &= \frac{\mathrm{i}}{\sqrt{2}}[\bar{\kappa}\Phi_{+2} + (\delta + 2\bar{\beta})\Phi_{+1} + (3\rho - 3\bar{\rho})\Phi_0 - (\bar{\delta} + 2\beta)\Phi_{-1} - \kappa\Phi_{-2}].
\end{aligned}$$

Finally, it should be pointed out that, in those cases where a (real) null direction is singled out, it is preferable to employ the connection symbols (5.43). From (5.43) and (6.8) one finds that the spinorial triad is related to the orthonormal basis $\{\partial_1, \partial_2, \partial_3\}$ by

$$\partial_{11} = \frac{1}{\sqrt{2}}(\partial_1 - \partial_3), \qquad \partial_{12} = \frac{1}{\sqrt{2}}\partial_2, \qquad \partial_{22} = -\frac{1}{\sqrt{2}}(\partial_1 + \partial_3),$$

in such a way that both ∂_{11} and ∂_{22} are null and real. Equations (6.11), (6.27), and (6.28) are then equivalent to those found in the triad formalism developed in Hall, Morgan and Perjés (1987).

6.4 Metric connections with torsion

The torsion of a connection ∇ is defined by

$$T(X, Y) \equiv \nabla_X Y - \nabla_Y X - [X, Y],$$

for any pair of vector fields, X, Y, on M. When the torsion does not vanish, the components of the second covariant derivatives of a vector field satisfy

$$(\nabla_a\nabla_b - \nabla_b\nabla_a)t^c = R^c{}_{dab}t^d - T^d_{ab}(\nabla_d t^c), \tag{6.108}$$

where the $T^a_{bc} = -T^a_{cb}$ are the components of the torsion tensor, $T(\partial_a, \partial_b) = T^c_{ab}\partial_c$. Assuming that the connection is compatible with the metric (*i.e.*, $\nabla_a g_{bc} = 0$), the curvature tensor of the connection has the symmetries $R_{abcd} = -R_{bacd} =$

$-R_{abdc}$, but $R_{abcd} + R_{acdb} + R_{adbc}$ may be different from zero and, therefore, R_{abcd} may not coincide with R_{cdab}. In the present case we find that

$$\det(g_{rs})\,\varepsilon^{abc} R^d{}_{abc} = \det(g_{rs})\,\varepsilon^{abc}(\nabla_a T^d_{bc} - T^d_{am} T^m_{bc}) \equiv 2\tau^d. \qquad (6.109)$$

Defining, as in Section 6.2,

$$G_{ab} \equiv -\tfrac{1}{4}\det(g_{rs})\,\varepsilon_{acd}\varepsilon_{bdf} R^{cdef}$$

we have $G^a{}_a = -\frac{1}{2}R^{cd}{}_{cd} = -\frac{1}{2}R$ and the components of the Ricci tensor are given by

$$R_{ab} \equiv R^c{}_{acb} = \tfrac{1}{2}Rg_{ab} + G_{ba}$$

i.e., $G_{ab} = R_{ba} - \frac{1}{2}Rg_{ba}$. Since R_{abcd} may not coincide with R_{cdab}, the Ricci tensor may not be symmetric. Making use of (6.109) it follows that

$$R_{ab} - R_{ba} = -\varepsilon_{abc}\tau^c.$$

By virtue of the antisymmetry $T^a_{bc} = -T^a_{cb}$, the spinor equivalent of T^a_{bc}, T^{AB}_{CDEF}, can be expressed as $T^{AB}_{CDEF} = \varepsilon_{CE}\Theta^{AB}{}_{DF} + \varepsilon_{DF}\Theta^{AB}{}_{CE}$, with $\Theta^{AB}{}_{CD} = \Theta^{(AB)}{}_{(CD)}$. Then from (6.108) we obtain $\Box_{AB} t^{CD} = -\frac{1}{2}G^C{}_{EAB} t^{ED} - \frac{1}{2}G^D{}_{EAB} t^{CE} + \Theta^{EF}{}_{AB}\nabla_{EF} t^{CD}$; therefore,

$$\begin{aligned}\Box_{AB}\psi_C &= -\tfrac{1}{2}G_{CDAB}\psi^D + \Theta^{DE}{}_{AB}\nabla_{DE}\psi_C \\ &= -\tfrac{1}{2}R_{ABCD}\psi^D - \tfrac{1}{8}R(\varepsilon_{AC}\psi_B + \varepsilon_{BC}\psi_A) + \Theta^{DE}{}_{AB}\nabla_{DE}\psi_C.\end{aligned} \qquad (6.110)$$

In particular, for a scalar function f, $\Box_{AB} f = \Theta^{CD}{}_{AB}\partial_{CD} f$. The torsion is real $(\overline{T^a_{bc}} = T^a_{bc})$ if and only if

$$\widehat{\Theta}_{ABCD} = \begin{cases} -\Theta_{ABCD} & \text{if } (g_{ab}) = \operatorname{diag}(1,1,1), \\ \Theta_{ABCD} & \text{if } (g_{ab}) = \operatorname{diag}(1,1,-1). \end{cases} \qquad (6.111)$$

Introducing the torsion 2-forms

$$\Theta_{AB} \equiv \Theta_{ABCD} S^{CD}, \qquad (6.112)$$

the first structural equations are now given by

$$\mathrm{d}\theta^{AB} = 2\Gamma^{C(A} \wedge \theta^{B)}{}_C + 2\Theta^{AB}. \qquad (6.113)$$

Therefore, the differentials of the basis 2-forms S^{AB} [(6.88)] are

$$\mathrm{d}S^{AB} = \mathrm{d}\theta^{C(A} \wedge \theta^{B)}{}_C = 2\Gamma^{C(A} \wedge S^{B)}{}_C + 2\Theta^{C(A} \wedge \theta^{B)}{}_C. \qquad (6.114)$$

Taking the differential of (6.113), using (6.113) and (6.89) we have

$$d\Theta^{AB} = 2\Gamma^{C(A} \wedge \Theta^{B)}{}_C - \mathcal{R}^{C(A} \wedge \theta^{B)}{}_C.$$

Making use of (6.87), (6.86), (6.112), and the relation

$$\theta^{AB} \wedge S^{CD} = \tfrac{1}{2}(\varepsilon^{AC}\varepsilon^{BD} + \varepsilon^{AD}\varepsilon^{BC})\theta^{11} \wedge \theta^{12} \wedge \theta^{22}, \tag{6.115}$$

we find

$$G^{C(A}{}_C{}^{B)} = -\nabla_{CD}\Theta^{ABCD} + 2\Theta^R{}_{CRD}\Theta^{ABCD},$$

which is equivalent to (6.109).

Similarly, taking the differential on both sides of (6.87) we obtain the Bianchi identities

$$d\mathcal{R}_{AB} = -2d\Gamma^C{}_{(A} \wedge \Gamma_{B)C} = 2\Gamma^C{}_{(A} \wedge \mathcal{R}_{B)C},$$

which, by virtue of (6.114), (6.87), and (6.115), reduce to

$$\nabla^{AB}G_{CDAB} - 2\Theta^{ARB}{}_R G_{CDAB} = 0. \tag{6.116}$$

If ∇ is a connection compatible with the metric and γ_{ABCD} is a spinor field such that $\gamma_{ABCD} = \gamma_{(AB)(CD)}$, the connection $\widetilde{\nabla}$ given by

$$\widetilde{\nabla}_{AB}\psi_C = \nabla_{AB}\psi_C - \gamma^D{}_{CAB}\psi_D, \tag{6.117}$$

for any one-index spinor field, is also a connection compatible with the metric. Conversely, given two connections compatible with the metric, ∇ and $\widetilde{\nabla}$, there exists a spinor field $\gamma_{ABCD} = \gamma_{(AB)(CD)}$ such that (6.117) holds. Then $\widetilde{\Box}_{AB}\psi_C \equiv \widetilde{\nabla}^R{}_{(A}\widetilde{\nabla}_{B)R}\psi_C$ is related to $\Box_{AB}\psi_C \equiv \nabla^R{}_{(A}\nabla_{B)R}\psi_C$ by

$$\begin{aligned}\widetilde{\Box}_{AB}\psi_C &= \Box_{AB}\psi_C + (\nabla^R{}_{(A}\gamma_{|DC|B)R} + \gamma^S{}_{CR(A}\gamma_{|DS|B)}{}^R)\psi^D \\ &\quad + (2\delta^{(R}_{(A}\gamma^{S)M}{}_{B)M} - \gamma^{RS}{}_{AB})\widetilde{\nabla}_{RS}\psi_C.\end{aligned}$$

Making use of (6.110) it follows that the curvature and torsion of the two connections are related by the formulas

$$\begin{aligned}-\tfrac{1}{2}\widetilde{G}_{CDAB} &= -\tfrac{1}{2}G_{CDAB} + \nabla^R{}_{(A}\gamma_{|DC|B)R} + \gamma^S{}_{CR(A}\gamma_{|DS|B)}{}^R \\ &\quad - \gamma_{CDRS}\Theta^{RS}{}_{AB}, \\ \widetilde{\Theta}^{RS}{}_{AB} &= \Theta^{RS}{}_{AB} - \gamma^{RS}{}_{AB} + 2\delta^{(R}_{(A}\gamma^{S)M}{}_{B)M}.\end{aligned}$$

In particular, if ∇ is the Levi-Civita connection (which is characterized by the condition $\Theta_{ABCD} = 0$) then the curvature and torsion of the connection $\widetilde{\nabla}$ defined by (6.117) are given by

$$\begin{aligned}-\tfrac{1}{2}\widetilde{G}_{CDAB} &= -\tfrac{1}{2}\Phi_{ABCD} - \tfrac{1}{24}R(\varepsilon_{AC}\varepsilon_{BD} + \varepsilon_{AD}\varepsilon_{BC}) + \nabla^R{}_{(A}\gamma_{|DC|B)R} \\ &\quad + \gamma^S{}_{CR(A}\gamma_{|DS|B)}{}^R,\end{aligned} \tag{6.118}$$

where Φ_{ABCD} is the spinor equivalent of the trace-free part of the Ricci tensor of the Levi-Civita connection and R is the corresponding scalar curvature, and

$$\widetilde{\Theta}^{RS}{}_{AB} = -\gamma^{RS}{}_{AB} + 2\delta^{(R}_{(A}\gamma^{S)M}{}_{B)M}. \tag{6.119}$$

From the last equation one obtains the expression

$$\gamma^{AB}{}_{CD} = 2\delta^{(A}_{(C}\widetilde{\Theta}^{B)M}{}_{D)M} - \widetilde{\Theta}^{AB}{}_{CD} - \tfrac{1}{4}\widetilde{\Theta}^{MS}{}_{MS}(\delta^A_C\delta^B_D + \delta^A_D\delta^B_C),$$

which allows one to find the metric connection with a given torsion. This formula shows explicitly that there is only one metric connection without torsion.

6.5 Congruences of curves

The spinor equivalent of any real vector field, $t^a\partial_a$, in a space with a positive definite metric is of the form

$$t_{AB} = \alpha_{(A}\widehat{\alpha}_{B)}. \tag{6.120}$$

When the metric is indefinite, (6.120) also holds if $t^a t_a < 0$, which is equivalent to the condition $\widehat{\alpha}^A\alpha_A \neq 0$. The spinor equivalent of a null vector field, $t^a t_a = 0$, is of the form (6.120), with $\widehat{\alpha}^A\alpha_A = 0$; however, in what follows it will be assumed that $\widehat{\alpha}_A$ and α_A form a basis for the one-index spinors, which amounts to the condition $\widehat{\alpha}^A\alpha_A \neq 0$ and therefore, the case where t_a is null will be excluded in this section. The spinor field α_A is not uniquely determined by the vector field $t^a\partial_a$; the components (6.120) are invariant under the transformation

$$\alpha_A \mapsto e^{i\chi/2}\alpha_A, \tag{6.121}$$

where χ is any real-valued function.

Proposition. Assuming that $\widehat{\alpha}^A\alpha_A \neq 0$, the vector field (6.120) is tangent to a geodesic if and only if

$$\alpha^A\widehat{\alpha}^B\alpha^C\nabla_{AB}\alpha_C = 0. \tag{6.122}$$

Note that condition (6.122) is invariant under the transformation (6.121) and that (6.122) is equivalent to

$$\alpha^A\widehat{\alpha}^B\widehat{\alpha}^C\nabla_{AB}\widehat{\alpha}_C = 0. \tag{6.123}$$

Proof. Using the identity (5.88) we see that for an arbitrary spinor ξ_A,

$$\xi_A = \frac{1}{\alpha^R\widehat{\alpha}_R}(\widehat{\alpha}_A\alpha^B\xi_B - \alpha_A\widehat{\alpha}^B\xi_B),$$

hence,

$$\alpha^A\widehat{\alpha}^B\nabla_{AB}\alpha_C = \frac{1}{\alpha^R\widehat{\alpha}_R}\left(\widehat{\alpha}_C\alpha^D\alpha^A\widehat{\alpha}^B\nabla_{AB}\alpha_D - \alpha_C\widehat{\alpha}^D\alpha^A\widehat{\alpha}^B\nabla_{AB}\alpha_D\right) \quad (6.124)$$

and, similarly,

$$\alpha^A\widehat{\alpha}^B\nabla_{AB}\widehat{\alpha}_C = \frac{1}{\alpha^R\widehat{\alpha}_R}\left(\widehat{\alpha}_C\alpha^D\alpha^A\widehat{\alpha}^B\nabla_{AB}\widehat{\alpha}_D - \alpha_C\widehat{\alpha}^D\alpha^A\widehat{\alpha}^B\nabla_{AB}\widehat{\alpha}_D\right).$$

Thus, the spinor equivalent of $t_a\nabla^a t_b$ is given by

$$\begin{aligned}
&-t_{AB}\nabla^{AB}t_{CD}\\
&= -\alpha_A\widehat{\alpha}_B(\alpha_{(C}\nabla^{AB}\widehat{\alpha}_{D)} + \widehat{\alpha}_{(D}\nabla^{AB}\alpha_{C)})\\
&= -\frac{1}{\alpha^R\widehat{\alpha}_R}\Big\{\alpha_{(C}\widehat{\alpha}_{D)}\alpha^E\alpha^A\widehat{\alpha}^B\nabla_{AB}\widehat{\alpha}_E - \alpha_C\alpha_D\widehat{\alpha}^E\alpha^A\widehat{\alpha}^B\nabla_{AB}\widehat{\alpha}_E\\
&\qquad + \widehat{\alpha}_D\widehat{\alpha}_C\alpha^E\alpha^A\widehat{\alpha}^B\nabla_{AB}\alpha_E - \widehat{\alpha}_{(D}\alpha_{C)}\widehat{\alpha}^E\alpha^A\widehat{\alpha}^B\nabla_{AB}\alpha_E\Big\}\\
&= -\frac{1}{\alpha^R\widehat{\alpha}_R}\Big\{t_{CD}\alpha^A\widehat{\alpha}^B\nabla_{AB}(\alpha^E\widehat{\alpha}_E) - \alpha_C\alpha_D\widehat{\alpha}^E\alpha^A\widehat{\alpha}^B\nabla_{AB}\widehat{\alpha}_E\\
&\qquad + \widehat{\alpha}_C\widehat{\alpha}_D\alpha^E\alpha^A\widehat{\alpha}^B\nabla_{AB}\alpha_E\Big\},
\end{aligned} \quad (6.125)$$

which is proportional to t_{CD} if and only if (6.122) and (6.123) are fulfilled. From (6.125) it also follows that (6.120) is tangent to an affinely parametrized geodesic (*i.e.*, $t_a\nabla^a t_b = 0$) if and only if, in addition to (6.122), $\alpha^A\widehat{\alpha}_A$ is constant along the geodesic.

Under the transformation (6.121) the function $\alpha^A\widehat{\alpha}^B\widehat{\alpha}^C\nabla_{AB}\alpha_C$ transforms according to

$$\begin{aligned}
\alpha^A\widehat{\alpha}^B\widehat{\alpha}^C\nabla_{AB}\alpha_C \quad &\mapsto \quad e^{-i\chi/2}\alpha^A\widehat{\alpha}^B\widehat{\alpha}^C\nabla_{AB}(e^{i\chi/2}\alpha_C)\\
&= \alpha^A\widehat{\alpha}^B\widehat{\alpha}^C\nabla_{AB}\alpha_C + \tfrac{1}{2}i\widehat{\alpha}^C\alpha_C\alpha^A\widehat{\alpha}^B\partial_{AB}\chi,
\end{aligned}$$

therefore, if χ is a solution of $t^a\partial_a\chi = -2i(\alpha^A\widehat{\alpha}^B\widehat{\alpha}^C\nabla_{AB}\alpha_C)/(\widehat{\alpha}^D\alpha_D)$, then the new spinor field α_A satisfies the condition $\alpha^A\widehat{\alpha}^B\widehat{\alpha}^C\nabla_{AB}\alpha_C = 0$. This last condition together with (6.122) are equivalent to $\alpha^A\widehat{\alpha}^B\nabla_{AB}\alpha_C = 0$; thus, if the vector field $t^a\partial_a$ is tangent to a geodesic, we can always find, locally, a spinor field such that $t_{AB} = \alpha_{(A}\widehat{\alpha}_{B)}$ and

$$\alpha^A\widehat{\alpha}^B\nabla_{AB}\alpha_C = 0$$

(*i.e.*, α_A is parallelly transported along the geodesic).

We shall assume in what follows that the metric is positive definite. Then given a congruence of curves (*i.e.*, a family of curves such that through each point there passes one curve in this family) we define a spinor field o_A such that

$$t_{AB} = o_{(A}\widehat{o}_{B)} \quad (6.126)$$

are the spinor components of a tangent vector to the congruence and

$$o^A \widehat{o}_A = 1. \tag{6.127}$$

Note that $t^a t_a = \frac{1}{2}$ [see (5.73)] and that (6.126) and (6.127) define o_A up to a factor of the form $e^{i\theta/2}$. Making use of the definitions

$$\begin{aligned} \kappa &= o^A o^B o^C \nabla_{AB} o_C, \\ \alpha &= o^A \widehat{o}^B o^C \nabla_{AB} o_C, \\ \beta &= o^A o^B \widehat{o}^C \nabla_{AB} o_C = o^A o^B o^C \nabla_{AB} \widehat{o}_C, \\ \rho &= o^A o^B \widehat{o}^C \nabla_{AB} \widehat{o}_C, \\ \varepsilon &= o^A \widehat{o}^B \widehat{o}^C \nabla_{AB} o_C = o^A \widehat{o}^B o^C \nabla_{AB} \widehat{o}_C, \end{aligned} \tag{6.128}$$

or, equivalently,

$$\begin{aligned} \overline{\kappa} &= -\widehat{o}^A \widehat{o}^B \widehat{o}^C \nabla_{AB} \widehat{o}_C, \\ \overline{\alpha} &= o^A \widehat{o}^B \widehat{o}^C \nabla_{AB} \widehat{o}_C, \\ \overline{\beta} &= \widehat{o}^A \widehat{o}^B o^C \nabla_{AB} \widehat{o}_C = \widehat{o}^A \widehat{o}^B \widehat{o}^C \nabla_{AB} o_C, \\ \overline{\rho} &= -\widehat{o}^A \widehat{o}^B o^C \nabla_{AB} o_C, \\ \overline{\varepsilon} &= -o^A \widehat{o}^B o^C \nabla_{AB} \widehat{o}_C = -o^A \widehat{o}^B \widehat{o}^C \nabla_{AB} o_C, \end{aligned}$$

which amount to (6.41)–(6.43) with

$$D = -o^A \widehat{o}^B \partial_{AB}, \qquad \delta = o^A o^B \partial_{AB}, \qquad \overline{\delta} = -\widehat{o}^A \widehat{o}^B \partial_{AB},$$

from (6.122) we see that $D = t^a \partial_a = -t^{AB} \partial_{AB}$ is tangent to a congruence of geodesics if and only if $\alpha = 0$.

Using (6.127) and (6.128) one finds that under the transformation $o_A \mapsto e^{i\theta/2} o_A$, where θ is a real function, which preserves conditions (6.126) and (6.127), the spin-coefficients (6.128) transform according to

$$\begin{aligned} &\kappa \mapsto e^{2i\theta} \kappa, \qquad \alpha \mapsto e^{i\theta} \alpha, \qquad \rho \mapsto \rho, \\ &\beta \mapsto e^{i\theta} \left(\beta - \tfrac{1}{2} i \delta\theta\right), \qquad \varepsilon \mapsto \varepsilon + \tfrac{1}{2} i D\theta, \end{aligned} \tag{6.129}$$

which are precisely (6.48) and (6.49). Therefore, choosing θ in such a way that $D\theta = 2i\varepsilon$ the new ε vanishes. In particular, if D is tangent to a congruence of geodesics, $\alpha = 0$ and we can always make $\varepsilon = 0$. Equations (6.124) and (6.128) show that α and ε vanish if and only if o_A (and hence $\widehat{o}_A$) is covariantly constant along the geodesics,

$$o^A \widehat{o}^B \nabla_{AB} o_C = 0.$$

This last condition implies that the triad $\{D, \delta, \overline{\delta}\}$ is parallelly transported along the geodesics.

Given a system of coordinates x^i ($i = 1, 2, 3$), the functions $x^i(u, v)$ define a one-parameter family of geodesics if for a given value of v, the curve $x^i(u) = x^i(u, v)$ is geodetic. The vector field $\zeta^i \equiv \partial x^i(u, v)/\partial v$ measures the displacement of neighboring geodesics and $t^i \equiv \partial x^i(u, v)/\partial u$ is tangent to the geodesics. Then, $t^i \partial\zeta^j/\partial x^i = \partial\zeta^j/\partial u = \partial^2 x^j/\partial u \partial v = \partial t^j/\partial v = \zeta^i \partial t^j/\partial x^i$ or, equivalently,

$$[t, \zeta] = 0, \tag{6.130}$$

where t and ζ are the differential operators (or vector fields) $t = t^i \partial/\partial x^i$, $\zeta = \zeta^i \partial/\partial x^i$. Any vector field ζ^a satisfying (6.130) is said to be a connecting vector of the congruence. (Equation (6.130) means that the Lie derivative of ζ^a with respect to t^a vanishes.)

Writing $t = D$ and $\zeta = fD + \overline{w}\delta + w\overline{\delta}$, where f is a real function and w is a complex function, making use of (6.46) and the properties of the commutator (or Lie bracket) one finds that

$$\begin{aligned}[t, \zeta] &= (Df + 2\alpha\overline{w} + 2\overline{\alpha}w)D + (D\overline{w} + (2\varepsilon - \rho)\overline{w} - \overline{\kappa}w)\delta \\ &\quad + (Dw + (-2\varepsilon - \overline{\rho})w - \kappa\overline{w})\overline{\delta},\end{aligned}$$

hence, assuming $\varepsilon = 0$, ζ is a connecting vector for a congruence of geodesics with tangent vector D if and only if

$$Df = 0 \tag{6.131}$$

and

$$Dw = \overline{\rho}w + \kappa\overline{w}. \tag{6.132}$$

Equation (6.131) implies that if ζ is orthogonal to D at some point of a geodesic, then it is orthogonal to D along that geodesic. (Note that $D = (1/\sqrt{2})\mathrm{d}/\mathrm{d}s$, where s is the arc length.) In what follows we set $f = 0$; therefore, ζ is orthogonal to the congruence of geodesics everywhere and we can write

$$\zeta = x\partial_1 + y\partial_2,$$

where $\partial_1 = (\delta + \overline{\delta})/\sqrt{2}$, $\partial_2 = \mathrm{i}(\overline{\delta} - \delta)/\sqrt{2}$ form an orthonormal basis of the normal planes to the geodesics and

$$w = \frac{1}{\sqrt{2}}(x + \mathrm{i}y). \tag{6.133}$$

In order to find the geometrical meaning of the functions $\Theta \equiv \operatorname{Re}\rho$, $\omega \equiv \operatorname{Im}\rho$ and κ, we consider first the case where $\kappa = 0$ and $\omega = 0$, then substituting (6.133) into (6.132) one finds that $Dx = \Theta x$, $Dy = \Theta y$, which means that as one moves along a geodesic, any connecting vector ζ orthogonal to D expands ($\Theta > 0$) or contracts ($\Theta < 0$), maintaining its orientation with respect to the axes ∂_1 and ∂_2, *i.e.*, the congruence is expanding ($\Theta > 0$) or contracting ($\Theta < 0$). In fact, using, *e.g.*, (6.62) it follows that $\operatorname{div} D = 2\Theta$.

When $\kappa = 0$ and $\Theta = 0$, (6.132) and (6.133) give $Dx = \omega y$, $Dy = -\omega x$, which corresponds to a rigid rotation of the connecting vector relative to the axes ∂_1 and ∂_2. If $\rho = 0$ and κ is real, from (6.132) and (6.133) we get $Dx = \kappa x$, $Dy = -\kappa y$, which correspond to an area-preserving shear with principal axes ∂_1 and ∂_2. When κ is complex, then, at a given point, κ is of the form $\kappa = |\kappa_0|\, e^{i\chi_0}$ and from (6.129) one finds that under the transformation $o_A \mapsto e^{-i\chi_0/4} o_A$ (which preserves the condition $\varepsilon = 0$ and corresponds to a rotation through an angle $-\chi_0/2$ about D), $\kappa \mapsto |\kappa_0|$ at that point. Therefore, (6.132) with $\rho = 0$ and κ complex corresponds to an area-preserving shear with principal axes that form an angle $-(\arg\kappa)/2$ with respect to ∂_1 and ∂_2. Θ, ω and κ will be called the expansion, twist and shear of the congruence, respectively.

In a similar way, one finds that in the case where t is tangent to a congruence of geodesics in a space with indefinite metric [of signature $(+ + -)$] with $t^a t_a < 0$, making $t = D$ and $\varepsilon = 0$, the meaning of ρ and κ is that found in the case where the metric is positive definite, with ∂_1 and ∂_2 interchanged.

Thus, D is tangent to a shear-free congruence of geodesics if and only if $\alpha = \kappa = 0$ which, according to (6.128), is equivalent to the condition

$$o^A o^C \nabla_{AB} o_C = 0. \tag{6.134}$$

Similarly, the vector field (6.120) is tangent to a shear-free congruence of geodesics if and only if

$$\alpha^A \alpha^C \nabla_{AB} \alpha_C = 0,$$

even if $\alpha^A \widehat{\alpha}_A$ is not constant. Indeed, assuming that α_A is different from zero, we can define $o_A \equiv (\alpha^R \widehat{\alpha}_R)^{-1/2} \alpha_A$, which satisfies (6.127), then $\alpha^A \alpha^C \nabla_{AB} \alpha_C = (\alpha^R \widehat{\alpha}_R)^{3/2} o^A o^C \nabla_{AB} o_C = 0$, where we have made use of (6.134).

From (6.66) and (6.100) it follows that $\kappa = 0 = \alpha$ imply $\Phi_{+2} = 0$, *i.e.*, if $\alpha_{(A} \widehat{\alpha}_{B)}$ is tangent to a shear-free congruence of geodesics, then α_A and $\widehat{\alpha}_A$ are principal spinors of Φ_{ABCD}.

6.6 Applications

In this section we apply the spinor formalism to various fields in three-dimensional Euclidean space; the corresponding equations are written in a form that is manifestly covariant under spatial rotations only. Further applications, in curved space-times, are given in the next chapter.

Dirac's equation

The Dirac equation can be written in the standard form

$$\begin{aligned} i\hbar\partial_t u &= -i\hbar c\sigma_j \partial v/\partial x^j + Mc^2 u, \\ i\hbar\partial_t v &= -i\hbar c\sigma_j \partial u/\partial x^j - Mc^2 v, \end{aligned} \tag{6.135}$$

where u and v are two-component spinors and the x^i are Cartesian coordinates. Recalling that the elements of the Pauli matrices σ_j are $\sigma_j{}^A{}_B$ and using (6.8) it can be seen that (6.135) corresponds to the covariant expression

$$\begin{aligned} \frac{1}{c}\partial_t u^A &= -\sqrt{2}\,\nabla^A{}_B v^B - \frac{iMc}{\hbar}u^A, \\ \frac{1}{c}\partial_t v^A &= -\sqrt{2}\,\nabla^A{}_B u^B + \frac{iMc}{\hbar}v^A, \end{aligned}$$

which is equivalent to

$$\begin{aligned} \frac{1}{c}\partial_t u^1 &= -\sqrt{2}\,(D+\varepsilon+\overline{\rho})v^1 - \sqrt{2}\,(\overline{\delta}-\overline{\beta}+\overline{\alpha})v^2 - \frac{iMc}{\hbar}u^1, \\ \frac{1}{c}\partial_t u^2 &= -\sqrt{2}\,(\delta-\beta+\alpha)v^1 + \sqrt{2}\,(D-\varepsilon+\rho)v^2 - \frac{iMc}{\hbar}u^2, \\ \frac{1}{c}\partial_t v^1 &= -\sqrt{2}\,(D+\varepsilon+\overline{\rho})u^1 - \sqrt{2}\,(\overline{\delta}-\overline{\beta}+\overline{\alpha})u^2 + \frac{iMc}{\hbar}v^1, \\ \frac{1}{c}\partial_t v^2 &= -\sqrt{2}\,(\delta-\beta+\alpha)u^1 + \sqrt{2}\,(D-\varepsilon+\rho)u^2 + \frac{iMc}{\hbar}v^2. \end{aligned} \tag{6.136}$$

From (6.136) one can readily obtain the explicit form of the Dirac equation in any orthogonal coordinate system or in an arbitrary system of coordinates. For instance, substituting (6.54) and (6.55) into (6.136) one obtains (3.121). (For the case of orthogonal coordinates, alternative procedures are given in Ley-Koo and Wang (1988) and in Chapters 3 and 4.)

Weyl neutrino field

The Weyl neutrino equation for the two-component neutrino field in Cartesian coordinates is given by

$$\frac{1}{c}\partial_t \psi^A = \sigma_i{}^A{}_B \partial\psi^B/\partial x_i, \tag{6.137}$$

where the $\sigma_i = (\sigma_i{}^A{}_B)$ are the usual Pauli matrices (see, *e.g.*, Rose 1961, Bjorken and Drell 1964); therefore, with respect to an arbitrary spinorial triad, making use of the definition (6.8) and replacing the partial derivatives by covariant derivatives,

$$\sqrt{2}\,\nabla^A{}_B\psi^B = \frac{1}{c}\partial_t\psi^A. \tag{6.138}$$

Then, according to (6.14), the mate of the neutrino field satisfies

$$\sqrt{2}\,\nabla^A{}_B\widehat{\psi}^B = -\frac{1}{c}\partial_t\widehat{\psi}^A \tag{6.139}$$

(the field $\widehat{\psi}$ obeys the equation for the antineutrino, see, *e.g.*, Rose 1961, Bjorken and Drell 1964).

Equations (6.138) and (6.139) lead to the continuity equation

$$c\sqrt{2}\,\nabla_{AB}(\psi^A\widehat{\psi}^B) = \psi^A(-\partial_t\widehat{\psi}_A) + \widehat{\psi}^B\partial_t\psi_B = -\partial_t(\psi^A\widehat{\psi}_A),$$

which is of the form $\operatorname{div}\mathbf{J} + \partial_t\rho_{\mathrm{n}} = 0$, with

$$J_{AB} \equiv -c\sqrt{2}\,\psi_{(A}\widehat{\psi}_{B)}, \qquad \rho_{\mathrm{n}} \equiv \psi^A\widehat{\psi}_A. \tag{6.140}$$

(Using (5.73) we obtain $J_aJ^a = c^2{\rho_{\mathrm{n}}}^2$, *i.e.*, $|\mathbf{J}| = \rho_{\mathrm{n}}c$; hence, the four-vector $(\rho_{\mathrm{n}}c, \mathbf{J})$ is null.)

Looking for plane wave solutions of (6.137), which are of the form $\psi^A = \alpha^A e^{i(k_ix^i-\omega t)}$, where the x^i are Cartesian coordinates, k_i and α^A are constant, we obtain

$$\sqrt{2}\,k^A{}_B\alpha^B = -\frac{\omega}{c}\alpha^A, \tag{6.141}$$

where k_{AB} are the spinor components of k_i. Hence, $k_{AB}\alpha^A\alpha^B = 0$, which means that α^A is a principal spinor of k_{AB}. Since k_i is real, from (5.76) it follows that k_{AB} must be proportional to $\alpha_{(A}\widehat{\alpha}_{B)}$ and from (6.141) one finds

$$k_{AB} = -\sqrt{2}\left(\frac{\omega}{c}\right)\frac{\alpha_{(A}\widehat{\alpha}_{B)}}{\alpha^R\widehat{\alpha}_R}. \tag{6.142}$$

(The minus sign appearing in (6.142) corresponds to the fact that, for a neutrino with positive energy, the spin and the momentum are antiparallel.)

Electromagnetic field

The source-free Maxwell equations in vacuum can be written as $\nabla\times(\mathbf{E}+i\mathbf{B}) = (i/c)\partial_t(\mathbf{E}+i\mathbf{B})$ and $\nabla\cdot(\mathbf{E}+i\mathbf{B}) = 0$ or, equivalently,

$$\sqrt{2}\,\nabla^C{}_{(A}F_{B)C} = \frac{1}{c}\partial_tF_{AB}, \qquad \nabla^{AB}F_{AB} = 0, \tag{6.143}$$

where F_{AB} are the spinor components of the complex vector field $\mathbf{F} \equiv \mathbf{E} + i\mathbf{B}$. Since the anti-symmetric part of $\nabla^C{}_A F_{BC}$ is proportional to $\nabla^{CR} F_{RC}$, the Maxwell equations (6.143) can be rewritten as

$$\sqrt{2}\,\nabla^C{}_A F_{BC} = \frac{1}{c}\partial_t F_{AB}, \qquad \nabla^{AB} F_{AB} = 0. \tag{6.144}$$

Thus, the mate of the spinor field F_{AB} satisfies

$$\sqrt{2}\,\nabla^C{}_A \widehat{F}_{BC} = -\frac{1}{c}\partial_t \widehat{F}_{AB}, \qquad \nabla^{AB} \widehat{F}_{AB} = 0. \tag{6.145}$$

The conservation laws of energy and momentum of the free electromagnetic field can be obtained from (6.144) and (6.145) in the following way.

$$\begin{aligned} \partial_t (F^{AB} \widehat{F}_{AB}) &= \sqrt{2}\,c\,(-F^{AB} \nabla^C{}_A \widehat{F}_{BC} + \widehat{F}_{AB} \nabla^{CA} F^B{}_C) \\ &= \sqrt{2}\,c\,(-F^{AB} \nabla^C{}_A \widehat{F}_{BC} + \widehat{F}_{AB} \nabla^{CA} F^B{}_C) \\ &= \sqrt{2}\,c\,\nabla^{CA} (\widehat{F}_{AB} F^B{}_C). \end{aligned} \tag{6.146}$$

The contraction $F^{AB} \widehat{F}_{AB}$ is real and positive (in fact, $F^{AB} \widehat{F}_{AB} = \mathbf{E}^2 + \mathbf{B}^2$) and $\widehat{F}_{B(A} F^B{}_{C)}$ is the spinor equivalent of a real vector field since $(\widehat{F}_{B(A} F^B{}_{C)})\widehat{} = F_{B(A} \widehat{F}^B{}_{C)} = -\widehat{F}_{B(A} F^B{}_{C)}$. Equation (6.146) is equivalent to $\partial_t u + \operatorname{div} \mathbf{S} = 0$, where $u = (1/8\pi) F^{AB} \widehat{F}_{AB}$ is the energy density and

$$S_{AB} = \frac{c}{4\sqrt{2}\,\pi} \widehat{F}_{C(A} F^C{}_{B)} \tag{6.147}$$

is the spinor equivalent of the Poynting vector.

In an analogous manner we find that

$$\begin{aligned} &\partial_t (\widehat{F}_{C(A} F^C{}_{B)}) \\ &= \sqrt{2}\,c\,(\widehat{F}_{C(A} \nabla^{RC} F_{B)R} - F^C{}_{(A} \nabla^R{}_{B)} \widehat{F}_{CR}) \\ &= \sqrt{2}\,c\,\{\nabla^{RC} (\widehat{F}_{C(A} F_{B)R}) - F_{R(A} \nabla^{RC} \widehat{F}_{B)C} + F_{R(A} \nabla_{B)}{}^C \widehat{F}^R{}_C\} \\ &= \sqrt{2}\,c\,\{\nabla^{RC} (\widehat{F}_{C(A} F_{B)R}) - F_{R(A} \delta^R_{B)} \nabla^{SC} \widehat{F}_{SC}\} \\ &= \sqrt{2}\,c\,\nabla^{RC} (\widehat{F}_{C(A} F_{B)R}). \end{aligned}$$

$\widehat{F}_{C(A} F^C{}_{B)} / (4\sqrt{2}\,\pi c)$ is the spinor equivalent of the density of linear momentum of the electromagnetic field and

$$T_{AB}{}^{CD} = -\frac{1}{4\pi} \widehat{F}_{(A}{}^{(C} F_{B)}{}^{D)} \tag{6.148}$$

is the spinor equivalent of the Maxwell stress tensor $T_{ab} = (1/4\pi)[E_a E_b + B_a B_b - \frac{1}{2}(E_c E^c + B_c B^c) g_{ab}]$.

The symmetry of F_{AB} implies that

$$F_{AB} = \alpha_{(A}\beta_{B)} \tag{6.149}$$

[*cf.* (5.68)]. The principal spinors α_A and β_A of F_{AB} define the real vector fields

$$v_{AB} \equiv \alpha_{(A}\widehat{\alpha}_{B)}, \qquad w_{AB} \equiv \beta_{(A}\widehat{\beta}_{B)}. \tag{6.150}$$

It may be noticed that α_A and β_A are defined by (6.149) up to the transformation $\alpha_A \mapsto \lambda\alpha_A$, $\beta_A \mapsto \lambda^{-1}\beta_A$, which induces the transformation $v_a \mapsto |\lambda|^2 v_a$, $w_a \mapsto |\lambda|^{-2} w_a$, on the vector fields (6.150). This means that a nondegenerate electromagnetic field defines, at each point of the space, two real vectors whose direction and sense are uniquely defined. In general, the direction of v_a or w_a does not coincide with that of the electric or the magnetic field. Substituting (6.149) into (6.147), making use of (5.71) and (5.88), one finds that

$$S_{AB} = \frac{c}{8\pi\sqrt{2}}(\alpha^R\widehat{\alpha}_R\, w_{AB} + \beta^R\widehat{\beta}_R\, v_{AB}),$$

which, by virtue of (5.73), amounts to

$$\mathbf{S} = \frac{c}{8\pi}(|\mathbf{v}|\,\mathbf{w} + |\mathbf{w}|\,\mathbf{v}). \tag{6.151}$$

Thus, $\mathbf{S}$ is a linear combination of $\mathbf{v}$ and $\mathbf{w}$ and it makes equal angles with $\mathbf{v}$ and $\mathbf{w}$. On the other hand, from (5.71), (6.148) and (6.149) it follows that the spinor equivalent of the trace-free part of the Maxwell stress tensor, $T_{ab} - \frac{1}{3}T_c{}^c g_{ab}$, is given by

$$-\frac{1}{4\pi}\widehat{F}_{(AB}F_{CD)} = -\frac{1}{4\pi}\alpha_{(A}\beta_B\widehat{\alpha}_C\widehat{\beta}_{D)}$$

therefore [*cf.* (5.89) and (5.90)]

$$4\pi\left(T_{ab} - \tfrac{1}{3}T_c{}^c g_{ab}\right) = -v_{(a}w_{b)} + \tfrac{1}{3}v^c w_c\, g_{ab}$$

and

$$4\pi T_{ab} = -v_{(a}w_{b)} + \tfrac{1}{3}(4\pi T_c{}^c + v^c w_c) g_{ab}.$$

Using (5.73), (5.88), and (6.148)–(6.150) we obtain

$$\begin{aligned} 4\pi T_c{}^c + v^c w_c &= -\tfrac{1}{2}\widehat{F}_{AB}F^{AB} - v^{AB}w_{AB} = -\tfrac{3}{4}\alpha^A\beta_A\widehat{\alpha}^B\widehat{\beta}_B \\ &= -\tfrac{3}{4}(|\mathbf{v}||\mathbf{w}| - v^c w_c), \end{aligned}$$

therefore,

$$4\pi T_{ab} = -v_{(a}w_{b)} - \tfrac{1}{4}(|\mathbf{v}||\mathbf{w}| - v^c w_c) g_{ab}. \tag{6.152}$$

From (6.149) it follows that $F_a F^a = -F_{AB}F^{AB} = \frac{1}{2}(\alpha^A\beta_A)^2$. On the other hand, $F_a F^a = E_a E^a - B_a B^a + 2\mathrm{i}E_a B^a = \mathbf{E}^2 - \mathbf{B}^2 + 2\mathrm{i}\mathbf{E}\cdot\mathbf{B}$; therefore, $\beta_A = \lambda\alpha_A$ if and only if $\mathbf{E}^2 = \mathbf{B}^2$ and $\mathbf{E}\cdot\mathbf{B} = 0$. By absorbing the factor $\lambda^{1/2}$ into α_A, we obtain

$$F_{AB} = \alpha_A\alpha_B \qquad \Leftrightarrow \qquad \mathbf{E}^2 = \mathbf{B}^2 \text{ and } \mathbf{E}\cdot\mathbf{B} = 0, \tag{6.153}$$

where the principal spinor α_A is defined up to sign. The electromagnetic fields with $\mathbf{E}^2 = \mathbf{B}^2$ and $\mathbf{E}\cdot\mathbf{B} = 0$ are called degenerate, algebraically special, null or pure radiation fields.

In the case of a degenerate electromagnetic field [(6.153)], (6.151) and (6.152) reduce to

$$\mathbf{S} = \frac{c}{4\pi}|\mathbf{v}|\mathbf{v}, \qquad 4\pi T_{ab} = -v_a v_b.$$

If the null electromagnetic field (6.153) satisfies the source-free Maxwell equations, then the vector field $\mathbf{v}$ (and hence $\mathbf{S}$) is tangent to a shear-free congruence of geodesics. (This result is a special case of the Mariot–Robinson theorem (Mariot 1954, Robinson 1961), which applies to curved space-times.) Indeed, if $F_{AB} = \alpha_A\alpha_B$ is an algebraically special electromagnetic field that satisfies the source-free Maxwell equations, the only nonvanishing component of F_{AB} with respect to the triad defined by

$$o_A \equiv (\alpha^R\widehat{\alpha}_R)^{-1/2}\alpha_A$$

is F_{22}; then, making use of the Maxwell equations in the explicit form

$$\begin{aligned}
\sqrt{2}\{(D - 2\varepsilon + \rho)F_{+1} + (\delta + 2\alpha)F_0 - \kappa F_{-1}\} - \frac{1}{c}\partial_t F_{+1} &= 0,\\
\sqrt{2}\{\tfrac{1}{2}(\overline{\delta} - 2\overline{\beta})F_{+1} + (\rho - \overline{\rho})F_0 + \tfrac{1}{2}(\delta - 2\beta)F_{-1}\} - \frac{1}{c}\partial_t F_0 &= 0,\\
\sqrt{2}\{\overline{\kappa}F_{+1} + (\overline{\delta} + 2\overline{\alpha})F_0 - (D + 2\varepsilon + \overline{\rho})F_{-1}\} - \frac{1}{c}\partial_t F_{-1} &= 0,\\
(\overline{\delta} - 2\overline{\beta} + 2\overline{\alpha})F_{+1} - 2(D + \rho + \overline{\rho})F_0 - (\delta - 2\beta + 2\alpha)F_{-1} &= 0,
\end{aligned}$$

one obtains

$$\kappa = \alpha = 0, \tag{6.154}$$

and

$$\begin{aligned}
(\delta - 2\beta)F_{-1} &= 0,\\
(D + 2\varepsilon + \overline{\rho})F_{-1} &= -\frac{1}{\sqrt{2}c}\partial_t F_{-1},
\end{aligned} \tag{6.155}$$

where $F_{-1} \equiv F_{22}$ [see (6.61)]. Equations (6.154) imply that the congruence with tangent vector $v_{AB} = \alpha_{(A}\widehat{\alpha}_{B)}$ is shear-free and geodetic.

Conversely, given a shear-free congruence of geodesics, there exists locally an algebraically special solution of the source-free Maxwell equations $F_{AB} = \alpha_A \alpha_B$ such that $\alpha_{(A}\widehat{\alpha}_{B)}$ is tangent to the congruence. Indeed, by choosing the triad $\{D, \delta, \overline{\delta}\}$, in such a way that D is tangent to the congruence, $\kappa = \alpha = 0$. Hence, the source-free Maxwell equations for an electromagnetic field with $F_{11} = F_{12} = 0$ reduce to (6.155). Since $[D, \delta] = (2\varepsilon - \rho)\delta$ [see (6.46)], the integrability condition of (6.155) is

$$D(2\beta F_{-1}) - \delta\big((-2\varepsilon - \overline{\rho})F_{-1} - \frac{1}{\sqrt{2}c}\partial_t F_{-1}\big) = (2\varepsilon - \rho)2\beta F_{-1}.$$

Using (6.155), (6.67) and (6.68) one finds that this condition is satisfied identically. The solution of (6.155) is not unique; in fact, it contains an arbitrary complex function of two variables. For example, all the spin-coefficients for the triad

$$D = \frac{1}{\sqrt{2}}\partial_z, \qquad \delta = \frac{1}{\sqrt{2}}(\partial_x + \mathrm{i}\partial_y), \qquad \overline{\delta} = \frac{1}{\sqrt{2}}(\partial_x - \mathrm{i}\partial_y),$$

induced by the Cartesian coordinates (x, y, z), are equal to zero. Hence, the integral curves of D, which are straight lines parallel to the z-axis, form a shear-free congruence of geodesics (with vanishing expansion and twist). Then from (6.155) we have, $F_{-1} = f(z - ct, x + \mathrm{i}y)$, where f is an arbitrary function of two variables.

As a second example, the triad (6.55) induced by the spherical coordinates satisfies the conditions $\kappa = 0 = \alpha$ and therefore the vector field $D = (1/\sqrt{2})\partial_r$ is tangent to a shear-free congruence of geodesics, which are straight lines through the origin. Substituting (6.54) and (6.55) into (6.155) one finds that

$$F_{-1} = \frac{1}{r \sin\theta} f\big(r - ct, \mathrm{e}^{-\mathrm{i}\phi}\cot\tfrac{1}{2}\theta\big),$$

where f is an arbitrary function. However, in this case, for any choice of the function f, F_{-1} will diverge in some direction. This example shows that, even though (6.155) are locally integrable for a given shear-free congruence of geodesics, their solution may not be well behaved globally.

The massless free field equations for arbitrary spin

The massless free field equations for spin s are given by

$$\sqrt{2}\,\nabla^R{}_{(A}\phi_{B\ldots L)R} = \pm\frac{1}{c}\partial_t\phi_{AB\ldots L}, \tag{6.156}$$

and

$$\nabla^{AB}\phi_{AB\ldots L} = 0, \tag{6.157}$$

where $\phi_{AB...L}$ is a $2s$-index totally symmetric spinor and the sign on the right-hand side of (6.156) depends on the helicity of the field [*cf.* (6.138), (6.139) and (6.143)].

Equations (6.156) and (6.157) are equivalent to

$$\sqrt{2}\,\nabla^R{}_{(A}\widehat{\phi}_{B...L)R} = \mp\frac{1}{c}\partial_t\widehat{\phi}_{AB...L}, \qquad \nabla^{AB}\widehat{\phi}_{AB...L} = 0. \tag{6.158}$$

Hence, $\phi_{AB...L}$ and $\widehat{\phi}_{AB...L}$ have opposite helicities [*cf.* (6.138) and (6.139)]. From (6.156) and (6.158) one obtains the continuity equation

$$\partial_t(\phi^{AB...L}\widehat{\phi}_{AB...L}) = \pm\sqrt{2}\,c\,\nabla^{AR}(\phi_{(A}{}^{B...L}\widehat{\phi}_{R)B...L})$$

[*cf.* (6.140) and (6.147)]. Note that $\phi^{AB...L}\widehat{\phi}_{AB...L} \geqslant 0$ and that the components $\phi_{(A}{}^{B...L}\widehat{\phi}_{R)B...L}$ belong to a real vector field.

In Cartesian coordinates, a plane wave solution of (6.156) and (6.157) is of the form $\phi_{AB...L} = \chi_{AB...L}e^{i(k_ix^i-\omega t)}$, where $\chi_{AB...L}$ is constant and k_i is a real constant vector. Taking into account (5.76), from (6.156) and (6.157) it follows that

$$\chi_{AB...L} = \alpha_A\alpha_B\cdots\alpha_L,$$

for some α_A, and

$$k_{AB} = \pm\sqrt{2}\left(\frac{\omega}{c}\right)\frac{\alpha_{(A}\widehat{\alpha}_{B)}}{\alpha^R\widehat{\alpha}_R}$$

[*cf.* (6.142)].

Killing spinors

In recent decades it has been found that spinors, and the Dirac operator, are very useful tools in differential geometry and topology. The standard treatment is based on the use of Clifford algebras, which allows a unified treatment for manifolds of any dimension (see, *e.g.*, Lawson and Michelsohn 1989, Friedrich 2000). On the other hand, the two-component spinor formalism developed in this chapter, although applicable to three-dimensional manifolds only, involves shorter derivations and leads readily to stronger results. For instance, a spinor field, ψ_A, is a Killing spinor if there exists a constant, μ, such that (see, *e.g.*, Friedrich 2000)

$$\nabla_{AB}\psi_C = \mu\,\varepsilon_{C(A}\psi_{B)}. \tag{6.159}$$

This equation implies that a Killing spinor is an eigenspinor of the Dirac operator,

$$\nabla_A{}^B\psi_B = -\tfrac{3}{2}\mu\,\psi_A, \tag{6.160}$$

and, therefore, $\nabla_D{}^A\nabla_{AB}\psi_C = \frac{1}{4}\mu^2(\varepsilon_{BD}\psi_C + 4\varepsilon_{BC}\psi_D)$. Thus, a Killing spinor is also an eigenspinor of the Laplacian operator

$$\nabla^2\psi_A \equiv -\nabla^{BC}\nabla_{BC}\psi_A = \tfrac{3}{2}\mu^2\psi_A$$

and $\Box_{DB}\psi_C = \frac{1}{2}\mu^2(\varepsilon_{BC}\psi_D + \varepsilon_{DC}\psi_B)$. Comparing with (6.26) we conclude that

$$\Phi_{ABCD} = 0 \quad \text{and} \quad R = -12\mu^2,$$

i.e., the scalar curvature is constant and $R_{ab} = -4\mu^2 g_{ab}$. Furthermore, μ must be real or pure imaginary and, since the Cotton–York tensor vanishes [see (6.36)], the manifold must be locally conformally flat.

In general, the square of the Dirac operator is related to the Laplacian according to

$$\begin{aligned} \nabla^A{}_B \nabla^B{}_C \psi^C &= \nabla^{BA} \nabla_B{}^C \psi_C = \nabla^{B(A} \nabla_B{}^{C)} \psi_C + \tfrac{1}{2}\varepsilon^{AC} \nabla^{BD} \nabla_{BD} \psi_C \\ &= -\tfrac{1}{8} R \psi^A - \tfrac{1}{2} \nabla^2 \psi^A. \end{aligned}$$

A similar relation holds in any dimension and in all cases the only part of the curvature involved is the scalar curvature (see, *e.g.*, Petersen 1998, Friedrich 2000). (In the case of dimension 4, the corresponding formula can also be derived by means of the two-component spinor formalism making use of (A10), while in dimension 2 it follows from (6.72).)

If ψ_A and ϕ_A are two solutions of (6.159) with the same constant μ, then

$$\nabla_{AB}(\psi_C \phi_D) = -\mu(\varepsilon_{AC}\psi_{(B}\phi_{D)} + \varepsilon_{BD}\psi_{(A}\phi_{C)}), \tag{6.161}$$

which implies that $\psi^R \phi_R$ is a constant and that $\psi_{(A}\phi_{B)}$ is the spinor equivalent of a possibly complex Killing vector field, the real and imaginary parts of which are Killing vector fields.

Thus, in particular, if ψ_A is a Killing spinor, $\psi_A \psi_B$ is the spinor equivalent of a possibly complex Killing vector field; furthermore, from (6.14) it follows that if the metric is positive definite and μ is pure imaginary ($R \geqslant 0$), or if the metric is indefinite and μ is real ($R \leqslant 0$), then $\widehat{\psi}_A$ also satisfies (6.159), therefore $\psi_{(A}\widehat{\psi}_{B)}$ is the spinor equivalent of a real Killing vector field.

In the case where the metric is positive definite, making use of (6.161), one finds that the Lie bracket, or commutator, of the Killing vector field $\psi^A \psi^B \partial_{AB}$ and its complex conjugate is given by

$$\begin{aligned} &[\psi^A \psi^B \partial_{AB}, -\widehat{\psi}^C \widehat{\psi}^D \partial_{CD}] \\ &\qquad = \begin{cases} 0 & \text{if } \mu \text{ is real,} \\ 4\mu(\widehat{\psi}^R \psi_R)\psi^{(A}\widehat{\psi}^{B)}\partial_{AB} & \text{if } \mu \text{ is pure imaginary.} \end{cases} \end{aligned}$$

When μ is pure imaginary, using again (6.161) one finds that the Lie bracket of $\psi^A \psi^B \partial_{AB}$ and $\psi^{(A}\widehat{\psi}^{B)}\partial_{AB}$ is

$$[\psi^A \psi^B \partial_{AB}, \psi^{(C}\widehat{\psi}^{D)}\partial_{CD}] = -2\mu(\widehat{\psi}^R \psi_R)\, \psi^A \psi^B \partial_{AB};$$

thus, if μ is different from zero, the three mutually orthogonal real Killing vector fields Re $(\psi^A\psi^B\partial_{AB})$, Im $(\psi^A\psi^B\partial_{AB})$, and $\psi^A\widehat{\psi}^B\partial_{AB}$ generate a Lie algebra isomorphic to that of the rotation group SO(3). Finally, if there exists a nontrivial solution to (6.159) with μ real, the two real Killing vector fields, Re $(\psi^A\psi^B\partial_{AB})$ and Im $(\psi^A\psi^B\partial_{AB})$ are orthogonal to each other and commute.

Similar results can be derived assuming that the metric is indefinite; however, an important difference comes from the fact that now there can be nonzero spinors ψ_A such that $\widehat{\psi}^A\psi_A = 0$.

The equations for a Killing spinor [(6.159)] can also be expressed in the form ${}^{(\pm)}\nabla_{AB}\psi_C = 0$, where

$$^{(\pm)}\nabla_{AB}\psi_C \equiv \nabla_{AB}\psi_C \pm \frac{1}{4}\sqrt{-\frac{R}{3}}(\varepsilon_{DB}\varepsilon_{CA} + \varepsilon_{DA}\varepsilon_{CB})\psi^D \tag{6.162}$$

[*cf.* (6.117)]. Since ${}^{(\pm)}\gamma_{ABCD} \equiv \pm\frac{1}{4}\sqrt{-\frac{R}{3}}(\varepsilon_{AC}\varepsilon_{BD} + \varepsilon_{AD}\varepsilon_{BC})$ satisfies ${}^{(\pm)}\gamma_{ABCD} = {}^{(\pm)}\gamma_{(AB)(CD)}$, the connections ${}^{(\pm)}\nabla$ defined by (6.162) are compatible with the metric, have a nonvanishing torsion (according to (6.119) the torsion of ${}^{(\pm)}\nabla$ is given by ${}^{(\pm)}\Theta_{ABCD} = 2{}^{(\pm)}\gamma_{ABCD}$) but a *vanishing curvature*, as can be seen using (6.118). Thus, a Killing spinor is parallel to itself with respect to one of the connections defined by (6.162). It may be noticed that the torsion tensor of the connections ${}^{(\pm)}\nabla$ is real or pure imaginary (depending on the sign of R and on the signature of the metric) [see (6.111)].

Another example is provided by the harmonic spinors, which are the solutions to

$$\nabla^B{}_A\psi_B = 0. \tag{6.163}$$

By means of a straightforward computation, making use of (6.26), we find that, for two arbitrary spinor fields, ψ_A, λ_A,

$$\begin{aligned} 2(\nabla^{AB}\widehat{\lambda}_B)(\nabla^C{}_A\psi_C) + \nabla^{AB}(\widehat{\lambda}^C\nabla_{AB}\psi_C + 2\widehat{\lambda}_B\nabla_{AC}\psi^C) \\ = (\nabla^{AB}\widehat{\lambda}^C)\nabla_{AB}\psi_C + \tfrac{1}{4}R\psi^A\widehat{\lambda}_A, \end{aligned} \tag{6.164}$$

therefore, taking $\lambda_A = \psi_A$ and assuming that ψ_A obeys (6.163) we have

$$\nabla^{AB}(\widehat{\psi}^C\nabla_{AB}\psi_C) = (\nabla^{AB}\widehat{\psi}^C)\nabla_{AB}\psi_C + \tfrac{1}{4}R\psi^A\widehat{\psi}_A. \tag{6.165}$$

If, for instance, M is a compact manifold with positive definite metric, from (6.165) it follows that

$$\int_M (\nabla^{AB}\psi^C)(\nabla_{AB}\psi_C)^\frown \mathrm{d}v = -\tfrac{1}{4}\int_M R\psi^A\widehat{\psi}_A\,\mathrm{d}v,$$

where $\mathrm{d}v$ is the volume element defined by the metric of M. Using the fact that $\phi^{AB\ldots L}\widehat{\phi}_{AB\ldots L} \geqslant 0$ and $\phi^{AB\ldots L}\widehat{\phi}_{AB\ldots L} = 0$ only if $\phi^{AB\ldots L} = 0$ [see (5.67)], one concludes that if $R > 0$, there are no nontrivial solutions to (6.163).

When M is a three-dimensional compact manifold with positive definite metric, a lower bound for the eigenvalues of the Dirac operator can be obtained directly in the following manner. If ψ_A is an eigenspinor of the Dirac operator, $\nabla^A{}_B\psi^B = \mathrm{i}\lambda\psi^A$, with $\lambda \in \mathbb{R}$, then

$$\xi_{ABC} \equiv \nabla_{AB}\psi_C - \tfrac{2}{3}\mathrm{i}\lambda\varepsilon_{C(A}\psi_{B)}$$

satisfies $\xi_{AB}{}^B = 0$ and therefore the spinor field ξ_{ABC} is totally symmetric. Hence, $(\nabla^{AB}\widehat{\psi}^C)\nabla_{AB}\psi_C = \xi^{ABC}\widehat{\xi}_{ABC} + \frac{2}{3}\lambda^2\psi^A\widehat{\psi}_A$ and from (6.164) it follows that

$$\tfrac{4}{3}\lambda^2 \int_M \psi^A\widehat{\psi}_A \,\mathrm{d}v = \int_M \xi^{ABC}\widehat{\xi}_{ABC}\,\mathrm{d}v + \tfrac{1}{4}\int_M R\psi^A\widehat{\psi}_A\,\mathrm{d}v.$$

This last equation implies that $\lambda^2 \geqslant \frac{3}{16}R_0$, where $R_0 \equiv \min\{R(p) \mid p \in M\}$ and that if $\lambda^2 = \frac{3}{16}R_0$, then $\xi_{ABC} = 0$, which means that ψ_A is a Killing spinor, and R is constant (*cf.* Friedrich 2000).

The notions of spinor field, Killing spinor and Dirac operator can be defined on two-dimensional manifolds (and on manifolds of any dimension provided that certain conditions are satisfied, see, *e.g.*, Lawson and Michelsohn 1989, Friedrich 2000) and, in particular, on two-dimensional surfaces of a three-dimensional manifold with positive definite induced metric. Assuming, for instance, that δ and $\overline{\delta}$, defined by (6.45), are tangent to such a surface, an eigenspinor of the Dirac spinor has two components with spin weights $-1/2$ and $1/2$, and the Dirac operator can be taken as

$$\begin{pmatrix} 0 & -\overline{\eth} \\ \eth & 0 \end{pmatrix},$$

therefore the eigenspinors of the Dirac operator of the sphere and of the plane are given by (3.129) and (4.103), (4.106), (4.166), respectively.

7
Applications to General Relativity

The spinor formalism employed in the four-dimensional space-time of general relativity is more powerful and basic than the tensor formalism and, as the latter, can be applied to express a field equation in space-plus-time form in terms of covariant derivatives associated with the intrinsic geometry of spacelike or timelike hypersurfaces (see, *e.g.*, Sommers 1980, Sen 1982, Shaw 1983a, 1983b, Ashtekar 1987, 1991).

A spacelike or timelike hypersurface in space-time becomes a three-dimensional Riemannian manifold with the metric induced by the space-time metric and, therefore, it has its own spinor structure. In this chapter we show that a null tetrad of space-time induces a triad on any spacelike or timelike hypersurface and that there exists a simple relation between the connection coefficients of the induced triad and the self-dual part of the connection for the null tetrad, which leads to the expression for the covariant derivative of a spinor field on a spacelike hypersurface given by Sommers (1980) and Sen (1982). The curvature of the hypersurface is expressed in terms of the self-dual part of the space-time curvature.

In Section 7.1 we consider the case of spacelike hypersurfaces in a four-dimensional Riemannian manifold with Lorentzian signature and in Section 7.2 we consider the case of timelike hypersurfaces. In Section 7.3 it is shown that the timelike Killing vector field of a stationary space-time allows one to relate null tetrads of space-time with triads of the manifold of orbits of the Killing vector field (see also Perjés 1970).

The notation and conventions followed here for the spinors in four-dimensional manifolds will be, essentially, those of Plebański (1975) and the relevant information is summarized in the Appendix. Lower-case Latin indices $a, b, \ldots$, range over 1,2,3, lower-case Greek indices range over 0,1,2,3, capital Latin indices A, B, ..., range over 1,2 and the dotted indices $\dot{A}$, $\dot{B}$, ..., range over $\dot{1}$, $\dot{2}$.

7.1 Spacelike hypersurfaces

Let M be a four-dimensional Riemannian manifold with signature $(-+++)$. A null tetrad for M is a set of four vector fields $\partial_{A\dot{B}}$ such that

$$\tilde{g}(\partial_{A\dot{B}}, \partial_{C\dot{D}}) = -2\varepsilon_{AC}\varepsilon_{\dot{B}\dot{D}}, \tag{7.1}$$

where $\tilde{g}$ is the metric tensor of M, and

$$\overline{\partial_{A\dot{B}}} = \partial_{B\dot{A}}. \tag{7.2}$$

Let Σ be a spacelike hypersurface and $\partial_n \equiv n^\mu \partial_\mu$ be a normal vector field to Σ, such that

$$n^\mu n_\mu = -1. \tag{7.3}$$

Expressing ∂_n in the form

$$\partial_n = -\tfrac{1}{2} n^{A\dot{B}} \partial_{A\dot{B}}, \tag{7.4}$$

from (7.1), (7.3), and (7.4) it follows that

$$n^{A\dot{B}} n_{A\dot{B}} = 2, \tag{7.5}$$

which amounts to

$$n_A{}^{\dot{B}} n_{C\dot{B}} = \varepsilon_{AC}. \tag{7.6}$$

As in the preceding chapters, the spinor indices are raised and lowered according to the rules

$$\psi_A = \varepsilon_{AB}\psi^B, \qquad \psi^A = \psi_B \varepsilon^{BA} \tag{7.7}$$

and similarly for the dotted indices.

We now introduce the three vector fields

$$\partial_{AB} \equiv \frac{1}{\sqrt{2}} n_{(A}{}^{\dot{C}} \partial_{B)\dot{C}}, \tag{7.8}$$

which are tangent to Σ. Indeed, from (7.1), (7.4), (7.6), and (7.8) we have

$$\tilde{g}(\partial_{AB}, \partial_n) = -\frac{1}{2\sqrt{2}} n_{(A}{}^{\dot{C}} n^{R\dot{S}} \tilde{g}(\partial_{B)\dot{C}}, \partial_{R\dot{S}}) = \frac{1}{\sqrt{2}} n_{(A}{}^{\dot{C}} n_{B)\dot{C}} = 0.$$

Furthermore, from (7.8), (7.1), and (7.6), one finds that

$$\tilde{g}(\partial_{AB}, \partial_{CD}) = \tfrac{1}{2} n_{(A}{}^{\dot{R}} n_{(C}{}^{\dot{S}} \tilde{g}(\partial_{B)\dot{R}}, \partial_{D)\dot{S}}) = -\tfrac{1}{2}(\varepsilon_{AC}\varepsilon_{BD} + \varepsilon_{BC}\varepsilon_{AD}). \tag{7.9}$$

Thus, the vector fields ∂_{AB} constitute a spinorial triad for the hypersurface Σ, in the sense of Chapter 6. Making use of (7.6), it can be seen that

$$\partial_{A\dot{B}} = -\sqrt{2}\, n^C{}_{\dot{B}} \partial_{AC} - n_{A\dot{B}} \partial_n. \tag{7.10}$$

Using (7.2), (7.6), and (7.8), and the fact that $\overline{n_{A\dot{B}}} = n_{B\dot{A}}$, one finds that the complex conjugate of ∂_{AB} is given by

$$\overline{\partial_{AB}} = -n^{C}{}_{\dot{A}} n^{D}{}_{\dot{B}} \partial_{CD}. \tag{7.11}$$

Therefore, $t_{ABCD\ldots}$ is the spinor equivalent of a *real* n-index tensor on Σ with respect to the basis ∂_{AB}, if and only if

$$\overline{t_{ABCD\ldots}} = (-1)^n n^{R}{}_{\dot{A}} n^{S}{}_{\dot{B}} n^{T}{}_{\dot{C}} n^{U}{}_{\dot{D}} \cdots t_{RSTU\ldots}. \tag{7.12}$$

In Chapters 5 and 6 it was assumed that a spinorial triad for a three-dimensional Riemannian manifold with positive definite metric satisfies $\overline{\partial_{AB}} = -\partial^{AB}$, which is a special case of (7.11) corresponding to

$$(n^{A\dot{B}}) = \begin{pmatrix} -1 & 0 \\ 0 & -1 \end{pmatrix}. \tag{7.13}$$

(Assuming that ∂_n, $\partial_{1\dot{1}}$ and $\partial_{2\dot{2}}$ are future-pointing.) By means of a tetrad transformation [preserving (7.1) and (7.2)], one can always take $n^{A\dot{B}}$ to the form (7.13).

As shown in the preceding chapters, there exists an antilinear mapping of the spin space onto itself defined in any Riemannian three-dimensional manifold. The components of the mate of a spinor $\psi_{AB\ldots}$, denoted by $\widehat{\psi}_{AB\ldots}$, with respect to a basis such that $\overline{\partial_{AB}} = -\partial^{AB}$, are defined by $\widehat{\psi}_{AB\ldots} \equiv \overline{\psi^{AB\ldots}}$. Since in the present case the triad ∂_{AB} satisfies (7.11), the components of the mate of a spinor will be given by

$$\widehat{\psi}_{AB\ldots} = n_{A}{}^{\dot{R}} n_{B}{}^{\dot{S}} \cdots \overline{\psi_{RS\ldots}}. \tag{7.14}$$

Then, for an m-index spinor $\psi_{AB\ldots}$, $\widehat{\widehat{\psi}}_{AB\ldots} = (-1)^m \psi_{AB\ldots}$, $\psi^{AB\ldots}\widehat{\psi}_{AB\ldots} \geqslant 0$, and, if $t_{AB\ldots}$ are the spinor components of a real n-index tensor, $\widehat{t}_{AB\ldots} = (-1)^n t_{AB\ldots}$ [see (7.12) and (7.14)]. The group of spin transformations leaving invariant (7.11) and (7.14) is isomorphic to SU(2). The mate of a spinor defined by (7.14) is denoted in Perjés (1970) as $\psi^{\dagger}_{AB\ldots}$ and called the adjoint of $\psi_{AB\ldots}$; in order to avoid confusion with the usual definition of the adjoint of a matrix, as in the previous chapters, we will denote the mate of a spinor as $\widehat{\psi}_{AB\ldots}$. (See also Sommers 1980, Sen 1981, 1982, Ashtekar 1987, 1991.)

As in every Riemannian manifold, there exists a unique torsion-free connection compatible with the metric induced on Σ. Such a connection is represented by the functions Γ_{ABCD} defined by

$$\nabla_{AB} \partial_{CD} = \Gamma^{R}{}_{CAB} \partial_{RD} + \Gamma^{R}{}_{DAB} \partial_{CR}, \tag{7.15}$$

where ∇_{AB} denotes the covariant derivative with respect to ∂_{AB} and

$$\Gamma_{ABCD} = \Gamma_{(AB)(CD)}. \tag{7.16}$$

On the other hand, since the torsion of the connection vanishes,

$$\begin{aligned}[\partial_{AB}, \partial_{CD}] &= \nabla_{AB}\partial_{CD} - \nabla_{CD}\partial_{AB} \\ &= 2\Gamma^R{}_{(C|AB|}\partial_{D)R} - 2\Gamma^R{}_{(A|CD|}\partial_{B)R},\end{aligned} \tag{7.17}$$

where we have made use of (7.15). We substitute (7.8) into the left-hand side of (7.17), making use of the relation

$$[\partial_{A\dot{B}}, \partial_{C\dot{D}}] = \Gamma^R{}_{CA\dot{B}}\partial_{R\dot{D}} + \Gamma^{\dot{R}}{}_{\dot{D}\dot{B}A}\partial_{C\dot{R}} - \Gamma^R{}_{AC\dot{D}}\partial_{R\dot{B}} - \Gamma^{\dot{R}}{}_{\dot{B}\dot{D}C}\partial_{A\dot{R}}, \tag{7.18}$$

which follows from the fact that the connection $\widetilde{\nabla}$ of M is torsion-free, where $\Gamma_{ABC\dot{D}}$ are the components of the connection $\widetilde{\nabla}$ with respect to the null tetrad $\partial_{A\dot{B}}$ (see the Appendix), together with (7.6) and (7.10). Then one finds that

$$\begin{aligned}&[\partial_{AB}, \partial_{CD}] \\ &\quad = (K_{ABCD} - K_{CDAB})\partial_n + \sqrt{2}\,(K_{AB(C}{}^R + \Gamma^R{}_{(C(A|\dot{S}|}n_{B)}{}^{\dot{S}})\,\partial_{D)R} \\ &\qquad -\sqrt{2}\,(K_{CD(A}{}^R + \Gamma^R{}_{(A(C|\dot{S}|}n_{D)}{}^{\dot{S}})\,\partial_{B)R},\end{aligned} \tag{7.19}$$

where we have introduced

$$K_{ABCD} \equiv -\tfrac{1}{2}n_{(A}{}^{\dot{R}}(\widetilde{\nabla}_{B)\dot{R}}n_C{}^{\dot{S}})n_{D\dot{S}}. \tag{7.20}$$

It can be verified that $K_{ABCD} = \tilde{g}(\partial_{CD}, \frac{1}{\sqrt{2}}n_{(A}{}^{\dot{R}}\,\widetilde{\nabla}_{B)\dot{R}}\partial_n)$, and therefore K_{ABCD} are the spinor components of the extrinsic curvature of Σ. The spinor field K_{ABCD} possesses the symmetries

$$K_{ABCD} = K_{(AB)(CD)} \tag{7.21}$$

and, since the vector fields ∂_{AB} are tangent to Σ, their Lie bracket must be also tangent to Σ, therefore from (7.19) we see that

$$K_{ABCD} = K_{CDAB}. \tag{7.22}$$

Comparing (7.17) and (7.19), it follows that

$$\Gamma_{ABCD} = \frac{1}{\sqrt{2}}(\Gamma_{AB(C|\dot{S}|}n_{D)}{}^{\dot{S}} + K_{ABCD}). \tag{7.23}$$

Thus, the connection coefficients for the triad ∂_{AB} are given in terms of the self-dual part of the connection of M and the extrinsic curvature of Σ.

Using (7.8) and (7.23) one can find the components of the covariant derivative of a spinor field on Σ in terms of the covariant differentiation of M. For instance,

$$\begin{aligned}\nabla_{AB}\psi_C &= \partial_{AB}\psi_C - \Gamma^R{}_{CAB}\psi_R \\ &= \frac{1}{\sqrt{2}}[n_{(A}{}^{\dot{R}}\partial_{B)\dot{R}}\psi_C - \Gamma^R{}_{C(A|\dot{S}|}n_{B)}{}^{\dot{S}}\psi_R - K^R{}_{CAB}\psi_R] \\ &= \frac{1}{\sqrt{2}}[n_{(A}{}^{\dot{R}}\,\widetilde{\nabla}_{B)\dot{R}}\psi_C - K_{ABC}{}^R\psi_R].\end{aligned} \tag{7.24}$$

This last expression is equivalent to the spatial covariant derivative defined in Sommers (1980) and Sen (1982).

The covariant derivative of the mate of a spinor field satisfies the relation

$$\nabla_{AB}\widehat{\psi}_{D\ldots}^{C\ldots} = -(\nabla_{AB}\psi_{D\ldots}^{C\ldots})^{\frown}, \tag{7.25}$$

as can be seen making use of (7.24), (7.14), (7.20), and the fact that K_{ABCD} are the spinor components of a real tensor (*i.e.*, $\widehat{K}_{ABCD} = K_{ABCD}$) [*cf.* (6.14)].

The intrinsic curvature

The curvature of the connection of Σ can be obtained from the relation [see (6.26)]

$$\Box_{AB}\psi_C = -\tfrac{1}{2}[\Phi_{ABCD} + \tfrac{1}{12}R(\varepsilon_{AC}\varepsilon_{BD} + \varepsilon_{BC}\varepsilon_{AD})]\psi^D, \tag{7.26}$$

where $\Box_{AB} \equiv \nabla^R{}_{(A}\nabla_{B)R}$, $\Phi_{ABCD} = \Phi_{(ABCD)}$ are the spinor components of the trace-free part of the Ricci tensor, $\Phi_{ab} \equiv R_{ab} - \frac{1}{3}Rg_{ab}$, $R_{ab} = R^c{}_{acb}$ and R are the Ricci tensor and the scalar curvature of Σ, respectively. Making use of (7.24), (7.20), (7.6), and (A10) one readily finds that

$$\begin{aligned} 2\Box_{AB}\psi_C &= \tfrac{1}{2}\widetilde{\Box}_{AB}\psi_C + \tfrac{1}{2}n_A{}^{\dot R}n_B{}^{\dot S}\widetilde{\Box}_{\dot R\dot S}\psi_C \\ &\quad + \big[\sqrt{2}\,\nabla^R{}_{(A}K_{B)RCD} + K^R{}_{(A|D}{}^S K_{R|B)SC}\big]\psi^D \\ &= \big[-C_{ABCD} - \tfrac{1}{24}\tilde{R}(\varepsilon_{AC}\varepsilon_{BD} + \varepsilon_{BC}\varepsilon_{AD}) - \tfrac{1}{4}n_A{}^{\dot R}n_B{}^{\dot S}C_{CD\dot R\dot S} \\ &\quad + \sqrt{2}\,\nabla^R{}_{(A}K_{B)RCD} + K^R{}_{(A|C}{}^S K_{R|B)DS}\big]\psi^D, \end{aligned} \tag{7.27}$$

where $\widetilde{\Box}_{AB} \equiv \widetilde{\nabla}_{(A}{}^{\dot R}\,\widetilde{\nabla}_{B)\dot R}$, $\widetilde{\Box}_{\dot A\dot B} \equiv \widetilde{\nabla}^R{}_{(\dot A}\,\widetilde{\nabla}_{|R|\dot B)}$, $C_{ABCD} = C_{(ABCD)}$ and $C_{AB\dot C\dot D} = C_{(AB)(\dot C\dot D)}$ are the spinor components of the self-dual part of the conformal curvature and of the trace-free part of the Ricci tensor of M, respectively, and $\tilde{R}$ is the scalar curvature (see the Appendix). Hence, from (7.26) and (7.27), one obtains

$$\begin{aligned} &\Phi_{ABCD} + \tfrac{1}{12}R(\varepsilon_{AC}\varepsilon_{BD} + \varepsilon_{BC}\varepsilon_{AD}) \\ &\quad = C_{ABCD} + \tfrac{1}{24}\tilde{R}(\varepsilon_{AC}\varepsilon_{BD} + \varepsilon_{BC}\varepsilon_{AD}) + \tfrac{1}{4}n_A{}^{\dot R}n_B{}^{\dot S}C_{CD\dot R\dot S} \\ &\quad\quad - \sqrt{2}\,\nabla^R{}_{(A}K_{B)RCD} - K^R{}_{(A|C}{}^S K_{R|B)DS}. \end{aligned} \tag{7.28}$$

The contraction of (7.28) with ε^{AC} yields

$$\begin{aligned} R\varepsilon_{BD} &= \tfrac{1}{2}\tilde{R}\varepsilon_{BD} + n^{A\dot R}n_B{}^{\dot S}C_{AD\dot R\dot S} - 2\sqrt{2}\,\nabla^{RA}K_{BRAD} - \sqrt{2}\,\partial_{DB}K \\ &\quad - \tfrac{1}{2}K^2\varepsilon_{BD} - 2K_A{}^{SR}{}_B K^A{}_{RSD}, \end{aligned}$$

where K is the trace of K_{ab} ($K = K_a{}^a = -K_{AB}{}^{AB}$). (Note that, by virtue of (7.22), $K_A{}^R{}_{BR} = -(K/2)\varepsilon_{AB}$.) The symmetric and anti-symmetric parts of this equation give

$$\frac{1}{2\sqrt{2}} n^{A\dot{R}} n_{(B}{}^{\dot{S}} C_{D)A\dot{R}\dot{S}} = \nabla^{RA} K_{BRAD} + \tfrac{1}{2} \partial_{BD} K$$
$$= \nabla^{RA} K_{RABD} + \partial_{BD} K, \qquad (7.29)$$

and

$$R = \tfrac{1}{2}\tilde{R} + \tfrac{1}{2} n^{R\dot{R}} n^{S\dot{S}} C_{RS\dot{R}\dot{S}} + K_{ABCD} K^{ABCD} - K^2, \qquad (7.30)$$

respectively. The totally symmetric part of (7.28) is

$$\Phi_{ABCD} = C_{ABCD} + \tfrac{1}{4} n_{(A}{}^{\dot{R}} n_B{}^{\dot{S}} C_{CD)\dot{R}\dot{S}} - \sqrt{2}\, \nabla^R{}_{(A} K_{BCD)R}$$
$$- K^R{}_{(AB}{}^S K_{|R|CD)S}, \qquad (7.31)$$

which is equivalent to

$$\Phi_{ABCD} = \widehat{C}_{ABCD} + \tfrac{1}{4} n_{(A}{}^{\dot{R}} n_B{}^{\dot{S}} C_{CD)\dot{R}\dot{S}} + \sqrt{2}\, \nabla^R{}_{(A} K_{BCD)R}$$
$$- K^R{}_{(AB}{}^S K_{|R|CD)S}. \qquad (7.32)$$

Making use of (7.12) one verifies that $\frac{1}{2}(C_{ABCD} + \widehat{C}_{ABCD})$ corresponds to a real tensor on Σ, which turns out to be the electric part of the conformal curvature relative to Σ, defined by

$$E_{ab} \equiv C_{a\mu b\nu} n^\mu n^\nu, \qquad (7.33)$$

as can be shown using (7.4), (7.8), (A5), and (A11). Similarly, one finds that $\frac{1}{2}(C_{ABCD} - \widehat{C}_{ABCD})$ are the spinor components of $\mathrm{i} B_{ab}$, where B_{ab} is the magnetic part of the conformal curvature tensor:

$$B_{ab} \equiv {}^*C_{a\mu b\nu} n^\mu n^\nu, \qquad {}^*C_{\mu\nu\rho\sigma} = \tfrac{1}{2}\sqrt{|\det(\tilde{g}_{\gamma\delta})|}\, \varepsilon_{\mu\nu\alpha\beta} C^{\alpha\beta}{}_{\rho\sigma}. \qquad (7.34)$$

Hence,

$$C_{ABCD} = E_{ABCD} + \mathrm{i} B_{ABCD}. \qquad (7.35)$$

Thus, (7.31) and (7.32) are equivalent to

$$\Phi_{ABCD} = E_{ABCD} + \tfrac{1}{4} n_{(A}{}^{\dot{R}} n_B{}^{\dot{S}} C_{CD)\dot{R}\dot{S}} - K^R{}_{(AB}{}^S K_{|R|CD)S}$$
$$= E_{ABCD} + \tfrac{1}{4} n_{(A}{}^{\dot{R}} n_B{}^{\dot{S}} C_{CD)\dot{R}\dot{S}} - K_{RS(AB} K^{RS}{}_{CD)}$$
$$- K\, K_{(ABCD)} \qquad (7.36)$$

and

$$B_{ABCD} = -\mathrm{i}\sqrt{2}\, \nabla^R{}_{(A} K_{BCD)R}. \qquad (7.37)$$

The Sen–Witten connections

The connections ${}^{(\pm)}D$ defined by

$$ {}^{(\pm)}D_{AB}\psi_C \equiv \nabla_{AB}\psi_C \pm \frac{1}{\sqrt{2}} K_{ABC}{}^R \psi_R \tag{7.38} $$

appear in the Ashtekar formulation of canonical gravity and can be employed to prove the positivity of total mass in general relativity following Witten's spinorial method. (The proof given by Witten (1981) makes use of Dirac four-component spinors.) Owing to (7.21), the connections ${}^{(\pm)}D$ are compatible with the metric of Σ [*cf.* (6.117)] but they do not coincide with the Levi-Civita connection (unless the extrinsic curvature K_{ABCD} vanishes). In fact, making use of (6.119) one finds that the connection ${}^{(\pm)}D$ behaves as if it had a nonzero torsion defined by ${}^{(\pm)}\Theta^{CD}{}_{AB} = \pm\frac{1}{\sqrt{2}}(K^{CD}{}_{AB} + K\,\delta^C_{(A}\delta^D_{B)})$, but ${}^{(\pm)}\Theta^{CD}{}_{AB}$ would correspond to a pure imaginary torsion tensor [see (6.111)].

From the definition (7.38) it follows that

$$ ({}^{(\pm)}D_{AB}\psi^{C\ldots}_{D\ldots})\widehat{\ } = -{}^{(\mp)}D_{AB}\widehat{\psi}^{C\ldots}_{D\ldots} \tag{7.39} $$

[*cf.* (7.25)] and

$$ {}^{(\pm)}D_{AB}t^{AB} = \nabla_{AB}t^{AB}, \tag{7.40} $$

for any two-index spinor field t^{AB}. According to (6.118) the spinor equivalent of the (nonsymmetric) Ricci tensor, ${}^{(\pm)}R_{ab} \equiv {}^{(\pm)}R^c{}_{acb}$, corresponding to the curvature of ${}^{(\pm)}D$ is given by

$$ \begin{aligned} &{}^{(\pm)}R_{ABCD} + \tfrac{1}{4}{}^{(\pm)}R(\varepsilon_{AC}\varepsilon_{BD} + \varepsilon_{BC}\varepsilon_{AD}) \\ &\quad = \Phi_{ABCD} + \tfrac{1}{12}R(\varepsilon_{AC}\varepsilon_{BD} + \varepsilon_{BC}\varepsilon_{AD}) \pm \sqrt{2}\,\nabla^R{}_{(A}K_{B)RCD} \\ &\qquad + K^R{}_{(A|C}{}^S K_{R|B)DS}, \end{aligned} $$

where ${}^{(\pm)}R = {}^{(\pm)}R^a{}_a$. Hence, using (7.28), it follows that

$$ \begin{aligned} &{}^{(+)}R_{ABCD} + \tfrac{1}{4}{}^{(+)}R(\varepsilon_{AC}\varepsilon_{BD} + \varepsilon_{BC}\varepsilon_{AD}) \\ &\quad = C_{ABCD} + \tfrac{1}{24}\tilde{R}(\varepsilon_{AC}\varepsilon_{BD} + \varepsilon_{BC}\varepsilon_{AD}) + \tfrac{1}{4}n_A{}^{\dot{R}}n_B{}^{\dot{S}}C_{CD\dot{R}\dot{S}}, \end{aligned} \tag{7.41} $$

therefore

$$ \begin{aligned} &{}^{(+)}D^R{}_{(A}{}^{(+)}D_{B)R}\psi_C \\ &\quad = -\tfrac{1}{2}\{C_{ABCD} + \tfrac{1}{24}\tilde{R}(\varepsilon_{AC}\varepsilon_{BD} + \varepsilon_{BC}\varepsilon_{AD}) + \tfrac{1}{4}n_A{}^{\dot{R}}n_B{}^{\dot{S}}C_{CD\dot{R}\dot{S}}\}\psi^D \\ &\qquad + \frac{1}{\sqrt{2}}(K^{RS}{}_{AB} + K\delta^R_{(A}\delta^S_{B)})\,{}^{(+)}D_{RS}\psi_C. \end{aligned} \tag{7.42} $$

Thus, even though the curvature of the Riemannian connection ∇ depends only on the metric induced on Σ, this intrinsic curvature expressed in terms of the curvature of M, involves the extrinsic curvature of Σ [see (7.28)]. By contrast, according to (7.41), the curvature of the connection ${}^{(+)}D$ can be expressed in terms of the curvature of M alone. This fact may be expected if one notes that (7.24) and (7.38) yield the expression

$$ {}^{(+)}D_{AB}\psi_C = \frac{1}{\sqrt{2}} n_{(A}{}^{\dot{R}}\, \tilde{\nabla}_{B)\dot{R}}\psi_C, $$

which does not involve the extrinsic curvature of Σ.

Making use of (7.38), (7.39), and (7.42) one obtains the Witten–Sen identity:

$$ \begin{aligned} 2({}^{(+)}D^{AB}\lambda_B)\widehat{\ }\,({}^{(+)}D_{CA}\psi^C) &+ {}^{(+)}D^{AB}\big(\widehat{\lambda}^{C}\,{}^{(+)}D_{AB}\psi_C + 2\widehat{\lambda}_B\,{}^{(+)}D_{AC}\psi^C\big) \\ &= ({}^{(+)}D^{AB}\lambda^C)\,({}^{(+)}D_{AB}\psi_C)\widehat{\ } - \tfrac{1}{4} n^{A\dot{R}} n_B{}^{\dot{S}} G_{AC\dot{R}\dot{S}} \widehat{\lambda}^B \psi^C, \end{aligned} \qquad (7.43) $$

where $G_{AB\dot{C}\dot{D}}$ is the spinor equivalent of the Einstein tensor $G_{\mu\nu} \equiv \tilde{R}_{\mu\nu} - \frac{1}{2}\tilde{R}\,\tilde{g}_{\mu\nu}$ [*cf.* (6.164)]. This identity can be employed to prove the positivity of total mass in the following manner (a detailed discussion can be found in Walker 1983, Penrose and Rindler 1986, Stewart 1990). Assuming that λ_A is a solution of the Sen–Witten equation

$$ {}^{(+)}D_{AB}\lambda^A = 0, \qquad (7.44) $$

and taking $\psi_A = \lambda_A$ from (7.43) and (7.40) one obtains

$$ \nabla^{AB}\big(\widehat{\lambda}^{C}\,{}^{(+)}D_{AB}\lambda_C\big) = ({}^{(+)}D^{AB}\lambda^C)({}^{(+)}D_{AB}\psi_C)\widehat{\ } + G_{\mu\nu}n^{\mu}k^{\nu}, \qquad (7.45) $$

where k^{ν} is the real, future pointing, null vector corresponding to $-\lambda^A\lambda^{\dot{B}}$, with $\lambda^{\dot{B}} \equiv \overline{\lambda^B}$. If the Einstein field equations hold ($G_{\mu\nu} = (8\pi G/c^4)T_{\mu\nu}$) and the energy-momentum tensor satisfies the dominant energy condition (see, *e.g.*, Penrose and Rindler 1984), the last term on the right-hand side of (7.45) is nonnegative. Since the first term on the right-hand side of (7.45) is also nonnegative, it follows that

$$ \nabla^{AB}\big(\widehat{\lambda}^{C}\,{}^{(+)}D_{AB}\lambda_C\big) \geqslant 0. \qquad (7.46) $$

On the other hand, Reula (1982) has shown that if the metric and the extrinsic curvature of Σ are asymptotically flat, then for any asymptotically constant spinor field λ_0^A there exists a solution of (7.44) that tends to λ_0^A at infinity. Furthermore, using Gauss's theorem,

$$ \int_{\Sigma} \nabla^{AB}\big(\widehat{\lambda}^{C}\,{}^{(+)}D_{AB}\lambda_C\big)\, dv = \frac{G}{2c^4}(E\delta^A_B + \sqrt{2}\,cP^A{}_B)\lambda_0^B\widehat{\lambda}_{0A}, \qquad (7.47) $$

where E and P_a are the Arnowitt–Deser–Misner (ADM) energy and momentum of the gravitational field measured at spatial infinity. Since λ_0^A is arbitrary, from (7.46) and (7.47) one concludes that $E \geqslant (P_a P^a)^{1/2}$.

Finally, it may be pointed out that the second term on the left-hand side of (7.43) can also be written as $\nabla^{AB}(\widehat{\lambda}^C\ {}^{(+)}D_{AB}\psi_C + 2\widehat{\lambda}_B\ {}^{(+)}D_{AC}\psi^C) = \nabla^{AB}(\widehat{\lambda}^C\ {}^{(+)}D_{CB}\psi_A + \widehat{\lambda}_A\ {}^{(+)}D_{AC}\psi^C)$ and in the case where $\psi_C = \lambda_C$, by means of a straightforward computation one finds that

$$\nabla^{AB}(\widehat{\lambda}^C\ {}^{(+)}D_{CB}\lambda_A + \widehat{\lambda}_A\ {}^{(+)}D_{AC}\lambda^C) = \frac{1}{\sqrt{2}}\nabla^{AB}(\phi_{AB} - \widehat{\phi}_{AB}),$$

where

$$\phi_{AB} \equiv \tfrac{1}{2}\big(\lambda_{(A}\widetilde{\nabla}_{B)}{}^{\dot{C}}\overline{\lambda_C} - \overline{\lambda_C}\widetilde{\nabla}_{(A}{}^{\dot{C}}\lambda_{B)}\big)$$

is a spinor field employed in the proof of the positivity of the total gravitational energy at retarded times (see, *e.g.*, Walker 1983, Penrose and Rindler 1986, Stewart 1990 and the references cited therein).

Integrability conditions for the neutrino zero modes

In the study of neutrino "zero modes" in vacuum space-times, Sen (1981) found that such a neutrino field, λ_A, must satisfy

$$\nabla_{AB}\lambda_C = -\frac{1}{\sqrt{2}}K_{ABC}{}^D\lambda_D. \tag{7.48}$$

Making use of the definitions (7.38), (7.48) can be written as ${}^{(+)}D_{AB}\lambda_C = 0$. Then, if the Einstein vacuum field equations hold (*i.e.*, $C_{AB\dot{C}\dot{D}} = 0$ and $\tilde{R} = 0$), from (7.42) one readily obtains

$$C_{ABCD}\lambda^D = 0,$$

which implies that the space-time is of type N or flat (*cf.* Sen 1981).

The Sen–Witten connection also appears in the three-surface twistor equation (Tod 1984). A three-surface twistor on Σ is a solution of

$${}^{(+)}D_{(AB}\omega_{C)} = 0. \tag{7.49}$$

There exists a four-parameter family of solutions of (7.49) if and only if Σ can be embedded in a conformally flat space-time with the same metric and extrinsic curvature (Tod 1984).

7.2 Timelike hypersurfaces

In the case of a timelike hypersurface Σ in a Lorentzian four-manifold M, the spinor components $n^{A\dot{B}}$ of a unit normal vector field $\partial_n = n^\mu \partial_\mu$ satisfy

$$n_A{}^{\dot{B}} n_{C\dot{B}} = -\varepsilon_{AC}. \tag{7.50}$$

A spinorial triad for Σ, which is a Riemannian manifold with a signature $(++-)$, can be defined by

$$\partial_{AB} = \frac{\mathrm{i}}{\sqrt{2}} n_{(A}{}^{\dot{C}} \partial_{B)\dot{C}}. \tag{7.51}$$

These vector fields are tangent to Σ and also satisfy

$$\tilde{g}(\partial_{AB}, \partial_{CD}) = -\tfrac{1}{2}(\varepsilon_{AC}\varepsilon_{BD} + \varepsilon_{BC}\varepsilon_{AD}), \tag{7.52}$$

as in the previous case [see (7.9)]. Hence, the tetrad vectors are given by

$$\partial_{A\dot{B}} = -\mathrm{i}\sqrt{2}\, n^C{}_{\dot{B}} \partial_{AC} + n_{A\dot{B}} \partial_n \tag{7.53}$$

[*cf.* (7.10)]. Under complex conjugation, the triad vectors are related as in (7.11),

$$\overline{\partial_{AB}} = -n^C{}_{\dot{A}} n^D{}_{\dot{B}} \partial_{CD} \tag{7.54}$$

and, therefore, the reality conditions for the spinor components of a tensor are given again by (7.12). The mate of an m-index spinor is now defined by

$$\widehat{\psi}_{AB\ldots} = \mathrm{i} n_{A\dot{R}} \mathrm{i} n_{B\dot{S}} \cdots \overline{\psi^{RS\ldots}}, \tag{7.55}$$

hence, $\widehat{\widehat{\psi}}_{AB\ldots} = \psi_{AB\ldots}$ and $\widehat{\varepsilon}_{AB} = \varepsilon_{AB}$. The group of the spin transformations preserving (7.54) and (7.55) is isomorphic to SU(1,1) or, equivalently, to SL(2,$\mathbb{R}$).

Following the same steps as in Section 7.1, one finds that the connection coefficients for the triad ∂_{AB} are

$$\Gamma_{ABCD} = \frac{\mathrm{i}}{\sqrt{2}}(\Gamma_{AB(C|\dot{S}|} n_{D)}{}^{\dot{S}} + K_{ABCD}), \tag{7.56}$$

where the components of the extrinsic curvature of Σ, defined by $K_{ABCD} = \tilde{g}(\partial_{CD}, \frac{1}{\sqrt{2}} n_{(A}{}^{\dot{R}}\, \widetilde{\nabla}_{B)\dot{R}} \partial_n)$, are

$$K_{ABCD} = \tfrac{1}{2} n_{(A}{}^{\dot{R}} (\widetilde{\nabla}_{B)\dot{R}} n_C{}^{\dot{S}}) n_{D\dot{S}}. \tag{7.57}$$

Hence,

$$\nabla_{AB}\psi_C = \frac{\mathrm{i}}{\sqrt{2}}[n_{(A}{}^{\dot{R}}\, \widetilde{\nabla}_{B)\dot{R}} \psi_C - K_{ABC}{}^R \psi_R] \tag{7.58}$$

[*cf.* (7.20) and (7.24)], and

$$\nabla_{AB}\widehat{\psi}^{C\dots}_{D\dots} = (\nabla_{AB}\psi^{C\dots}_{D\dots})^{\widehat{}}. \tag{7.59}$$

Using (7.50), (7.57), and (7.58) one finds that

$$\begin{aligned} 2\square_{AB}\psi_C &= \tfrac{1}{2}\widetilde{\square}_{AB}\psi_C - \tfrac{1}{2}n_A{}^{\dot{R}}n_B{}^{\dot{S}}\,\widetilde{\square}_{\dot{R}\dot{S}}\psi_C \\ &\quad + \left[\mathrm{i}\sqrt{2}\,\nabla^R{}_{(A}K_{B)RCD} - K^R{}_{(A|D}{}^S K_{R|B)SC}\right]\psi^D, \end{aligned}$$

therefore, the curvature of Σ is given by

$$\begin{aligned} &\Phi_{ABCD} + \tfrac{1}{12}R(\varepsilon_{AC}\varepsilon_{BD} + \varepsilon_{BC}\varepsilon_{AD}) \\ &\quad = C_{ABCD} + \tfrac{1}{24}\tilde{R}(\varepsilon_{AC}\varepsilon_{BD} + \varepsilon_{BC}\varepsilon_{AD}) - \tfrac{1}{4}n_A{}^{\dot{R}}n_B{}^{\dot{S}}C_{CD\dot{R}\dot{S}} \\ &\quad\quad - \mathrm{i}\sqrt{2}\,\nabla^R{}_{(A}K_{B)RCD} + K^R{}_{(A|C}{}^S K_{R|B)DS}. \end{aligned} \tag{7.60}$$

As in the case of (7.28), one can decompose this last relation into irreducible parts. We obtain

$$\frac{\mathrm{i}}{2\sqrt{2}}n^{A\dot{R}}n_{(B}{}^{\dot{S}}C_{D)A\dot{R}\dot{S}} = \nabla^{RA}K_{RABD} + \partial_{BD}K, \tag{7.61}$$

$$R = \tfrac{1}{2}\tilde{R} - \tfrac{1}{2}n^{R\dot{R}}n^{S\dot{S}}C_{RS\dot{R}\dot{S}} - K_{ABCD}K^{ABCD} + K^2, \tag{7.62}$$

together with

$$\begin{aligned} &\Phi_{ABCD} \\ &\quad = E_{ABCD} - \tfrac{1}{4}n_{(A}{}^{\dot{R}}n_B{}^{\dot{S}}C_{CD)\dot{R}\dot{S}} + K^R{}_{(AB}{}^S K_{|R|CD)S} \\ &\quad = E_{ABCD} - \tfrac{1}{4}n_{(A}{}^{\dot{R}}n_B{}^{\dot{S}}C_{CD)\dot{R}\dot{S}} + K_{RS(AB}K^{RS}{}_{CD)} + K\,K_{(ABCD)} \end{aligned} \tag{7.63}$$

and

$$B_{ABCD} = \sqrt{2}\,\nabla^R{}_{(A}K_{BCD)R}, \tag{7.64}$$

where E_{ABCD} and B_{ABCD} are the spinor equivalents of the electric and magnetic parts of the conformal curvature relative to Σ, respectively.

7.3 Stationary space-times

A three-plus-one decomposition analogous to that given in the preceding sections can be obtained if, instead of the normal vector to a hypersurface, one makes use of a Killing vector field. In this section we shall assume that the space-time metric admits a timelike Killing vector field (*i.e.*, the space-time is stationary)

$$K = K^\mu\partial_\mu = -\tfrac{1}{2}K^{A\dot{B}}\partial_{A\dot{B}}, \tag{7.65}$$

hence $f \equiv -K_\mu K^\mu > 0$ and $\widetilde{\nabla}_\mu K_\nu + \widetilde{\nabla}_\nu K_\mu = 0$ or, equivalently,

$$\widetilde{\nabla}_{A\dot{B}} K_{C\dot{D}} = L_{AC}\varepsilon_{\dot{B}\dot{D}} + L_{\dot{B}\dot{D}}\varepsilon_{AC}, \tag{7.66}$$

where L_{AB} and $L_{\dot{A}\dot{B}}$ are symmetric (with $L_{\dot{A}\dot{B}} = \overline{L_{AB}}$) [*cf.* (6.16)].

Locally there exist coordinate systems, x^μ, such that $K = \partial/\partial x^0$; then $f = -\tilde{g}_{00}$ and $\partial \tilde{g}_{\mu\nu}/\partial x^0 = 0$. The metric of the space-time can be written as

$$\begin{aligned} ds^2 &= \tilde{g}_{00}\left[\left(dx^0 + \frac{\tilde{g}_{0i}}{\tilde{g}_{00}}dx^i\right)^2 - \left(\frac{\tilde{g}_{0i}\tilde{g}_{0j} - \tilde{g}_{00}\tilde{g}_{ij}}{(\tilde{g}_{00})^2}\right)dx^i dx^j\right] \\ &= f\left[\gamma_{ij}dx^i dx^j - (dx^0 - \mathcal{A}_i dx^i)^2\right], \end{aligned} \tag{7.67}$$

where $\mathcal{A}_i \equiv -\tilde{g}_{0i}/\tilde{g}_{00}$ and $\gamma_{ij} \equiv (\tilde{g}_{00})^{-2}(\tilde{g}_{0i}\tilde{g}_{0j} - \tilde{g}_{00}\tilde{g}_{ij})$ (see Geroch 1971, Heusler 1996, Beig and Schmidt 2000). Let ∂_{AB} be a spinorial triad with respect to the metric $d\sigma^2 \equiv \gamma_{ij}dx^i dx^j$, such that $\overline{\partial_{AB}} = -\partial^{AB}$ or, equivalently,

$$\overline{\partial_{AB}} = -n^C{}_{\dot{A}} n^D{}_{\dot{B}} \partial_{CD} \tag{7.68}$$

with

$$(n^{A\dot{B}}) \equiv \begin{pmatrix} -1 & 0 \\ 0 & -1 \end{pmatrix} \tag{7.69}$$

and let $\mathcal{A}_{AB}$ be the spinor equivalent of $\mathcal{A}_i$ with respect to ∂_{AB} ($\mathcal{A}_{AB} = \mathcal{A}_i dx^i(\partial_{AB}) = \mathcal{A}_i \partial_{AB} x^i$). Then

$$\partial_{A\dot{B}} = f^{-1/2}\{-\sqrt{2}\,n^R{}_{\dot{B}}(\partial_{AR} + \mathcal{A}_{AR}\partial_0) - n_{A\dot{B}}\partial_0\} \tag{7.70}$$

is a null tetrad for the metric (7.67), satisfying $\overline{\partial_{A\dot{B}}} = \partial_{B\dot{A}}$ [*cf.* (7.10)]. Since

$$n^B{}_{\dot{A}} n_B{}^{\dot{C}} = -\delta_{\dot{A}}^{\dot{C}}, \qquad n_A{}^{\dot{B}} n^C{}_{\dot{B}} = -\delta_A^C, \tag{7.71}$$

(7.70) implies that $\partial_0 = -\frac{1}{2}\sqrt{f}\,n^{A\dot{B}}\partial_{A\dot{B}}$, hence, with respect to the null tetrad (7.70),

$$K^{A\dot{B}} = \sqrt{f}\,n^{A\dot{B}}. \tag{7.72}$$

A decomposition of the space-time metric similar to (7.67) and the corresponding spinor formulation have been considered in Perjés (1970). One of the advantages of the decomposition (7.67) is that the Maxwell equations can be written in a form similar to the one they have in flat space-time (Torres del Castillo and Mercado-Pérez 1999, *cf.* also Sonego and Abramowicz 1998 and the references cited therein); however, in some cases, it is more convenient to employ, instead of $d\sigma^2$, a metric conformally equivalent to $d\sigma^2$ (see below).

Taking into account the fact that the triad ∂_{AB} satisfies the reality conditions (5.39), the mate of a spinor will be defined as in Section 5.3, namely, $\widehat{\psi}_{AB\ldots} = n_A{}^{\dot R} n_B{}^{\dot S} \cdots \overline{\psi_{RS\ldots}}$ or, by virtue of (7.72),

$$\widehat{\psi}_{AB\ldots} = f^{-1/2} K_A{}^{\dot R} f^{-1/2} K_B{}^{\dot S} \cdots \overline{\psi_{RS\ldots}}. \tag{7.73}$$

The vector field K is (locally) hypersurface orthogonal if and only if

$$\omega_\mu \equiv \sqrt{|\det(\tilde g_{\alpha\beta})|}\, \varepsilon_{\mu\nu\rho\sigma} K^\nu \widetilde\nabla^\rho K^\sigma,$$

vanishes, in which case the space-time is said to be static. Making use of (A7), one finds that the spinor equivalent of ω_μ is given by

$$\omega_{A\dot B} = \tfrac{1}{2}\mathrm{i}(K^{C\dot D} \widetilde\nabla_{C\dot B} K_{A\dot D} - K^{C\dot D} \widetilde\nabla_{A\dot D} K_{C\dot B}).$$

Therefore, from (7.66), (7.71), and (7.73) it follows that

$$\omega_{A\dot B} = \mathrm{i} K^R{}_{\dot B} (L_{RA} + \widehat{L}_{RA}). \tag{7.74}$$

In order to find the components of the Levi-Civita connection of the space-time metric with respect to the null tetrad (7.70) in terms of the components of the connection of the metric $\mathrm{d}\sigma^2$ with respect to ∂_{AB}, we compute the commutator $[\partial_{A\dot B}, \partial_{C\dot D}]$. Making use of (7.70), (7.17), (7.71), and the identity

$$\alpha_{A\dot B}\beta_{C\dot D} - \alpha_{C\dot D}\beta_{A\dot B} = \varepsilon_{AC}\alpha^R{}_{(\dot B}\beta_{|R|\dot D)} + \varepsilon_{\dot B\dot D}\alpha_{(A}{}^{\dot R}\beta_{C)\dot R}$$

we obtain

$$[\partial_{A\dot B}, \partial_{C\dot D}] = \gamma^R{}_{CA\dot B}\partial_{R\dot D} + \gamma^{\dot R}{}_{\dot D\dot BA}\partial_{C\dot R} - \gamma^R{}_{AC\dot D}\partial_{R\dot B} - \gamma^{\dot R}{}_{\dot B\dot DC}\partial_{A\dot R}, \tag{7.75}$$

where

$$\gamma^R{}_{CA\dot B} = \frac{1}{\sqrt{2f}}\big[-2\Gamma^R{}_{CAS} n^S{}_{\dot B} + \mathrm{i}\mathcal{B}_{AC} n^R{}_{\dot B} + \tfrac{1}{2}\delta^R_C n^S{}_{\dot B}\partial_{AS}\ln f\big] \tag{7.76}$$

and $\mathcal{B}_{AB}$ is the curl of $\mathcal{A}_{AB}$ [*cf.* (6.63)]

$$\mathcal{B}_{AB} \equiv \sqrt{2}\,\mathrm{i}\nabla^C{}_{(A}\mathcal{A}_{B)C} \tag{7.77}$$

(note that $\overline{\Gamma_{ABCD}} = -n^M{}_{\dot A} n^N{}_{\dot B} n^R{}_{\dot C} n^S{}_{\dot D}\Gamma_{MNRS}$). The functions $\gamma_{ABC\dot D}$ are not symmetric in their first two indices and, therefore, are not the components of the connection $\Gamma_{ABC\dot D}$. However, under the replacement of $\gamma^A{}_{BC\dot D}$ by $\gamma^A{}_{BC\dot D} + \delta^A_B\lambda_{C\dot D} + \delta^A_C\mu_{B\dot D}$, the right-hand side of (7.75) is unchanged if $\lambda_{A\dot B} = \overline{-\lambda_{B\dot A}}$ and $\mu_{A\dot B} = \overline{\mu_{B\dot A}}$. Furthermore, $\gamma^A{}_{AC\dot D} + \delta^A_A\lambda_{C\dot D} + \delta^A_C\mu_{A\dot D} = (\mathrm{i}\mathcal{B}_{CS}n^S{}_{\dot D} + n^S{}_{\dot D}\partial_{CS}\ln f)/\sqrt{2f} + 2\lambda_{C\dot D} + \mu_{C\dot D}$; hence, choosing $\lambda_{C\dot D} = -\mathrm{i}\mathcal{B}_{CS}n^S{}_{\dot D}/(2\sqrt{2f})$

and $\mu_{C\dot{D}} = -n^S{}_{\dot{D}}(\partial_{CS} \ln f)/\sqrt{2f}$, we find that $\Gamma_{ABC\dot{D}} = \gamma_{ABC\dot{D}} + \varepsilon_{AB}\lambda_{C\dot{D}} + \varepsilon_{AC}\mu_{B\dot{D}}$, *i.e.*,

$$\Gamma_{ABC\dot{D}} = \frac{1}{\sqrt{2f}} n^S{}_{\dot{D}}\big[-2\Gamma_{ABCS} + \mathrm{i}\mathcal{B}_{C(A}\varepsilon_{B)S} + \varepsilon_{C(A}\partial_{B)S} \ln f\big]. \tag{7.78}$$

It is a remarkable fact that the relations (7.78) are much simpler than the corresponding relations in the tensor formalism (see, *e.g.*, Torres del Castillo and Mercado-Pérez 1999).

With the aid of (7.70) and (7.78), any spinor equation involving the space-time metric can be written in terms of the three-dimensional metric $\mathrm{d}\sigma^2$, the torsion-free connection compatible with $\mathrm{d}\sigma^2$, and the objects f, $\mathcal{A}_i$ and $\mathcal{B}_i$, which can be considered as fields defined on the three-dimensional manifold formed by the orbits of K. For instance, from (7.66) it follows that $L_{AB} = \frac{1}{2}\widetilde{\nabla}_{(A}{}^{\dot{R}} K_{B)\dot{R}}$. Using (7.70), (7.72), and (7.78) one obtains

$$L_{AB} = -\frac{1}{\sqrt{2}}(\partial_{AB} \ln f + \mathrm{i}\mathcal{B}_{AB}), \tag{7.79}$$

thus,

$$\widehat{L}_{AB} = \frac{1}{\sqrt{2}}(\partial_{AB} \ln f - \mathrm{i}\mathcal{B}_{AB}) \tag{7.80}$$

and substituting into (7.74) we find that

$$\omega_{A\dot{B}} = \sqrt{2}\, K^C{}_{\dot{B}} \mathcal{B}_{AC}. \tag{7.81}$$

Thus, K is locally hypersurface orthogonal (*i.e.*, the space-time is static) if and only if $\mathcal{B}_{AC} = 0$.

According to (7.79) and (7.80), the components $\partial_{AB} \ln f$ and $\mathcal{B}_{AB}$ can be expressed in terms of L_{AB} and $\widehat{L}_{AB}$; in particular, from (7.71), (7.72), and (7.78)–(7.80) it follows that

$$\Gamma_{AB(C|\dot{R}|}K_{D)}{}^{\dot{R}} = \sqrt{2}\,\Gamma_{ABCD} + \tfrac{1}{2}\varepsilon_{AC}\widehat{L}_{BD} + \tfrac{1}{2}\varepsilon_{BD}\widehat{L}_{AC}. \tag{7.82}$$

(Note that, owing to (5.18), the last two terms are proportional to the dual of the vector equivalent of $\widehat{L}_{AB}$.) Similarly, one obtains $\Gamma_{ABC\dot{D}}K^{C\dot{D}} = -L_{AB}$.

Making use of (A9) and (7.66) one finds that the relation $\widetilde{\nabla}_\mu \widetilde{\nabla}_\nu K_\rho = \tilde{R}^\sigma{}_{\mu\nu\rho} K_\sigma$, which follows from the Killing equations, is equivalent to

$$\widetilde{\nabla}_{A\dot{B}} L_{CD} = 2C_{ACDR}K^R{}_{\dot{B}} + \tfrac{1}{2}C_{CD\dot{B}\dot{R}}K_A{}^{\dot{R}} - \tfrac{1}{6}\tilde{R}\varepsilon_{A(C}K_{D)\dot{B}}. \tag{7.83}$$

Then, by a contraction of indices, we obtain

$$\widetilde{\nabla}^A{}_{\dot{R}} L_{AB} = \tfrac{1}{2}C_{BC\dot{R}\dot{S}}K^{C\dot{S}} - \tfrac{1}{4}\tilde{R}\, K_{B\dot{R}}, \tag{7.84}$$

which, by virtue of (7.70), (7.72), and (7.78), amounts to

$$f^{-1}\nabla^{A}{}_{R}(fL_{AB})+\frac{\mathrm{i}}{2}\varepsilon_{BR}\mathcal{B}^{AC}L_{AC}=\frac{1}{4\sqrt{2}}(2C_{BC\dot{D}\dot{S}}K^{C\dot{S}}K_{R}{}^{\dot{D}}+\tilde{R}f\,\varepsilon_{BR}).$$

Substituting (7.79), the symmetric and the anti-symmetric parts of this last equation yield

$$\nabla^{A}{}_{(R}L_{B)A}+L_{A(R}\partial^{A}{}_{B)}\ln f=\frac{1}{2\sqrt{2}}K_{(R}{}^{\dot{D}}C_{B)C\dot{D}\dot{S}}K^{C\dot{S}},\qquad(7.85)$$

$$\nabla^{AB}L_{AB}+\sqrt{2}\,\widehat{L}^{AB}L_{AB}=\frac{1}{2\sqrt{2}}(C_{BC\dot{D}\dot{S}}K^{C\dot{S}}K^{B\dot{D}}-\tilde{R}f).\qquad(7.86)$$

Hence, by virtue of (7.79), we also have

$$\mathrm{i}f^{-1}\nabla^{A}{}_{(B}(f\mathcal{B}_{R)A})=-\tfrac{1}{2}K_{(B}{}^{\dot{D}}C_{R)C\dot{D}\dot{S}}K^{C\dot{S}}\qquad(7.87)$$

and

$$f^{-1}\nabla^{AB}\partial_{AB}f+\mathcal{B}^{AB}\mathcal{B}_{AB}=-\tfrac{1}{2}C_{BC\dot{D}\dot{S}}K^{B\dot{D}}K^{C\dot{S}}+\tfrac{1}{2}\tilde{R}f.\qquad(7.88)$$

Making use of (7.70) and (7.82) one finds that the irreducible components of the contraction of (7.83) with $K_{B\dot{B}}$ are (7.85), (7.86) and

$$\sqrt{2}\,\nabla_{(AB}L_{CD)}-\widehat{L}_{(AB}L_{CD)}=-2fC_{ABCD}+\tfrac{1}{2}K_{(A}{}^{\dot{R}}K_{B}{}^{\dot{S}}C_{CD)\dot{R}\dot{S}}.\qquad(7.89)$$

As a consequence of this relation, $\sqrt{2}\,\nabla_{(AB}(L_{CD)}+\widehat{L}_{CD)})=-2f(C_{ABCD}-\widehat{C}_{ABCD})$; therefore, from (7.79) and (7.80),

$$B_{ABCD}=\frac{1}{2f}\nabla_{(AB}\mathcal{B}_{CD)},\qquad(7.90)$$

which shows that if the space-time is static, then the magnetic part of the conformal curvature (relative to the Killing vector field K) vanishes and, in that case, C_{ABCD} must be of the form $\alpha_{(A}\beta_{B}\widehat{\alpha}_{C}\widehat{\beta}_{D)}$; hence, for a static space-time, C_{ABCD} must be of type D or G.

On the other hand, computing $\nabla^{R}{}_{(A}\nabla_{B)R}\psi_{C}$, where ψ_{C} are the components of a spinor field independent of x^{0}, we obtain

$$\begin{aligned}&2\Phi_{ABCD}+\tfrac{1}{6}R(\varepsilon_{AC}\varepsilon_{BD}+\varepsilon_{BC}\varepsilon_{AD})\\&\quad=2fC_{ABCD}+\tfrac{1}{12}\tilde{R}f(\varepsilon_{AC}\varepsilon_{BD}+\varepsilon_{BC}\varepsilon_{AD})+\tfrac{1}{2}K_{A}{}^{\dot{R}}K_{B}{}^{\dot{S}}C_{CD\dot{R}\dot{S}}\\&\qquad+\sqrt{2}\,\varepsilon_{C(A}\nabla^{R}{}_{B)}\widehat{L}_{RD}+\sqrt{2}\,\nabla_{D(A}\widehat{L}_{B)C}-\widehat{L}_{C(A}\widehat{L}_{B)D}-L_{AB}L_{CD}\\&\qquad-\widehat{L}_{AB}L_{CD}-\tfrac{1}{4}\widehat{L}^{RS}\widehat{L}_{RS}(\varepsilon_{AC}\varepsilon_{BD}+\varepsilon_{BC}\varepsilon_{AD}).\end{aligned}\qquad(7.91)$$

The combination of (7.89) and (7.91) leads to several useful relations. For instance, (7.89) and the totally symmetric part of (7.91) yield

$$\sqrt{2}\,\nabla_{(AB}[\widehat{L}_{CD)} - L_{CD)}] - L_{(AB}L_{CD)} - \widehat{L}_{(AB}\widehat{L}_{CD)} = 2\Phi_{ABCD} - K_{(A}{}^{\dot{R}}K_{B}{}^{\dot{S}}C_{CD)\dot{R}\dot{S}},$$

which, by virtue of (7.79) and (7.80), can be written as

$$\Phi_{ABCD} + f\nabla_{(AB}\nabla_{CD)}f^{-1} + L_{(AB}\widehat{L}_{CD)} = \tfrac{1}{2}K_{(A}{}^{\dot{R}}K_{B}{}^{\dot{S}}C_{CD)\dot{R}\dot{S}}. \qquad (7.92)$$

Similarly, the contraction of (7.91) with $\varepsilon^{AC}\varepsilon^{BD}$, together with (7.79), (7.80), and (7.86) give

$$R - 4f\nabla^{AB}\nabla_{AB}f^{-1} + 6f^2(\partial^{AB}f^{-1})(\partial_{AB}f^{-1}) - L^{AB}\widehat{L}_{AB} = \tfrac{3}{2}\tilde{R}f - \tfrac{1}{2}K^{A\dot{R}}K^{B\dot{S}}C_{AB\dot{R}\dot{S}}. \qquad (7.93)$$

Maxwell's equations

As pointed out at the beginning of this section, the decomposition of the space-time metric (7.67) allows us to write the Maxwell equations in a form similar to the one they have in flat space-time. The source-free Maxwell equations are given by

$$\widetilde{\nabla}^{A}{}_{\dot{R}}\varphi_{AB} = 0, \qquad (7.94)$$

where $\varphi_{AB} = \varphi_{BA} \equiv \frac{1}{2}F_{AB}{}^{\dot{C}}{}_{\dot{C}}$, and $F_{AB\dot{C}\dot{D}}$ is the spinor equivalent of the electromagnetic field tensor [see (A13)]. Following the same steps as in the case of (7.84), it follows that (7.94) amounts to

$$\mathcal{A}^{A}{}_{R}\partial_0\varphi_{AB} + \frac{1}{\sqrt{2}}\partial_0\varphi_{RB} + f^{-1}\nabla^{A}{}_{R}(f\varphi_{AB}) + \frac{\mathrm{i}}{2}\varepsilon_{BR}\mathcal{B}^{AC}\varphi_{AC} = 0. \qquad (7.95)$$

(It may be noticed that if the Einstein vacuum field equations are fulfilled, *i.e.*, $C_{AB\dot{C}\dot{D}} = 0$ and $\tilde{R} = 0$, then (7.84) implies that L_{AB} satisfies the source-free Maxwell equations.) By combining the symmetric and antisymmetric parts of (7.95) one finds that in a stationary space-time the Maxwell equations are given by

$$\partial_0(f\varphi_{AB} + \sqrt{2}\,\mathcal{A}^{R}{}_{(A}f\varphi_{B)R}) + \sqrt{2}\,\nabla^{R}{}_{(A}(f\varphi_{B)R}) = 0,$$
$$\nabla^{AB}(f\varphi_{AB} + \sqrt{2}\,\mathcal{A}^{R}{}_{A}f\varphi_{BR}) = 0. \qquad (7.96)$$

If we denote by $\sqrt{2}(E_a - \mathrm{i}H_a)$ the vector equivalent of $f\varphi_{AB}$, then the vector form of (7.96) is

$$\operatorname{div}\mathbf{D} = 0, \qquad \operatorname{curl}\mathbf{H} - \partial_0\mathbf{D} = 0,$$
$$\operatorname{div}\mathbf{B} = 0, \qquad \operatorname{curl}\mathbf{E} + \partial_0\mathbf{B} = 0, \qquad (7.97)$$

where

$$\mathbf{D} = \mathbf{E} - \boldsymbol{\mathcal{A}} \times \mathbf{H}, \qquad \mathbf{B} = \mathbf{H} + \boldsymbol{\mathcal{A}} \times \mathbf{E}. \tag{7.98}$$

Einstein vacuum field equations

If the Einstein vacuum field equations hold, from (7.87) it follows that locally there exists a real-valued function, ω, such that

$$\mathcal{B}_{AB} = f^{-1}\partial_{AB}\omega. \tag{7.99}$$

Then, from (7.79) we have

$$L_{AB} = -\frac{1}{\sqrt{2}} f^{-1}\partial_{AB}\chi, \tag{7.100}$$

with

$$\chi \equiv f + \mathrm{i}\omega \tag{7.101}$$

and $\widehat{L}_{AB} = (1/\sqrt{2})f^{-1}\partial_{AB}\overline{\chi}$. Furthermore, (7.81) gives $\omega_{A\dot{B}} = -\partial_{A\dot{B}}\omega$ (ω is, essentially, the so-called twist potential).

Substituting (7.100) into (7.86) one finds that

$$\nabla^{AB}(f\partial_{AB}\chi) - (\partial^{AB}\chi)\partial_{AB}\chi = 0 \tag{7.102}$$

while (7.92) and (7.93) give

$$\begin{aligned} f^{-2}\Phi_{ABCD} + f^{-1}\nabla_{(AB}\nabla_{CD)}f^{-1} &= \tfrac{1}{2}f^{-4}(\partial_{(AB}\chi)(\partial_{CD)}\overline{\chi}), \\ f^{-2}R - 4f^{-1}\nabla^{AB}\nabla_{AB}f^{-1} & \\ + 6(\partial^{AB}f^{-1})(\partial_{AB}f^{-1}) &= -\tfrac{1}{2}f^{-4}(\partial^{AB}\chi)(\partial_{AB}\overline{\chi}), \end{aligned}$$

or, introducing the conformally related triad

$$\partial'_{AB} \equiv f^{-1}\partial_{AB}, \tag{7.103}$$

we have

$$\begin{aligned} \Phi'_{ABCD} &= \tfrac{1}{2}f^{-2}(\partial'_{(AB}\chi)(\partial'_{CD)}\overline{\chi}), \\ R' &= -\tfrac{1}{2}f^{-2}(\partial'^{AB}\chi)(\partial'_{AB}\overline{\chi}), \end{aligned} \tag{7.104}$$

where Φ'_{ABCD} and R' are the components of the curvature of the metric tensor $f^2\mathrm{d}\sigma^2$ with respect to the triad (7.103) [see (6.32) and (6.33)].

In order to show explicitly certain symmetry of the equations for the stationary vacuum space-times, it is useful to consider f and ω as coordinates of an auxiliary two-dimensional manifold, P. Letting $(x^{(1)}, x^{(2)}) \equiv (\omega, f)$, one finds that (7.104) can be expressed in the form

$$\begin{aligned} \Phi'_{ABCD} &= \tfrac{1}{2} h_{(i)(j)} (\partial'_{(AB} x^{(i)})(\partial'_{CD)} x^{(j)}) \\ R' &= -\tfrac{1}{2} h_{(i)(j)} (\partial'^{AB} x^{(i)})(\partial'_{AB} x^{(j)}), \end{aligned} \tag{7.105}$$

where

$$(h_{(i)(j)}) \equiv f^{-2} \mathrm{diag}\,(1, 1),$$

which is the metric tensor of the Poincaré half-plane. The only nonvanishing Christoffel symbols corresponding to the metric $(h_{(i)(j)})$ are given by $\Gamma^{(1)}_{(1)(2)} = -f^{-1}$, $\Gamma^{(2)}_{(1)(1)} = f^{-1}$, $\Gamma^{(2)}_{(2)(2)} = -f^{-1}$; hence, (7.102) amounts to the pair of equations

$$\nabla'^{AB} \partial'_{AB} x^{(i)} + \Gamma^{(i)}_{(j)(k)} (\partial'^{AB} x^{(j)})(\partial'_{AB} x^{(k)}) = 0. \tag{7.106}$$

These equations mean that the functions $x^{(i)}$ give the local expression of a harmonic map of the manifold formed by the orbits of K with the metric tensor $f^2 \mathrm{d}\sigma^2$ into P with the metric tensor $h_{(i)(j)} \mathrm{d}x^{(i)} \mathrm{d}x^{(j)}$.

Equations (7.105) and (7.106) are invariant under the replacement of $x^{(i)}$ by $F^{(i)}(x^{(1)}, x^{(2)})$ if this mapping is an isometry of $h_{(i)(j)} \mathrm{d}x^{(i)} \mathrm{d}x^{(j)}$, with the metric tensor of N, $f^2 \mathrm{d}\sigma^2$, fixed. As is well known, the orientation-preserving isometries of the Poincaré half-plane can be expressed as

$$x^{(1)} + \mathrm{i} x^{(2)} \mapsto \frac{a(x^{(1)} + \mathrm{i} x^{(2)}) + b}{c(x^{(1)} + \mathrm{i} x^{(2)}) + d}, \tag{7.107}$$

with $\begin{pmatrix} a & b \\ c & d \end{pmatrix} \in \mathrm{SL}(2, \mathbb{R})$ (see, *e.g.*, Section 1.4). Thus, given f and ω corresponding to an exact solution of the Einstein vacuum field equations, by means of (7.107) one obtains the functions f and ω corresponding to another stationary vacuum space-time, with the metric $f^2 \mathrm{d}\sigma^2$ fixed (Geroch 1971, see also Beig and Schmidt 2000).

Axisymmetric solutions of the Einstein vacuum field equations

Equation (7.102), written in terms of the metric tensor $f^2 \mathrm{d}\sigma^2$, takes the form

$$(\mathrm{Re}\, \chi) \nabla'^{AB} \partial'_{AB} \chi - (\partial'^{AB} \chi)(\partial'_{AB} \chi) = 0. \tag{7.108}$$

Even though this equation involves the three-dimensional metric $f^2 \mathrm{d}\sigma^2$, when the space-time metric admits a spacelike Killing vector in addition to $K = \partial_0$, that commutes with K, one can replace in (7.108) the differential operators corresponding to $f^2 \mathrm{d}\sigma^2$ by those of a three-dimensional flat metric. In fact, under these assumptions, there exist coordinates (ρ, ϕ, z) such that $f^2 \mathrm{d}\sigma^2 = e^{2\gamma}(\mathrm{d}\rho^2 + \mathrm{d}z^2) + \rho^2 \mathrm{d}\phi^2$, where γ is a function of ρ and z only (Lewis 1932, Papapetrou 1963). Then, if $\partial\chi/\partial\phi = 0$, one finds that (7.108) reduces to

$$(\mathrm{Re}\,\chi)\left(\frac{1}{\rho}\partial_\rho(\rho\partial_\rho\chi) + \partial_z^2\chi\right) - (\partial_\rho\chi)^2 - (\partial_z\chi)^2 = 0,$$

which does not involve γ. This last equation is known as the Ernst equation (Ernst 1968a) and χ is the Ernst potential. Given a solution of the Ernst equation, γ is determined by (7.104).

Einstein–Maxwell equations

We shall consider now stationary solutions of the Einstein–Maxwell equations. Assuming that $\partial_0\varphi_{AB} = 0$, from the first equation in (7.96) it follows that there exists locally a complex-valued function Φ such that

$$f\varphi_{AB} = c^2\sqrt{2/G}\,\partial_{AB}\Phi, \tag{7.109}$$

where the constant factor is introduced for later convenience, and from the second equation in (7.96) and (7.77) it follows that

$$\nabla^{AB}(f\varphi_{AB}) - \mathrm{i}\mathcal{B}^{AB}f\varphi_{AB} = 0. \tag{7.110}$$

(Equation (7.109) is equivalent to $\partial_{A\dot{B}}\Phi = -(\sqrt{G}/c^2)K^{R\dot{S}}\varphi_{AR}\varepsilon_{\dot{B}\dot{S}}$, which involves only the spinor components of the self-dual part of the electromagnetic field, $\varphi_{AC}\varepsilon_{\dot{B}\dot{D}}$.)

On the other hand, using the Einstein field equations, $\tilde{R}_{\mu\nu} - \frac{1}{2}\tilde{R}\,\tilde{g}_{\mu\nu} = (8\pi G/c^4)T_{\mu\nu}$, and (A14), we have $C_{AB\dot{C}\dot{D}} = (2G/c^4)\varphi_{AB}\varphi_{\dot{C}\dot{D}}$. Hence, (7.87) and (7.73) yield

$$\mathrm{i}f^{-1}\nabla^A{}_{(B}(f\mathcal{B}_{R)A}) = \frac{G}{c^4}f\varphi^A{}_{(B}\widehat{\varphi}_{R)A}.$$

Making use of (7.109) we obtain

$$\mathrm{i}\nabla^A{}_{(B}(f\mathcal{B}_{R)A}) = \nabla^A{}_{(B}\big(\overline{\Phi}\,\partial_{R)A}\Phi - \Phi\,\partial_{R)A}\overline{\Phi}\big),$$

which implies the local existence of a real-valued function ω such that

$$\mathrm{i}f\mathcal{B}_{AB} = \overline{\Phi}\,\partial_{AB}\Phi - \Phi\,\partial_{AB}\overline{\Phi} + \mathrm{i}\partial_{AB}\omega. \tag{7.111}$$

Thus,

$$L_{AB} = -\frac{1}{\sqrt{2}}f^{-1}(\partial_{AB}\chi + 2\overline{\Phi}\,\partial_{AB}\Phi), \tag{7.112}$$

with

$$\chi \equiv f - \overline{\Phi}\Phi + \mathrm{i}\omega.$$

Substituting (7.109) and (7.111) into (7.110) we obtain

$$\nabla^{AB}(f\partial_{AB}\Phi) - (\partial^{AB}\chi + 2\overline{\Phi}\,\partial^{AB}\Phi)\partial_{AB}\Phi = 0. \tag{7.113}$$

Similarly, the substitution of (7.112) and $C_{AB\dot{C}\dot{D}} = (2G/c^4)\varphi_{AB}\varphi_{\dot{C}\dot{D}}$ into (7.86) yields

$$\nabla^{AB}(f\partial_{AB}\chi) - (\partial^{AB}\chi + 2\overline{\Phi}\,\partial^{AB}\Phi)\partial_{AB}\chi = 0. \tag{7.114}$$

In terms of the triad (7.103) corresponding to the metric $f^2\mathrm{d}\sigma^2$, (7.113) and (7.114) are

$$\begin{aligned}(\mathrm{Re}\,\chi + \overline{\Phi}\Phi)\nabla'^{AB}(\partial'_{AB}\Phi) - (\partial'^{AB}\chi + 2\overline{\Phi}\,\partial'^{AB}\Phi)\partial'_{AB}\Phi &= 0,\\ (\mathrm{Re}\,\chi + \overline{\Phi}\Phi)\nabla'^{AB}(\partial'_{AB}\chi) - (\partial'^{AB}\chi + 2\overline{\Phi}\,\partial'^{AB}\Phi)\partial'_{AB}\chi &= 0.\end{aligned} \tag{7.115}$$

As in the case of the Einstein vacuum field equations, when there exists a space-like Killing vector field that commutes with K, the differential operators appearing in (7.115) can be replaced by those corresponding to a flat three-dimensional space (Ernst 1968b).

Appendix
Spinors in the Four-Dimensional Space-Time

The spinor equivalent of a tensor $t_{\mu\nu\ldots}$ on a four-dimensional Riemannian manifold of signature $(-+++)$, M, is defined as

$$t_{AB\ldots\dot{C}\dot{D}\ldots} \equiv \sigma^{\mu}{}_{A\dot{C}}\sigma^{\nu}{}_{B\dot{D}}\cdots t_{\mu\nu\ldots}, \tag{A1}$$

where the $\sigma^{\mu}{}_{A\dot{B}}$ are Infeld–van der Waerden symbols, which in this case satisfy

$$\tilde{g}_{\mu\nu}\sigma^{\mu}{}_{A\dot{B}}\sigma^{\nu}{}_{C\dot{D}} = -2\varepsilon_{AC}\varepsilon_{\dot{B}\dot{D}}, \tag{A2}$$

where $\tilde{g}_{\mu\nu}$ are the components of the metric tensor of M, and

$$\overline{\sigma^{\mu}{}_{A\dot{B}}} = \sigma^{\mu}{}_{B\dot{A}} \tag{A3}$$

[*cf.* (2.63)]. Hence, the inverse relation to (A1) is

$$t_{\mu\nu\ldots} = (-\tfrac{1}{2}\sigma_{\mu}{}^{A\dot{C}})(-\tfrac{1}{2}\sigma_{\nu}{}^{B\dot{D}})\cdots t_{AB\ldots\dot{C}\dot{D}\ldots} \tag{A4}$$

and

$$t_{\ldots\mu\ldots}s^{\ldots\mu\ldots} = -\tfrac{1}{2}t_{\ldots A\dot{B}\ldots}s^{\ldots A\dot{B}\ldots}. \tag{A5}$$

The tangent vectors $\partial_{A\dot{B}} \equiv \sigma^{\mu}{}_{A\dot{B}}\partial_{\mu}$ form a null tetrad.

If $t_{\mu\nu}$ is anti-symmetric then its spinor equivalent satisfies $t_{AB\dot{C}\dot{D}} = -t_{BA\dot{D}\dot{C}}$; therefore,

$$\begin{aligned} t_{AB\dot{C}\dot{D}} &= \tfrac{1}{2}(t_{AB\dot{C}\dot{D}} + t_{BA\dot{C}\dot{D}}) + \tfrac{1}{2}(t_{AB\dot{C}\dot{D}} - t_{BA\dot{C}\dot{D}}) \\ &= \tfrac{1}{2}(t_{AB\dot{C}\dot{D}} - t_{AB\dot{D}\dot{C}}) + \tfrac{1}{2}(t_{AB\dot{C}\dot{D}} - t_{BA\dot{C}\dot{D}}) \\ &= \tfrac{1}{2}t_{AB}{}^{\dot{R}}{}_{\dot{R}}\varepsilon_{\dot{C}\dot{D}} + \tfrac{1}{2}t^{R}{}_{R\dot{C}\dot{D}}\varepsilon_{AB}, \end{aligned}$$

with the two-index spinors $t_{AB}{}^{\dot{R}}{}_{\dot{R}}$ and $t^{R}{}_{R\dot{C}\dot{D}}$ being symmetric (for instance, $t_{BA}{}^{\dot{R}}{}_{\dot{R}} = -t_{AB\dot{R}}{}^{\dot{R}} = t_{AB}{}^{\dot{R}}{}_{\dot{R}}$). Therefore, the spinor equivalent of an antisymmetric two-index tensor $t_{\mu\nu}$ is of the form

$$t_{AB\dot{C}\dot{D}} = \tau_{AB}\varepsilon_{\dot{C}\dot{D}} + \tau_{\dot{C}\dot{D}}\varepsilon_{AB}, \tag{A6}$$

with τ_{AB} and $\tau_{\dot{A}\dot{B}}$ symmetric. By virtue of (A3), $t_{\mu\nu}$ is real if and only if $\tau_{\dot{A}\dot{B}} = \overline{\tau_{AB}}$.

If the orientation of the tetrad is chosen in such a way that

$$e_{ACEG\dot{B}\dot{D}\dot{F}\dot{H}} = 4\mathrm{i}(\varepsilon_{AE}\varepsilon_{CG}\varepsilon_{\dot{B}\dot{H}}\varepsilon_{\dot{D}\dot{F}} - \varepsilon_{AG}\varepsilon_{CE}\varepsilon_{\dot{B}\dot{F}}\varepsilon_{\dot{D}\dot{H}}) \tag{A7}$$

are the spinor components of $e_{\mu\nu\rho\sigma} \equiv \sqrt{|\det(\tilde{g}_{\alpha\beta})|}\,\varepsilon_{\mu\nu\rho\sigma}$, the spinor equivalent of the dual of $t_{\mu\nu}$, defined by ${}^*t_{\mu\nu} = \frac{1}{2}e_{\mu\nu\rho\sigma}t^{\rho\sigma}$, is

$$^*t_{AB\dot{C}\dot{D}} = -\mathrm{i}\tau_{AB}\varepsilon_{\dot{C}\dot{D}} + \mathrm{i}\tau_{\dot{C}\dot{D}}\varepsilon_{AB}.$$

The spinor fields $\tau_{AB}\varepsilon_{\dot{C}\dot{D}}$ and $\tau_{\dot{C}\dot{D}}\varepsilon_{AB}$ correspond to the self-dual and the anti-self-dual parts of $t_{\mu\nu}$, respectively.

The spinor equivalent of the covariant derivative of a vector field, $\widetilde{\nabla}_{\mu}t_{\nu}$, is given by

$$\begin{aligned}\widetilde{\nabla}_{A\dot{B}}t_{C\dot{D}} &= \sigma^{\mu}{}_{A\dot{B}}\sigma^{\nu}{}_{C\dot{D}}(\partial_{\mu}t_{\nu} - \Gamma^{\rho}_{\nu\mu}t_{\rho})\\ &= \partial_{A\dot{B}}(\sigma^{\nu}{}_{C\dot{D}}t_{\nu}) - (\partial_{A\dot{B}}\sigma^{\nu}{}_{C\dot{D}})t_{\nu} - \sigma^{\mu}{}_{A\dot{B}}\sigma^{\nu}{}_{C\dot{D}}\Gamma^{\rho}_{\nu\mu}t_{\rho}\\ &= \partial_{A\dot{B}}t_{C\dot{D}} + \tfrac{1}{2}[(\partial_{A\dot{B}}\sigma^{\nu}{}_{C\dot{D}})\sigma_{\nu}{}^{E\dot{F}} + \sigma^{\mu}{}_{A\dot{B}}\sigma^{\nu}{}_{C\dot{D}}\sigma_{\rho}{}^{E\dot{F}}\Gamma^{\rho}_{\nu\mu}]t_{E\dot{F}}.\end{aligned} \tag{A8}$$

As a consequence of the relation

$$\begin{aligned}(\partial_{A\dot{B}}\sigma^{\nu}{}_{C\dot{D}})\sigma_{\nu}{}^{E\dot{F}} &= \partial_{A\dot{B}}(\sigma^{\nu}{}_{C\dot{D}}\sigma_{\nu}{}^{E\dot{F}}) - (\partial_{A\dot{B}}\sigma_{\nu}{}^{E\dot{F}})\sigma^{\nu}{}_{C\dot{D}}\\ &= -(\partial_{A\dot{B}}\tilde{g}_{\nu\rho}\sigma^{\rho E\dot{F}})\sigma^{\nu}{}_{C\dot{D}}\\ &= -(\partial_{A\dot{B}}\sigma^{\rho E\dot{F}})\tilde{g}_{\nu\rho}\sigma^{\nu}{}_{C\dot{D}} - \sigma^{\mu}{}_{A\dot{B}}(\partial_{\mu}\tilde{g}_{\nu\rho})\sigma^{\rho E\dot{F}}\sigma^{\nu}{}_{C\dot{D}}\\ &= -(\partial_{A\dot{B}}\sigma^{\rho E\dot{F}})\sigma_{\rho C\dot{D}}\\ &\quad - \sigma^{\mu}{}_{A\dot{B}}(\Gamma^{\lambda}_{\nu\mu}\tilde{g}_{\rho\lambda} + \Gamma^{\lambda}_{\rho\mu}\tilde{g}_{\lambda\nu})\sigma^{\rho E\dot{F}}\sigma^{\nu}{}_{C\dot{D}},\end{aligned}$$

the expression between brackets in the last line of (A8) can be written in the form $2(\Gamma_{C}{}^{E}{}_{A\dot{B}}\varepsilon_{\dot{D}}{}^{\dot{F}} + \Gamma_{\dot{D}}{}^{\dot{F}}{}_{\dot{B}A}\varepsilon_{C}{}^{E})$, where $\Gamma_{ABC\dot{D}} = \Gamma_{(AB)C\dot{D}}$ and $\Gamma_{\dot{A}\dot{B}\dot{C}D} = \overline{\Gamma_{ABC\dot{D}}}$ [*cf.* (A6)], hence

$$\widetilde{\nabla}_{A\dot{B}}t_{C\dot{D}} = \partial_{A\dot{B}}t_{C\dot{D}} - \Gamma^{E}{}_{CA\dot{B}}t_{E\dot{D}} - \Gamma^{\dot{F}}{}_{\dot{D}\dot{B}A}t_{C\dot{F}}.$$

Similarly, in the case of an arbitrary spinor, $\psi^{A\ldots\dot{B}\ldots}_{C\ldots\dot{D}\ldots}$, the components of the covariant derivative with respect to $\partial_{R\dot{S}}$ are given by

$$\begin{aligned}\tilde{\nabla}_{R\dot{S}}\psi^{A\ldots\dot{B}\ldots}_{C\ldots\dot{D}\ldots} &= \partial_{R\dot{S}}\psi^{A\ldots\dot{B}\ldots}_{C\ldots\dot{D}\ldots} + \Gamma^{A}{}_{MR\dot{S}}\psi^{M\ldots\dot{B}\ldots}_{C\ldots\dot{D}\ldots} + \cdots + \Gamma^{\dot{B}}{}_{\dot{M}\dot{S}R}\psi^{A\ldots\dot{M}\ldots}_{C\ldots\dot{D}\ldots} \\ &\quad + \cdots - \Gamma^{M}{}_{CR\dot{S}}\psi^{A\ldots\dot{B}\ldots}_{M\ldots\dot{D}\ldots} - \cdots - \Gamma^{\dot{M}}{}_{\dot{D}\dot{S}R}\psi^{A\ldots\dot{B}\ldots}_{C\ldots\dot{M}\ldots} - \cdots.\end{aligned}$$

In particular, $\tilde{\nabla}_{A\dot{B}}\varepsilon_{CD} = 0$. The covariant derivatives of the vector fields $\partial_{A\dot{B}}$ are given by $\tilde{\nabla}_{A\dot{B}}\partial_{C\dot{D}} = \Gamma^{M}{}_{CA\dot{B}}\partial_{M\dot{D}} + \Gamma^{\dot{M}}{}_{\dot{D}\dot{B}A}\partial_{C\dot{M}}$ and therefore the spinor components of the connection are determined by

$$[\partial_{A\dot{B}}, \partial_{C\dot{D}}] = \Gamma^{R}{}_{CA\dot{B}}\partial_{R\dot{D}} + \Gamma^{\dot{R}}{}_{\dot{D}\dot{B}A}\partial_{C\dot{R}} - \Gamma^{R}{}_{AC\dot{D}}\partial_{R\dot{B}} - \Gamma^{\dot{R}}{}_{\dot{B}\dot{D}C}\partial_{A\dot{R}}.$$

If the curvature tensor is defined by $(\tilde{\nabla}_\alpha\tilde{\nabla}_\beta - \tilde{\nabla}_\beta\tilde{\nabla}_\alpha)t_\gamma = -\tilde{R}^{\mu}{}_{\gamma\alpha\beta}t_\mu$, then its spinor equivalent can be expressed as

$$\begin{aligned}R_{ABCD\dot{E}\dot{F}\dot{G}\dot{H}} &= 4C_{ABCD}\varepsilon_{\dot{E}\dot{F}}\varepsilon_{\dot{G}\dot{H}} + 4C_{\dot{E}\dot{F}\dot{G}\dot{H}}\varepsilon_{AB}\varepsilon_{CD} \\ &\quad + C_{AB\dot{G}\dot{H}}\varepsilon_{CD}\varepsilon_{\dot{E}\dot{F}} + C_{CD\dot{E}\dot{F}}\varepsilon_{AB}\varepsilon_{\dot{G}\dot{H}} \\ &\quad + \tfrac{1}{6}\tilde{R}\big((\varepsilon_{AC}\varepsilon_{BD} + \varepsilon_{AD}\varepsilon_{BC})\varepsilon_{\dot{E}\dot{F}}\varepsilon_{\dot{G}\dot{H}} \\ &\qquad + (\varepsilon_{\dot{E}\dot{G}}\varepsilon_{\dot{F}\dot{H}} + \varepsilon_{\dot{E}\dot{H}}\varepsilon_{\dot{F}\dot{G}})\varepsilon_{AB}\varepsilon_{CD}\big), \qquad \text{(A9)}\end{aligned}$$

where C_{ABCD} and $C_{\dot{A}\dot{B}\dot{C}\dot{D}}$ represent the conformal curvature, $C_{AB\dot{C}\dot{D}}$ is the spinor equivalent of the trace-free part of the Ricci tensor, $C_{AB\dot{C}\dot{D}} = \sigma^{\mu}{}_{A\dot{C}}\sigma^{\nu}{}_{B\dot{D}}(\tilde{R}_{\mu\nu} - \frac{1}{4}\tilde{R}\,\tilde{g}_{\mu\nu})$, $\tilde{R}_{\mu\nu} \equiv \tilde{R}^{\rho}{}_{\mu\rho\nu}$ is the Ricci tensor and $\tilde{R}$ is the scalar curvature. The spinor equivalent of the commutator of covariant derivatives, $\tilde{\nabla}_\alpha\tilde{\nabla}_\beta - \tilde{\nabla}_\beta\tilde{\nabla}_\alpha$, can be expressed as $\tilde{\nabla}_{A\dot{B}}\tilde{\nabla}_{C\dot{D}} - \tilde{\nabla}_{C\dot{D}}\tilde{\nabla}_{A\dot{B}} = \varepsilon_{AC}\tilde{\Box}_{\dot{B}\dot{D}} + \varepsilon_{\dot{B}\dot{D}}\tilde{\Box}_{AC}$, where $\tilde{\Box}_{AB} \equiv \tilde{\nabla}_{(A}{}^{\dot{R}}\tilde{\nabla}_{B)\dot{R}}$ and $\tilde{\Box}_{\dot{A}\dot{B}} \equiv \tilde{\nabla}^{R}{}_{(\dot{A}}\tilde{\nabla}_{|R|\dot{B})}$; then

$$\begin{aligned}\tilde{\Box}_{AB}\psi_C &= [-2C_{ABCD} - \tfrac{1}{12}\tilde{R}(\varepsilon_{AC}\varepsilon_{BD} + \varepsilon_{BC}\varepsilon_{AD})]\psi^D, \\ \tilde{\Box}_{AB}\psi_{\dot{C}} &= -\tfrac{1}{2}C_{AB\dot{C}\dot{D}}\psi^{\dot{D}}.\end{aligned} \qquad \text{(A10)}$$

According to (A9), the spinor equivalent of the conformal curvature tensor is

$$C_{ABCD\dot{E}\dot{F}\dot{G}\dot{H}} = 4C_{ABCD}\varepsilon_{\dot{E}\dot{F}}\varepsilon_{\dot{G}\dot{H}} + 4C_{\dot{E}\dot{F}\dot{G}\dot{H}}\varepsilon_{AB}\varepsilon_{CD}, \qquad \text{(A11)}$$

therefore, the left dual of the conformal curvature, ${}^*C_{\mu\nu\rho\sigma} = \frac{1}{2}e_{\mu\nu\alpha\beta}C^{\alpha\beta}{}_{\rho\sigma}$, corresponds to

$${}^*C_{ABCD\dot{E}\dot{F}\dot{G}\dot{H}} = -4\mathrm{i}C_{ABCD}\varepsilon_{\dot{E}\dot{F}}\varepsilon_{\dot{G}\dot{H}} + 4\mathrm{i}C_{\dot{E}\dot{F}\dot{G}\dot{H}}\varepsilon_{AB}\varepsilon_{CD}. \qquad \text{(A12)}$$

The Maxwell equations are given by

$$\tilde{\nabla}_\mu F^{\mu\nu} = -\frac{4\pi}{c}J^\nu, \qquad \tilde{\nabla}_\mu{}^*F^{\mu\nu} = 0,$$

where $F_{\mu\nu} = -F_{\nu\mu}$ is the electromagnetic field tensor and J^{μ} is the four-current density. Hence, the spinor equivalents of the Maxwell equations are

$$-\frac{1}{2}\tilde{\nabla}_{A\dot{B}}(\varphi^{AC}\varepsilon^{\dot{B}\dot{D}} + \varphi^{\dot{B}\dot{D}}\varepsilon^{AC}) = -\frac{4\pi}{c}J^{C\dot{D}},$$
$$-\frac{1}{2}\tilde{\nabla}_{A\dot{B}}(-\mathrm{i}\varphi^{AC}\varepsilon^{\dot{B}\dot{D}} + \mathrm{i}\varphi^{\dot{B}\dot{D}}\varepsilon^{AC}) = 0,$$

where $\varphi_{AB} = \varphi_{BA} \equiv \frac{1}{2}F_{AB}{}^{\dot{C}}{}_{\dot{C}}$, and $F_{AB\dot{C}\dot{D}}$ is the spinor equivalent of the electromagnetic field tensor, $F_{\mu\nu}$, hence

$$\tilde{\nabla}_A{}^{\dot{D}}\varphi^{AC} = \frac{4\pi}{c}J^{C\dot{D}}. \tag{A13}$$

The spinor equivalent of the energy-momentum tensor of the electromagnetic field, $T_{\mu\nu} = (4\pi)^{-1}(F_{\mu\rho}F_{\nu}{}^{\rho} - \frac{1}{4}g_{\mu\nu}F_{\rho\sigma}F^{\rho\sigma})$, is

$$T_{AB\dot{C}\dot{D}} = \frac{1}{4\pi}\varphi_{AB}\varphi_{\dot{C}\dot{D}}. \tag{A14}$$

References

Arfken, G. (1985). *Mathematical Methods for Physicists*, 3rd ed. (Academic Press, San Diego).

Ashtekar, A. (1987). *Phys. Rev.* D **36**, 1587.

Ashtekar, A. (1991). *Lectures on Non-perturbative Canonical Gravity* (World Scientific, Singapore).

Bander, M. and Itzykson, C. (1966). *Rev. Mod. Phys.* **38**, 330.

Beig, R. and Schmidt, B. (2000). In *Einstein's Field Equations and Their Physical Implications*, ed. B.G. Schmidt (Springer-Verlag, Berlin).

Bjorken, J.D. and Drell, S.D. (1964). *Relativistic Quantum Mechanics* (McGraw-Hill, New York).

Brink, D.M. and Satcher, G.R. (1993). *Angular Momentum*, 3rd ed. (Oxford University Press, Oxford).

Burn, R.P. (1985). *Groups: a path to geometry* (Cambridge University Press, Cambridge).

Campbell, W.B. and Morgan, T. (1971). *Physica* **53**, 264.

Cartan, E. (1966). *The theory of spinors* (Hermann, Paris) (Dover, New York, reprinted 1981).

Chandrasekhar, S. and Kendall, P.C. (1957). *Astrophys. J.* **126**, 457.

Cortés-Cuautli, L.C. (1997). *Rev. Mex. Fís.* **43**, 527.

Davis, Jr., L. (1939). *Phys. Rev.* **56**, 186.

Davydov, A.S. (1988). *Quantum Mechanics*, 2nd ed. (Pergamon, Oxford).

Dirac, P.A.M. (1931). *Proc. R. Soc. London* A **133**, 60.

Dirac, P.A.M. (1948). *Phys. Rev.* **74**, 817.

Eastwood, M. and Tod, P. (1982). *Proc. Camb. Phil. Soc.* **92**, 317.

Ernst, F.J. (1968a). *Phys. Rev.* **167**, 1175.

Ernst, F.J. (1968b). *Phys. Rev.* **168**, 1415.

Eyges, L. (1972). *The Classical Electromagnetic Field* (Addison-Wesley, Reading, Mass.) (Dover, New York, reprinted 1980).

Feynman, R.P. (1987). In Feynman, R.P. and Weinberg, S., *Elementary Particles and the Laws of Physics* (Cambridge University Press, Cambridge).

Friedrich, Th. (2000). *Dirac Operators in Riemannian Geometry* (American Mathematical Society, Providence, Rhode Island).

Fung, Y.C. (1965). *Foundations of Solid Mechanics* (Prentice-Hall, Englewood Cliffs, N.J.).

Geroch, R.P. (1971). *J. Math. Phys.* **12**, 918.

Geroch, R.P., Held, A., and Penrose, R. (1973). *J. Math. Phys.* **14**, 874.

Goldberg, J.N., Macfarlane, A.J., Newman, E.T., Rohrlich, F., and Sudarshan, E.C.G. (1967). *J. Math. Phys.* **8**, 2155.

Goldstein, H. (1980). *Classical Mechanics*, 2nd ed. (Addison-Wesley, Reading, Mass.), Chap. 4.

Hall, G.S. and Capocci, M.S. (1999). *J. Math. Phys.* **40**, 1466.

Hall, G.S., Morgan, T., and Perjés, Z. (1987). *Gen. Rel. Grav.* **19**, 1137.

Heusler, M. (1996). *Black Hole Uniqueness Theorems* (Cambridge University Press, Cambridge).

Hill, E.H. (1954). *Am. J. Phys.* **22**, 211.

Hochstadt, H. (1971). *The Functions of Mathematical Physics* (Wiley, New York) (Dover, New York, reprinted 1986).

Jackson, J.D. (1975). *Classical Electrodynamics*, 2nd ed. (Wiley, New York).

Knopp, K. (1952). *Elements of the Theory of Functions* (Dover, New York).

Landau, L.D. and Lifshitz, E.M. (1975). *Theory of Elasticity*, 2nd ed. (Pergamon, Oxford).

Lawson, H.B. and Michelsohn, M.-L. (1989). *Spin Geometry* (Princeton University Press, Princeton, N.J.).

Lebedev, N.N. (1965). *Special Functions and their Applications* (Prentice-Hall, Englewood Cliffs, N.J.) (Dover, New York, reprinted 1972).

Lewis, T. (1932). *Proc. R. Soc. London* A **136**, 176.

Ley-Koo, E. and Wang, R.C. (1988). *Rev. Mex. Fís.* **34**, 296.

Mariot, L. (1954). *C. R. Acad. Sci. Paris* **238**, 2055.

Messiah, A. (1962). *Quantum Mechanics*, Vol. II, (North Holland, Amsterdam).

Miller, Jr., W. (1977). *Symmetry and Separation of Variables* (Addison-Wesley, Reading, Mass.).

Misner, C.W., Thorne, K.S., and Wheeler, J.A. (1973). *Gravitation* (Freeman, San Francisco), Chap. 41.

Morse, P.M. and Feshbach, H. (1953). *Methods of Theoretical Physics* (McGraw-Hill, New York), Chap. 11.

Newman, E.T. and Penrose, R. (1966). *J. Math. Phys.* **7**, 863.

Newman, E.T. and Penrose, R. (1968). *Proc. R. Soc. London* A **305**, 175.

Papapetrou, A. (1963). *C. R. Acad. Sci. Paris* **257**, 2797.

Payne, W.T. (1952). *Am. J. Phys.* **20**, 253.

Penrose, R. (1960). *Ann. Phys.* **10**, 171.

Penrose, R. (1994). *Shadows of the Mind* (Oxford University Press, New York), Chap. 5.

Penrose, R. and Rindler, W. (1984). *Spinors and Space-Time*, Vol. 1, (Cambridge University Press, Cambridge).

Penrose, R. and Rindler, W. (1986). *Spinors and Space-Time*, Vol. 2, (Cambridge University Press, Cambridge).

Perjés, Z. (1970). *J. Math. Phys.* **11**, 3383.

Perjés, Z. (1993). *Nucl. Phys.* B **403**, 809.

Petersen, P. (1998). *Riemannian Geometry* (Springer-Verlag, New York).

Plebański, J. (1975). *J. Math. Phys.* **16**, 2395.

Reitz, J.R., Milford, F.J., and Christy, R.W. (1993). *Foundations of Electromagnetic Theory*, 4th ed. (Addison-Wesley, Reading, Mass.).

Robinson, I. (1961). *J. Math. Phys.* **2**, 290.

Rose, M.E. (1961). *Relativistic Electron Theory* (Wiley, New York).

Reula, O. (1982). *J. Math. Phys.* **23**, 810.

Sakurai, J.J. (1994). *Modern Quantum Mechanics* (Addison-Wesley, Reading, Mass.).

Sattinger, D.H. and Weaver, O.L. (1986). *Lie Groups and Algebras with Applications to Physics, Geometry, and Mechanics* (Springer-Verlag, New York).

Schiff, L.I. (1968). *Quantum Mechanics*, 3rd ed. (McGraw-Hill, New York).

Schouten, J.A. (1921). *Math. Zeit.* **11**, 58.

Sen, A. (1981). *J. Math. Phys.* **22**, 1781.

Sen, A. (1982). *Int. J. Theor. Phys.* **21**, 1.

Shaw, W.T. (1983a). *Gen. Rel. Grav.* **15**, 1163.

Shaw, W.T. (1983b). *Proc. R. Soc. London* A **390**, 191.

Sokolnikoff, I.S. (1956). *Mathematical Theory of Elasticity*, 2nd ed. (McGraw-Hill, New York).

Sommers, P. (1980). *J. Math. Phys.* **21**, 2567.

Sonego, S. and Abramowicz, M.A. (1998). *J. Math. Phys.* **39**, 3158.

Stewart, J. (1990). *Advanced General Relativity* (Cambridge University Press, Cambridge).

Stillwell, J. (1992). *Geometry of Surfaces* (Springer-Verlag, New York).

Tamm, Ig. (1931). *Z. Phys.* **71**, 141.

Timoshenko, S.P. and Goodier, J.N. (1970). *Theory of Elasticity*, 3rd ed. (McGraw-Hill, Tokyo).

Tod, K.P. (1984). *Gen. Rel. Grav.* **16**, 435.

Torres del Castillo, G.F. (1990a). *Rev. Mex. Fís.* **36**, 446.

Torres del Castillo, G.F. (1990b). *Rev. Mex. Fís.* **36**, 510.

Torres del Castillo, G.F. (1992a). *Rev. Mex. Fís.* **38**, 863.

Torres del Castillo, G.F. (1992b). *Rev. Mex. Fís.* **38**, 19.

Torres del Castillo, G.F. (1992c). *Rev. Mex. Fís.* **38**, 753.

Torres del Castillo, G.F. (1993). *J. Math. Phys.* **34**, 3856.

Torres del Castillo, G.F. (1994a). *Rev. Mex. Fís.* **40**, 195.

Torres del Castillo, G.F. (1994b). *Rev. Mex. Fís.* **40**, 713.

Torres del Castillo, G.F. (1996). *J. Math. Phys.* **37**, 5684.

Torres del Castillo, G.F. and Cartas-Fuentevilla, R. (1994). *Rev. Mex. Fís.* **40**, 833.

Torres del Castillo, G.F. and Cortés-Cuautli, L.C. (1997). *J. Math. Phys.* **38**, 2996.

Torres del Castillo, G.F. and Hernández-Guevara, A. (1995). *Rev. Mex. Fís.* **41**, 139.

Torres del Castillo, G.F. and Hernández-Moreno, F.J. (2002). *J. Math. Phys.* **43**, 5172.

Torres del Castillo, G.F. and Mercado-Pérez, J. (1999). *J. Math. Phys.* **40**, 2882.

Torres del Castillo, G.F. and Quintero-Téllez, G. (1999). *Rev. Mex. Fís.* **45**, 557.

Torres del Castillo, G.F. and Rojas-Marcial, J.E. (1993). *Rev. Mex. Fís.* **39**, 32.

Tung, W.-K. (1985). *Group Theory in Physics* (World Scientific, Singapore).

Vilenkin, N. Ja. (1968). *Special Functions and the Theory of Group Representations* (American Mathematical Society, Providence, Rhode Island).
Villalba, V.M. (1990). *J. Math. Phys.* **31**, 2702.
Virchenko, N. and Fedotova, I. (2001). *Generalized Associated Legendre Functions and their Applications* (World Scientific, Singapore).
Wald, R.M. (1984). *General Relativity* (University of Chicago Press, Chicago).
Walker, M. (1983). In *Gravitational Radiation*, ed. N. Deruelle and T. Piran (North-Holland, Amsterdam).
Witten, E. (1981). *Commun. Math. Phys.* **80**, 381.
Wu, T.T. and Yang, C.N. (1976). *Nucl. Phys.* B **107**, 365.
York, J.W. (1971). *Phys. Rev. Lett.* **26**, 1656.
Yoshida, Z. (1992). *J. Math. Phys.* **33**, 1252.

Index

Progress in Mathematical Physics

Progress in Mathematical Physics is a book series encompassing all areas of mathematical physics. It is intended for mathematicians, physicists and other scientists, as well as graduate students in the above related areas.

This distinguished collection of books includes authored monographs and textbooks, the latter primarily at the senior undergraduate and graduate levels. Edited collections of articles on important research developments or expositions of particular subject areas may also be included.

This series is reasonably priced and is easily accessible to all channels and individuals through international distribution facilities.

Preparation of manuscripts is preferable in LaTeX. The publisher will supply a macro package and examples of implementation for all types of manuscripts.

Proposals should be sent directly to the series editors:

Anne Boutet de Monvel
Mathématiques, case 7012
Université Paris VII Denis Diderot
2, place Jussieu
F-75251 Paris Cedex 05
France

Gerald Kaiser
The Virginia Center for Signals and Waves
1921 Kings Road
Glen Allen, VA 23059
U.S.A.

or to the Publisher:

Birkhäuser Boston
675 Massachusetts Avenue
Cambridge, MA 02139
U.S.A.
Attn: Ann Kostant

Birkhäuser Verlag
40-44 Viadukstrasse
CH-4010 Basel
Switzerland
Attn: Thomas Hempfling

1 COLLET/ECKMANN. Iterated Maps on the Interval as Dynamical Systems
ISBN 3-7643-3510-6

2 JAFFE/TAUBES. Vortices and Monopoles, Structure of Static Gauge Theories
ISBN 3-7643-3025-2

3 MANIN. Mathematics and Physics
ISBN 3-7643-3027-9

4 ATWOOD/BJORKEN/BRODSKY/STROYNOWSKI. Lectures on Lepton Nucleon Scattering and Quantum Chromodynamics
ISBN 3-7643-3079-1

5 DITA/GEORGESCU/PURICE. Gauge Theories: Fundamental Interactions and Rigorous Results
ISBN 3-7643-3095-3

6 FRAMPTON/GLASHOW/VAN DAM. Third Workshop on Grand Unification, 1982
ISBN 3-7643-3105-4

7 FRÖHLICH. Scaling and Self-Similarity in Physics: Renormalization in Statistical Mechanics and Dynamics
ISBN 3-7643-3168-2

8 MILTON/SAMUEL. Workshop on Non-Perturbative Quantum Chromodynamics
ISBN 3-7643-3127-5

9 LANGACKER/STEINHARDT/WELDON. Fourth Workshop on Grand Unification
ISBN 3-7643-3169-0

10 FRITZ/JAFFE/SZÁSZ. Statistical Physics and Dynamical Systems: Rigorous Results
ISBN 3-7643-3300-6

11 CEAUSESCU/COSTACHE/GEORGESCU. Critical Phenomena: 1983 Brasov School Conference
ISBN 3-7643-3289-1

12 PIGUET/SIBOLD. Renormalized Supersymmetry: The Perturbation Theory of N=1 Supersymmetric Theories in Flat Space-Time
ISBN 3-7643-3346-4

13 HABA/SOBCZYK. Functional Integration, Geometry and Strings: Proceedings of the XXV Karpacz Winter School of Theoretical Physics
ISBN 3-7643-2387-6

14 SMIRNOV. Renormalization and Asymptotic Expansions
ISBN 3-7643-2640-9

15 Leznov/Saveliev. Group-Theoretical Methods for Integration of Nonlinear Dynamical Systems
ISBN 3-7643-2615-8

16 MASLOV. The Complex WKB Method for Nonlinear Equations I: Linear Theory
ISBN 3-7643-5088-1

17 BAYLIS. Electrodynamics: A Modern Geometric Approach
ISBN 0-8176-4025-8

18 ABŁAMOWICZ/FAUSER. Clifford Algebras and their Applications in Mathematical Physics, Volume 1: Algebra and Physics
ISBN 0-8176-4182-3

19 RYAN/SPRÖßIG. Clifford Algebras and their Applications in Mathematical Physics, Volume 2: Clifford Analysis
ISBN 0-8176-4183-1

20 STOLLMANN. Caught by Disorder: Bound States in Random Media
ISBN 0-8176-4210-2

21 PETTERS/LEVINE/WAMBSGANSS. Singularity Theory and Gravitational Lensing
ISBN 0-8176-3668-4

22 CERCIGNANI. The Relativistic Boltzmann Equation: Theory and Applications
ISBN 3-7643-6693-1

23 KASHIWARA/MIWA. MathPhys Odyssey 2001: Integrable Models and Beyond—*In Honor of Barry M. McCoy*
ISBN 0-8176-4260-9

24 CNOPS. An Introduction to Dirac Operators on Manifolds
ISBN 0-8176-4298-6

25 KLAINERMAN/NICOLÒ. The Evolution Problem in General Relativity
ISBN 0-8176-4254-4

26 BLANCHARD/BRÜNING. Mathematical Methods in Physics
ISBN 0-8176-4228-5

27 WILLIAMS. Topics in Quantum Mechanics
ISBN 0-8176-4311-7

28 OBOLASHVILI. Higher Order Partial Differential Equations in Clifford Analysis
ISBN 0-8176-4286-2

29 CORDANI. The Kepler Problem: Group Theoretical Apects, Regularization and Quantization, with Applications to the Study of Perturbations
ISBN 3-7643-6902-7

30 DUPLANTIER/RIVASSEAU. Poincaré Seminar 2002: Vacuum Energy-Renormalization
ISBN 3-7643-0579-7

31 RAKOTOMANANA. A Geometrical Approach to Thermomechanics of Dissipating Continua
ISBN 0-8176-4283-8

32 TORRES DEL CASTILLO. 3-D Spinors, Spin-Weighted Functions and their Applications
ISBN 0-8176-3249-2

33 HEHL/OBUKHOV. Foundations of Classical Electrodynamics: Charge, Flux, and Metric
ISBN 3-7643-4222-6